AF689189

# Mathematical Modeling in Bioscience

# Advanced Studies in Complex Systems

**Series Editors:** Valentina Emilia Balas, Dumitru Baleanu and Hemen Dutta

# Mathematical Modeling in Bioscience

## Theory and Applications

Edited by

Hemen Dutta

ELSEVIER

ACADEMIC PRESS

An imprint of Elsevier

Academic Press is an imprint of Elsevier
125 London Wall, London EC2Y 5AS, United Kingdom
525 B Street, Suite 1650, San Diego, CA 92101, United States
50 Hampshire Street, 5th Floor, Cambridge, MA 02139, United States

Copyright © 2025 Elsevier Inc. All rights are reserved, including those for text and data mining, AI training, and similar technologies.

For accessibility purposes, images in electronic versions of this book are accompanied by alt text descriptions provided by Elsevier. For more information, see https://www.elsevier.com/about/accessibility.

Publisher's note: Elsevier takes a neutral position with respect to territorial disputes or jurisdictional claims in its published content, including in maps and institutional affiliations.

MATLAB® is a trademark of The MathWorks, Inc. and is used with permission.
The MathWorks does not warrant the accuracy of the text or exercises in this book.
This book's use or discussion of MATLAB® software or related products does not constitute endorsement or sponsorship by The MathWorks of a particular pedagogical approach or particular use of the MATLAB® software.

No part of this publication may be reproduced or transmitted in any form or by any means, electronic or mechanical, including photocopying, recording, or any information storage and retrieval system, without permission in writing from the publisher. Details on how to seek permission, further information about the Publisher's permissions policies and our arrangements with organizations such as the Copyright Clearance Center and the Copyright Licensing Agency, can be found at our website: www.elsevier.com/permissions.

This book and the individual contributions contained in it are protected under copyright by the Publisher (other than as may be noted herein).

**Notices**

Knowledge and best practice in this field are constantly changing. As new research and experience broaden our understanding, changes in research methods, professional practices, or medical treatment may become necessary.

Practitioners and researchers must always rely on their own experience and knowledge in evaluating and using any information, methods, compounds, or experiments described herein. In using such information or methods they should be mindful of their own safety and the safety of others, including parties for whom they have a professional responsibility.

To the fullest extent of the law, neither the Publisher nor the authors, contributors, or editors, assume any liability for any injury and/or damage to persons or property as a matter of products liability, negligence or otherwise, or from any use or operation of any methods, products, instructions, or ideas contained in the material herein.

ISBN: 978-0-443-15445-4

For information on all Academic Press publications
visit our website at https://www.elsevier.com/books-and-journals

*Publisher:* Mara Conner
*Acquisitions Editor:* Chris Katsaropoulos
*Editorial Project Manager:* Teddy A Lewis
*Production Project Manager:* Omer Mukthar
*Cover Designer:* Greg Harris

Typeset by VTeX

Working together to grow libraries in developing countries

ELSEVIER Book Aid International

www.elsevier.com • www.bookaid.org

# Contents

# Contributors

**Maricel Agop**
Physics Department, "Gheorghe Asachi" Technical University, Iaşi, Romania
Academy of Romanian Scientists, Bucureşti, Romania

**Fahad Al Basir**
Department of Mathematics, Asansol Girls' College, Asansol, West Bengal, India

**Raghda A.M. Attia**
School of Medical Informatics and Engineering, Xuzhou Medical University, Xuzhou, Jiangsu Province, PR China

**Laid Boudjellal**
Centre of Mathematics, University of Minho, Braga, Portugal

**Amar Nath Chatterjee**
Department of Mathematics, K.L.S. College, Nawada, Magadh University, Bodh Gaya, Bihar, India

**F. Chipepa**
Department of Mathematics and Statistical Sciences, Botswana International University of Science and Technology, Palapye, Botswana

**Watcharaporn Cholamjiak**
Department of Mathematics, School of Science, University of Phayao, Phayao, Thailand

**Shilpa Das**
Department of Mathematics, R.G. Baruah College, Fatasil Ambari, Guwahati, India

**Lucian Eva**
Nicolae Oblu Hospital, Iaşi, Romania

**M. Gabanakgosi**
Department of Mathematics and Statistical Sciences, Botswana International University of Science and Technology, Palapye, Botswana

**Alina Gavriluţ**
Faculty of Mathematics, Alexandru Ioan Cuza University from Iaşi, Iaşi, Romania

**Behzad Ghanbari**
Department of Engineering Science, Kermanshah University of Technology, Kermanshah, Iran

**Vlad Ghizdovăţ**
Biophysics and Medical Physics Department, "Grigore T. Popa" University of Medicine and Pharmacy, Iaşi, Romania

**Eduardo González-Olivares**
Pontificia Universidad Católica de Valparaíso, Valparaíso, Chile

**Sonal Jain**
School of Technology, Woxsen University, Hyderabad, Telangana, India

**Muhammad Altaf Khan**
Institute of Ground Water Studies, Faculty of Natural and Agricultural Sciences, University of the Free State, Bloemfontein, South Africa

**Mostafa M.A. Khater**
School of Medical Informatics and Engineering, Xuzhou Medical University, Xuzhou, Jiangsu Province, PR China
Department of Basic Science, The Higher Institute for Engineering & Technology, Cairo, Egypt
Institute of Digital Economy, Ugra State University, Khanty-Mansiysk, Russia

**Nicole Martínez-Jeraldo**
Doctorado en Modelamiento Matemático Aplicado, Universidad Católica del Maule, Talca, Chile

**T. Moakofi**
Department of Mathematics and Statistical Sciences, Botswana International University of Science and Technology, Palapye, Botswana

**B. Oluyede**
Department of Mathematics and Statistical Sciences, Botswana International University of Science and Technology, Palapye, Botswana

**Kolade M. Owolabi**
Department of Mathematical Sciences, Federal University of Technology Akure, Akure, Ondo State, Nigeria

**Edson Pindza**
Department of Decision Sciences, College of Economic and Management Sciences, University of South Africa (UNISA), Pretoria, South Africa

**Arunee Promsri**
Department of Physical Therapy, School of Allied Health Sciences, University of Phayao, Phayao, Thailand

**Elizabeth Rozas-Torres**
Centro de Docencia de Ciencias Básicas para Ingeniería, Facultad de Ciencias de la Ingeniería, Universidad Austral de Chile, Valdivia, Chile

**Cristiana J. Silva**
Department of Mathematics, ISTA, Iscte – Instituto Universitário de Lisboa, Lisboa, Portugal
Center for Research & Development in Mathematics and Applications (CIDMA), University of Aveiro, Aveiro, Portugal

**Paulo C. Tintinago-Ruiz**
Universidad del Quindío, Armenia, Colombia

**Aleksander Vokhmintsev**
Institute of Digital Economy, Ugra State University, Khanty-Mansiysk, Russia
Institute of Information Technology, Chelyabinsk State University, Chelyabinsk, Russia

**D. Wanduku**
Department of Mathematical Sciences, Georgia Southern University, Statesboro, GA, United States

**Watcharaporn Yajai**
Department of Mathematics, School of Science, University of Phayao, Phayao, Thailand

1

# Analysis of the impact of time delay incorporation in mathematical models of cellular population dynamics

Laid Boudjellal[a] and Cristiana J. Silva[b,c]

[a]Centre of Mathematics, University of Minho, Braga, Portugal [b]Department of Mathematics, ISTA, Iscte – Instituto Universitário de Lisboa, Lisboa, Portugal [c]Center for Research & Development in Mathematics and Applications (CIDMA), University of Aveiro, Aveiro, Portugal

## 1.1 Introduction

There are many types of autoimmune diseases. In general terms, autoimmune diseases can de described as heterogeneous conditions characterized by the disruption of tolerogenic mechanisms and subsequent activation of self-reactive immune cells [1]. In other words, autoimmune diseases occur when the immune system attacks and destroys its own healthy cells, tissues, and organs, as a consequence of an interruption of their mechanisms. Due to its complexity, autoimmune diseases constitute a very interesting topic for mathematicians, and mathematical models have been successfully used to study this complex dynamics.

Over the years, many mathematicians have focused their work on this topic and several mathematical approaches have been used to investigate the dynamical behavior of the various cellular populations in the context of autoimmune diseases. See, for instance, kinetic theory of active particles [2–4], optimal control theory [5], dynamical system theory [6], among others. Mathematical models of autoimmune diseases can have a significant role in understanding the underlying mechanisms, predicting disease progression, and also evaluating potential treatment strategies.

In this work, we study the effect of considering a discrete time delay in the dynamical biological model proposed in [7]. The model in [7] aims to study the cellular interactions and the global population behavior in autoimmune diseases. It is based on a kinetic theory description of the cellular dynamics and possesses a macroscopic counterpart derived on the hydrodynamical limit of the kinetic system, which is able to describe the global behavior of the cellular populations. The macroscopic analogue is a system of ordinary differential equations (ODEs), which constitutes an autonomous dynamical system. Three cellular populations are involved, namely the population of self-antigen presenting cells

Mathematical Modeling in Bioscience. https://doi.org/10.1016/B978-0-44-315445-4.00007-X

Copyright © 2025 Elsevier Inc. All rights are reserved, including those for text and data mining, AI training, and similar technologies.

(SAPCs), the population of self-reactive $T$ cells (SRTCs), and the population of immunosuppressive cells (ISCs). The immunosuppressive cells, such as Treg and natural killer cells, are essential components of immune homoeostasis since they provide a therapeutic opportunity for active immune regulation, which can have a long time effect in controlling the development of autoimmunity.

We are interested in the role of ISCs in controlling autoimmunity, and we consider a time delay, describing the cumulative duration of the processes of viral attachment and interaction between the ISCs and SAPCs. This time delay is associated with various aspects that result from both a) internal effects, such as the ability of spread of immune response, and b) external effects, such as the age and gender. This represents the innovative aspect of our model.

For the delayed model proposed here, we study the local asymptotic stability of its equilibrium states. Then, we study the global stability of a biologically meaningful disease-free equilibrium. The extension of the global stability analysis to the other equilibrium points of the time delayed model is not a trivial task and is a challenge for a future work.

After this Introduction, our work is organized as follows: In Section 1.2, we recall the non-delayed model from [7] and present the time delayed model, explaining their main assumptions and computing the equilibrium points. In Section 1.3, we study the local asymptotical stability of the equilibrium states. In Section 1.4, we prove the global stability of the disease-free equilibrium. In Section 1.5, we determine numerically the solution of both the non-delayed model and the delayed model, and compare their behavior to appraise the effect of the time delay on the solution. The simulations illustrate the stability results proved in the previous sections. Finally, in Section 1.6, we summarize our results and present new perspectives in connection with our work.

## 1.2 The mathematical model for autoimmunity with time delay

In this section, we describe the mathematical model considered in our study. Such model is derived from a previous model without time delay, which has been proposed in [7]. Hereby, we present first the original model, and then we introduce our delayed model.

### 1.2.1 The non-delayed model

The macroscopic model proposed in [7] considers three cellular populations, namely the population $p_1$ of self-antigen presenting cells (SAPCs), the population $p_2$ of self-reactive $T$ cells (SRTCs), and the population $p_3$ of immunosuppressive cells (ISCs), mainly constituted by Treg cells and natural killer cells. We represent by $n_i(t)$, with $i = 1, 2, 3$, the expected number of cells of each population $p_i$ at time $t \geq 0$. In the model of paper [7], the density $n_i(t)$ is defined by

$$n_i(t) = \int_0^1 f_i(t,u)du, \quad i = 1, 2, 3,$$

where $f_i(t,u)$ is the distribution function associated to the population $p_i$, which denotes the number of cells of the population $p_i$ with biological activity $u \in [0,1]$, at time $t \geq 0$. The biological activity $u$ describes the functional state expressed by each cellular population and is explained in [7]. Functions $f_i(t,u)$ are the unknowns of the kinetic model considered in [7], and their behavior is specified by a system of nonlinear partial differential equations (PDEs), involving integral terms of quadratic type, which describe binary interactions among cells. See paper [7] and the references cited therein for a detailed account of the kinetic models for biological processes.

The effective interactions among the populations $p_i$ occur between a pair of cells, and are instantaneous and homogeneous in space. At the macroscopic scale, the admissible interactions are of proliferative or destructive type and are the following: proliferative interactions between SAPCs and SRTCs, which increase both populations; and between SAPCs and ISCs, which increase the population of ISCs; destructive interactions between SAPCs and ISCs, which decrease the population of SAPCs; and between SRTCs and ISCs, which decrease the population of SRTCs. Both the proliferative and the destructive rates are assumed to be constant, and are here denoted by $\alpha, \gamma, \delta$ and by $\beta, \rho$, respectively. Moreover, the model considers a source term representing a constant input of SAPCs, due to the effect of certain environmental factors and unhealthy dietary habits. Such recruitment term is denoted here by $\Lambda$. Finally, the model considers the natural death rates of all populations, here denoted by $\mu_i$, $i = 1, 2, 3$. Omitting here other details, for the sake of brevity, we introduce the ODE system of paper [7] in the form

$$\begin{cases} \dot{n_1}(t) = \Lambda + \alpha n_1(t)n_2(t) - \beta n_1(t)n_3(t) - \mu_1 n_1(t) \\ \dot{n_2}(t) = \gamma n_2(t)n_1(t) - \rho n_2(t)n_3(t) - \mu_2 n_2(t) \\ \dot{n_3}(t) = \delta n_3(t)n_1(t) - \mu_3 n_3(t), \end{cases} \tag{1.1}$$

where all parameters are strictly positive. Model (1.1) is studied in detail in paper [7], for what concerns the equilibrium states, asymptotical stability, and bifurcations. An interesting result of paper [7] is the capability of the model in capturing the recurrent dynamics describing the relapse-remission pattern typical of many autoimmune diseases, in terms of long-term periodic oscillations.

## 1.2.2 The model with time delay

The mathematical model with time delay considered in the present work constitutes a simple variant of model (1.1), in the sense that we introduce a delay only in one destructive interaction. Specifically, we assume that there exists a latent period for the effectiveness of the destructive interactions among SAPCs and ISCs. Such period is represented by a time delay $\tau > 0$, which is introduced in the term describing the above-mentioned interactions. Accordingly, the new model becomes

$$\begin{cases} \dot{n_1}(t) = \Lambda + \alpha n_1(t)n_2(t) - \beta n_1(t-\tau)n_3(t-\tau) - \mu_1 n_1(t) \\ \dot{n_2}(t) = \gamma n_2(t)n_1(t) - \rho n_2(t)n_3(t) - \mu_2 n_2(t) \\ \dot{n_3}(t) = \delta n_3(t-\tau)n_1(t-\tau) - \mu_3 n_3(t), \end{cases} \tag{1.2}$$

where $\tau \geq 0$. In the particular case in which $\tau = 0$, system (1.2) reduces to system (1.1). Again, all parameters that figure in system (1.2) are strictly positive, that is, $\Lambda, \alpha, \beta, \gamma, \rho, \delta, \mu_1, \mu_2, \mu_3 > 0$.

The initial conditions for model (1.2) describe the past history of each population density in the time interval $[-\tau, 0]$, and are given by

$$n_1(\theta) = \Phi_1(\theta) > 0,\ n_2(\theta) = \Phi_2(\theta) > 0,\ n_3(\theta) = \Phi_3(\theta) > 0, \quad \text{for} \quad \theta \in [-\tau, 0], \tag{1.3}$$

where $\Phi_1$, $\Phi_2$, $\Phi_3$ are continuous functions from the interval $[-\tau, 0]$ into $\mathbb{R}_+$.

### 1.2.3 Equilibrium states of the model with delay

To investigate the effects of the time delay introduced in our model, we first determine the equilibrium states of the model. Models (1.1) and (1.2) have the same equilibria, as expected. We restrict our analysis to the set of biologically significant solutions to the Cauchy problem (1.2)–(1.3), namely

$$\mathcal{D} = \left\{ (n_1(t), n_2(t), n_3(t)) \in \mathbb{R}^3 : n_1(t) \geq 0,\ n_2(t) \geq 0,\ n_3(t) \geq 0 \right\}. \tag{1.4}$$

Our result is the following:

**Theorem 1.2.1** (Equilibrium states). *The model* (1.2) *has four equilibrium states in* $\mathcal{D}$, *which are given by*

**a)** $E_0 = \left(\dfrac{\Lambda}{\mu_1}, 0, 0\right)$, *representing an equilibrium state without both SRTC and ISC cells;*

**b)** $E_1 = \left(\dfrac{\mu_3}{\delta}, 0, \dfrac{\Lambda\delta - \mu_1\mu_3}{\beta\mu_3}\right)$, *representing an equilibrium state without SRTC cells, which exists if* $\Lambda\delta \geq \mu_1\mu_3$;

**c)** $E_2 = \left(\dfrac{\mu_3}{\delta}, \dfrac{-\Lambda\delta^2\rho + \gamma\beta\mu_3^2 - \delta\beta\mu_2\mu_3 + \delta\rho\mu_1\mu_3}{\alpha\delta\rho\mu_3}, \dfrac{\gamma\mu_3 - \delta\mu_2}{\delta\rho}\right)$ *representing an equilibrium state of co-existence of all population's cells, which exists if* $\gamma\mu_3 \geq \delta\mu_2$ *and* $\Lambda \leq \dfrac{\mu_1\mu_3}{\delta} + \dfrac{\mu_3\gamma - \mu_2\delta}{\delta\rho}\dfrac{\mu_3\beta}{\delta}$;

**d)** $E_3 = \left(\dfrac{\mu_2}{\gamma}, \dfrac{\mu_1\mu_2 - \Lambda\gamma}{\alpha\mu_2}, 0\right)$, *representing an equilibrium state without ISC cells, that exists if* $\mu_1\mu_2 \geq \Lambda\gamma$.

We note that equilibrium $E_2$, with strict inequalities in its existence conditions, is the unique with meaningful significance in the context of our study since the other equilibria predict the extinction of some populations. Anyway, all other equilibria have their mathematical importance and are relevant for our study.

The following sections in this work will focus on stability and asymptotic behavior of the equilibria of the model (1.2). In Section 1.3, we study the local asymptotic stability of the equilibria; Section 1.4 is devoted to the global stability. Finally, in Section 1.5, we study, numerically, the time behavior of the solution to our system (1.2) and investigate the asymptotic behavior of the equilibria.

## 1.3 Local asymptotical stability of the equilibrium states

The stability analysis developed here for system (1.2) is based on a general result for non-linear delay equations, see [8] and [9]. For the sake of completeness, we introduce this result below.

Let us consider the following system of $n$ delay differential equations:

$$\dot{x} = f(x, x_\tau), \tag{1.5}$$

where $f$ is a continuously differentiable function as many times as we need; $\tau > 0$ represents the time delay, and $x, x_\tau \in \mathbb{R}^n$. Let us assume that the system (1.5) can be written in the form

$$\dot{x} = f_1(x) + f_2(x_\tau), \tag{1.6}$$

where $f_1$, $f_2$ is continuously differentiable functions as many times as we need. Then, we have the following result, based on the linearization of system (1.6), as usual, see [8], [9] and [10]:

**Theorem 1.3.1** (Stability of the delayed system). *Assume that $\bar{x}$ is an equilibrium solution of system* (1.6), *and denote by $A_1$ and $A_2$ the Jacobian matrices of $f_1$ and $f_2$, respectively, with respect to $x$ and $x_\tau$, that is $A_1 = D_x f_1(x)$ and $A_2 = D_{x_\tau} f_2(x_\tau)$. For $k \in \mathbb{N}$, $k \leq n$, suppose that the characteristic equation of system* (1.6),

$$\det(\lambda I_n - A_1 - e^{-\lambda\tau} A_2) = 0,$$

*evaluated at the equilibrium state $\bar{x}$, has $k$ different solutions $\lambda_1, \cdots, \lambda_k$. Therefore he following hold:*

- **i)** *the equilibrium solution $\bar{x}$ is locally asymptotically stable if* $\mathrm{Re}(\lambda_i) < 0$, *for all $i = 1, \cdots, k$;*
- **ii)** *the equilibrium solution $\bar{x}$ is unstable if* $\mathrm{Re}(\lambda_i) > 0$ *for some $i = 1, \cdots, k$.*

Having Theorem 1.3.1 in mind, we start by re-writing our system (1.2) in a more compact form, similar to that of the general system (1.6). Let us introduce $n_i^\tau(t) = n_i(t - \tau)$, for

$i = 1, 2, 3$, and

$$N(t) = \begin{pmatrix} n_1(t) \\ n_2(t) \\ n_3(t) \end{pmatrix}, \quad N^\tau(t) = \begin{pmatrix} n_1^\tau(t) \\ n_2^\tau(t) \\ n_3^\tau(t) \end{pmatrix}.$$

System (1.2) can be written as

$$\dot{N}(t) = F_1(N(t)) + F_2(N^\tau(t)), \tag{1.7}$$

where

$$F_1(N(t)) = \begin{pmatrix} \Lambda + \alpha n_1(t)n_2(t) - \mu_1 n_1(t) \\ \gamma n_2(t)n_1(t) - \rho n_2(t)n_3(t) - \mu_2 n_2(t) \\ -\mu_3 n_3(t) \end{pmatrix} \tag{1.8}$$

and

$$F_2(N^\tau(t)) = \begin{pmatrix} -\beta n_1^\tau(t)n_3^\tau(t) \\ 0 \\ \delta n_3^\tau(t)n_1^\tau(t) \end{pmatrix}. \tag{1.9}$$

Then, we define the Jacobian matrices of functions $F_1$ and $F_2$ with respect to $t$ and $\tau$, given by

$$A_1 = J_{F_1}(N(t)) = \begin{pmatrix} \alpha n_2(t) - \mu_1 & \alpha n_1(t) & 0 \\ \gamma n_2(t) & \gamma n_1(t) - \rho n_3(t) - \mu_2 & -\rho n_2(t) \\ 0 & 0 & -\mu_3 \end{pmatrix}$$

and

$$A_2 = J_{F_2^\tau}(N^\tau(t)) = \begin{pmatrix} -\beta n_3^\tau(t) & 0 & -\beta n_1^\tau(t) \\ 0 & 0 & 0 \\ \delta n_3^\tau(t) & 0 & \delta n_1^\tau(t) \end{pmatrix}.$$

The characteristic equation of model (1.2) is then given by

$$\det\left(\lambda I_3 - A_1 - e^{-\lambda\tau} A_2\right) = 0. \tag{1.10}$$

### 1.3.1 Local stability of $E_0$

Consider the equilibrium state $E_0$ determined in Theorem 1.2.1, which is specified by

$$n_1(t) = \frac{\Lambda}{\mu_1}, \quad n_2(t) = n_3(t) = 0.$$

The characteristic equation of system (1.2), referred to the equilibrium state $E_0$, is given by

$$\frac{1}{\mu_1^2}\left[(\lambda + \mu_1)(-\Lambda\gamma + \lambda\mu_1 + \mu_1\mu_2)(-\Lambda\delta e^{-\lambda\tau} + \lambda\mu_1 + \mu_1\mu_3)\right] = 0, \tag{1.11}$$

which possesses the two real roots given by

$$\lambda_1 = -\mu_1, \qquad \lambda_2 = \frac{\Lambda\gamma - \mu_1\mu_2}{\mu_1} \tag{1.12}$$

and a third real root $\lambda_3$ such that

$$-\Lambda\delta e^{-\lambda_3\tau} + \lambda_3\mu_1 + \mu_1\mu_3 = 0. \tag{1.13}$$

Using Theorem 1.3.1, we state the following result:

**Theorem 1.3.2** (Local stability of the equilibrium state $E_0$). *Let $\tau \geq 0$. The equilibrium state $E_0$ is locally asymptotically stable if*

$$\Lambda\gamma < \mu_1\mu_2 \qquad \textit{and} \qquad \Lambda\delta < \mu_1\mu_3,$$

*and it is unstable if*

$$\Lambda\gamma > \mu_1\mu_2 \qquad \textit{or} \qquad \Lambda\delta > \mu_1\mu_3.$$

*Proof.* (a) Let $\tau = 0$. The roots (1.12) and (1.13) of the characteristic equation (1.11) reduce to

$$\lambda_1 = -\mu_1, \quad \lambda_2 = \frac{\Lambda\gamma - \mu_1\mu_2}{\mu_1}, \quad \lambda_3 = \frac{\Lambda\delta - \mu_1\mu_3}{\mu_1},$$

and using Theorem 1.3.1, the proof of the local stability of $E_0$ follows immediately.

(b) Let us consider $\tau > 0$. The first and second roots of the characteristic equation (1.11), namely $\lambda_1$ and $\lambda_2$, are not affected by the value of $\tau$. We just need to study the third root, namely $\lambda_3$. We consider the auxiliary transcendental function defined by

$$g(\lambda) = -\Lambda\delta\, e^{-\lambda\tau} + \lambda\mu_1 + \mu_1\mu_3, \ \lambda \in \mathbb{R}$$

such that

$$g'(\lambda) = \tau\Lambda\delta\, e^{-\lambda\tau} + \mu_1 > 0, \ \lambda \in \mathbb{R}, \qquad \text{and} \qquad g'(\lambda) > 0, \ \lambda \in \mathbb{R}.$$

Therefore $g$ is a strictly increasing function such that

$$\begin{cases} g(0) = -\Lambda\delta + \mu_1\mu_3 \\ \lim\limits_{\lambda \to +\infty} g(\lambda) = +\infty. \end{cases}$$

If $\Lambda\delta > \mu_1\mu_3$, then function $g$ has a positive real root so that $\lambda_3 > 0$ and $E_0$ is unstable.

If $\Lambda\delta < \mu_1\mu_3$, then function $g$ has a negative real root so that $\lambda_3 < 0$. The conclusion about the local stability of $E_0$ follows immediately. The proof of Theorem 1.3.2 is then completed. □

### 1.3.2 Local stability of $E_1$

The second equilibrium point $E_1$ has a biological meaning if and only if $\Lambda\delta \geq \mu_1\mu_3$. In this case, $E_1$ is given by

$$n_1(t) = \frac{\mu_3}{\delta}, \quad n_2(t) = 0, \quad n_3(t) = \frac{\Lambda\delta - \mu_1\mu_3}{\beta\mu_3}.$$

At the equilibrium point $E_1$, the characteristic equation of the model (1.2) is given by

$$(\lambda + K)\left\{\lambda^2 + \lambda\left[e^{-\lambda\tau}\left(\frac{\Lambda\delta}{\mu_3} - \mu_1 - \mu_3\right) + \mu_1 + \mu_3\right] + e^{-\lambda\tau}(\Lambda\delta - 2\mu_1\mu_3) + \mu_1\mu_3\right\} = 0,$$

where

$$K = \frac{\Lambda\delta^2\rho + \beta\delta\mu_2\mu_3 - \beta\gamma\mu_3^2 - \delta\mu_1\mu_3\rho}{\beta\delta\mu_3}.$$

**Theorem 1.3.3** (Local stability of the equilibrium state $E_1$).

**(a)** *Let $\tau = 0$. The equilibrium state $E_1$ is locally asymptotically stable if $\Lambda > \frac{\beta\mu_3}{\rho\delta^2}(\gamma\mu_3 - \delta\mu_2) + \frac{\mu_1\mu_3}{\delta}$, and it is unstable if $\Lambda < \frac{\beta\mu_3}{\rho\delta^2}(\gamma\mu_3 - \delta\mu_2) + \frac{\mu_1\mu_3}{\delta}$.*

**(b)** *Let $\tau > 0$, and assume that $\Lambda\delta < 2\mu_1\mu_3$. Then, the equilibrium state $E_1$ is locally asymptotically stable.*

*Proof.* (a) Let $\tau = 0$. Then the characteristic equation (1.10), computed at the equilibrium state $E_1$, is given by

$$(\lambda + K)\left[\lambda^2 + \lambda\frac{\Lambda\delta}{\mu_3} + \Lambda\delta - \mu_1\mu_3\right] = 0, \tag{1.14}$$

which has three different roots, namely

- $\lambda_1 = \frac{1}{2\mu_3}\left(-\Lambda\delta + \sqrt{\Lambda^2\delta^2 - 4\mu_3^2(\Lambda\delta - \mu_1\mu_3)}\right)$,
- $\lambda_2 = \frac{1}{2\mu_3}\left(-\Lambda\delta - \sqrt{\Lambda^2\delta^2 - 4\mu_3^2(\Lambda\delta - \mu_1\mu_3)}\right)$,
- $\lambda_3 = -K$, that is, $\lambda_3 = -\frac{1}{\beta\delta\mu_3}\left[\rho\delta(\Lambda\delta - \mu_1\mu_3) - \beta\mu_3(\gamma\mu_3 - \delta\mu_2)\right]$.

If $\lambda_1, \lambda_2$ are real eigenvalues, that is, if $\Lambda^2\delta^2 > 4\mu_3^2(\Lambda\delta - \mu_1\mu_3)$, then, it is immediate that $\lambda_1 < 0$ and $\lambda_2 < 0$ whenever $E_1$ exists, namely when $\Lambda\delta \geq \mu_1\mu_3$. Moreover, if $\lambda_1$ and $\lambda_2$ are complex eigenvalues, then $\mathrm{Re}(\lambda_1) = \mathrm{Re}(\lambda_2) = \frac{-\Lambda\delta}{2\mu_3} < 0$, where Re is the real part of the complex number. Therefore the equilibrium state $E_1$ is stable if $\lambda_3 < 0$, that is, if

$$\rho\delta(\Lambda\delta - \mu_1\mu_3) - \beta\mu_3(\gamma\mu_3 - \delta\mu_2) > 0 \Leftrightarrow \rho\delta(\Lambda\delta - \mu_1\mu_3) > \beta\mu_3(\gamma\mu_3 - \delta\mu_2)$$

or, equivalently, if

$$\Lambda > \frac{\beta\mu_3}{\rho\delta^2}(\gamma\mu_3 - \delta\mu_2) + \frac{\mu_1\mu_3}{\delta}. \tag{1.15}$$

This concludes the proof for non-delayed model.

(b) Let $\tau > 0$. Then the characteristic equation (1.10) computed at the equilibrium point $E_1$ is given by

$$(\lambda + K)\,g(\lambda) = 0,$$

where $g(\lambda)$ is the transcendental function, defined by

$$g(\lambda) = \lambda^2 + (\mu_1 + \mu_3)\lambda + \mu_1\mu_3 + e^{-\lambda\tau}\left[\left(\frac{\Lambda\delta}{\mu_3} - \mu_1 - \mu_3\right)\lambda + \Lambda\delta - 2\mu_1\mu_3\right].$$

First, we notice that the eigenvalue $\lambda_3 = -K$ is not affected by $\tau$, and has been studied in the previous case $\tau = 0$ so that $\lambda_3 < 0$ if condition (1.15) holds.

Concerning the transcendental function $g(\lambda)$, let us assume the existence condition for $E_1$, namely $\Lambda\delta \geq \mu_1\mu_3$. Our aim is to prove that the equilibrium $E_1$ is locally stable for any positive time delay $\tau$. Following Rouché's theorem [11], if instability occurs for a particular time delay $\tau$, then the characteristic equation evaluated at $E_1$ must have purely imaginary roots; see, *e.g.*, [12–15] for analogous stability proofs. Accordingly, in the following, we prove that function $g$ does not have any pure imaginary roots. Suppose that $\lambda = \pm bi$ are pure imaginary roots of $g$. Let $m = \mu_1 + \mu_3$ and $n = \mu_1\mu_3$. Therefore

$$g(bi) = -b^2 + mbi + n + (\cos(b\tau) - i\sin(b\tau))\left[\left(\frac{\Lambda\delta}{\mu_3} - m\right)bi + \Lambda\delta - 2n\right].$$

Then,

$$g(bi) = 0 \Leftrightarrow \begin{cases} -b^2 + n + \big(\Lambda\delta - 2n\big)\cos(b\tau) + b\left(\dfrac{\Lambda\delta}{\mu_3} - m\right)\sin(b\tau) = 0 \\ mb + b\left(\dfrac{\Lambda\delta}{\mu_3} - m\right)\cos(b\tau) - \sin(b\tau)\big(\Lambda\delta - 2n\big) = 0. \end{cases}$$

Let $A = \Lambda\delta - 2n$ and $B = \frac{\Lambda\delta}{\mu_3} - m$. Therefore

$$\begin{cases} -b^2 + n = -A\cos(b\tau) - bB\sin(b\tau) \\ mb = -bB\cos(b\tau) + A\sin(b\tau). \end{cases}$$

Squaring both equations, and then summing, we get

$$(-b^2 + n)^2 + (mb)^2 = A^2 + b^2B^2,$$

that is,

$$b^4 + \left(m^2 - 2n - B^2\right) b^2 + n^2 - A^2 = 0.$$

Introducing $m = \mu_1 + \mu_3$ and $n = \mu_1\mu_3$, we obtain

$$b^4 + (\mu_1^2 + \mu_3^2 - B^2)b^2 + \mu_1^2\mu_3^2 - A^2 = 0,$$

which means that $u = b^2$ is a positive root of the polynomial

$$f(u) = u^2 + (\mu_1^2 + \mu_3^2 - B^2)u + \mu_1^2\mu_3^2 - A^2.$$

In view of using the Descarte's rule of signs, we also consider

$$f(-u) = u^2 - (\mu_1^2 + \mu_3^2 - B^2)u + \mu_1^2\mu_3^2 - A^2,$$

and analyze the coefficient of $u$ in $f(-u)$,

$$\begin{aligned}
\mu_1^2 + \mu_3^2 - B^2 &= \mu_1^2 + \mu_3^2 - \left[\frac{\Lambda\delta}{\mu_3} - (\mu_1 + \mu_3)\right]^2 \\
&= -\left(\frac{\Lambda\delta}{\mu_3}\right)^2 + 2\frac{\Lambda\delta}{\mu_3}(\mu_1 + \mu_3) - 2\mu_1\mu_3 \\
&= 2(\Lambda\delta - \mu_1\mu_3) - \frac{\Lambda\delta}{\mu_3^2}(\Lambda\delta - 2\mu_1\mu_3),
\end{aligned}$$

where $\Lambda\delta - \mu_1\mu_3 \geq 0$ whenever $E_1$ exists. We see that if $\Lambda\delta < 2\mu_1\mu_3$, then $\mu_1^2 + \mu_3^2 - B^2 > 0$ and $f(u)$ has one negative root, at least, of the form $u = b^2$, which is a contradiction.

Thus we conclude that $g$ does not have any purely imaginary root, and then $E_1$ is locally stable for any positive time delay $\tau$. The proof is then completed. □

### 1.3.3 Local stability of $E_3$

Consider now the equilibrium state $E_3$ without ISC cells, which has been determined in Theorem 1.2.1 as

$$n_1(t) = \frac{\mu_2}{\gamma}, \quad n_2(t) = \frac{\mu_1\mu_2 - \Lambda\gamma}{\alpha\mu_2}, \quad n_3(t) = 0.$$

The characteristic equation of system (1.2), referring to the equilibrium state $E_3$, is given by

$$\mu_2\left(\lambda^2 + \frac{\Lambda\gamma}{\mu_2}\lambda + \Lambda\gamma - \mu_1\mu_2\right)(-\delta\mu_2 e^{-\lambda\tau} + \gamma\lambda + \gamma\mu_3) = 0. \tag{1.16}$$

**Theorem 1.3.4** (Local stability of the equilibrium state $E_3$). *The equilibrium state $E_3$ is unstable for any time delay $\tau \geq 0$.*

*Proof.* Independently of the delay $\tau$, the characteristic equation (1.16) possesses the roots

$$\lambda_1 = \frac{-\gamma\Lambda + \sqrt{\gamma^2\Lambda^2 + 4\mu_2^2(\mu_1\mu_2 - \gamma\Lambda)}}{2\mu_2}, \qquad \lambda_2 = \frac{-\gamma\Lambda - \sqrt{\gamma^2\Lambda^2 + 4\mu_2^2(\mu_1\mu_2 - \gamma\Lambda)}}{2\mu_2}, \tag{1.17}$$

resulting from the first factor in (1.16). The third eigenvalue, say $\lambda_3$, is also real and is given by the root of the last factor in (1.16) so that $\lambda_3$ depends on the delay $\tau$.

From Theorem 1.2.1, item 4), it is immediate that the discriminant in (1.17) is positive whenever $E_3$ exists. Consequently, independently of the delay $\tau$, the eigenvalues $\lambda_1, \lambda_2$ are real such that

$$\lambda_1 > 0 \quad \text{and} \quad \lambda_2 < 0,$$

and the equilibrium state $E_3$ is unstable whenever it exists for any time delay. □

## 1.4 Global stability of the equilibrium states

The global stability of the equilibrium states is an important topic when studying evolutionary systems. This is a difficult topic, in general, and we limit our analysis in this work to the global stability of the equilibrium state $E_0$; see the first item of Theorem 1.2.1 since this corresponds to the easiest case.

In view of using LaSalle's invariance principle [16], we construct a Lyapunov function associated to our differential system (1.2), for the equilibrium state $E_0$, $V : \mathcal{D} \subset \mathbb{R}^3 \longmapsto \mathbb{R}$, with $\mathcal{D}$ being the interest set given in (1.4). Then, we prove that

- $V\big(n_1(t), n_2(t), n_3(t)\big) \geq 0$, for all $\big(n_1(t), n_2(t), n_3(t)\big) \in \mathcal{D}$;
- $\dot{V}\big(n_1(t), n_2(t), n_3(t)\big) < 0$, for all $\big(n_1(t), n_2(t), n_3(t)\big) \neq E_0$;
- $\dot{V}\Big|_{sys(2)}(n_1(t), n_2(t), n_3(t)) = 0$, if and only if $\big(n_1(t), n_2(t), n_3(t)\big) = E_0$.

We state the global stability of the equilibrium $E_0$ under two assumptions, which prevent the blow up of the solution. Namely, we assume that the density of SAPC cells is bounded from above by the corresponding density of $E_0$, and also that the ratio between the death rates of SRTC cells and ISC cells dominates the ratio of their proliferative rates resulting from their interaction with SAPC cells. In the sequel, we consider the following assumptions, which are in agreement with the biological features of the model:

$$n_1(t) \leq \frac{\Lambda}{\mu_1}, \qquad \frac{\mu_2}{\mu_3} > \frac{\gamma}{\delta},$$

and our result is stated in the following theorem:

**Theorem 1.4.1** (Global stability of the equilibrium state $E_0$). *The equilibrium state $E_0$ is globally asymptotically stable if* $\dfrac{\delta\Lambda}{\mu_1\mu_3} < 1$.

*Proof.* Consider the continuously differentiable function $V : \mathcal{D} \subset \mathbb{R}^3 \longmapsto \mathbb{R}$ defined by

$$V\big(n_1(t), n_2(t), n_3(t)\big) = n_2(t) + n_3(t). \tag{1.18}$$

It is immediate that

$$V\big(n_1(t), n_2(t), n_3(t)\big) \geq 0, \text{ for all } \big(n_1(t), n_2(t), n_3(t)\big) \in \mathcal{D}.$$

Along the solutions of system (1.2), we have

$$\begin{aligned}\dot{V}(t) &= \Big[\gamma n_2(t)n_1(t) - \rho n_2(t)n_3(t) - \mu_2 n_2(t)\Big] + \Big[\delta n_3(t-\tau)n_1(t-\tau) - \mu_3 n_3(t)\Big] \\ &= \gamma n_2(t)n_1(t) - \mu_2 n_2(t) + \delta n_3(t-\tau)n_1(t-\tau) - \mu_3 n_3(t) - \rho n_2(t)n_3(t).\end{aligned}$$

For $t$ large enough, we have $n_3(t) = n_3(t-\tau)$. Using that $n_1(t) \leq \dfrac{\Lambda}{\mu_1} = \bar{n}_1$, we get

$$\begin{aligned}\dot{V}(t) &\leq \gamma n_2(t)\bar{n}_1 - \mu_2 n_2(t) + \delta n_3(t)\bar{n}_1 - \mu_3 n_3(t) - \rho n_2(t)n_3(t) \\ &\leq \gamma n_2(t)\bar{n}_1 - \mu_2 n_2(t) + \delta n_3(t)\bar{n}_1 - \mu_3 n_3(t) \\ &= \mu_2\left(\frac{\gamma \bar{n}_1}{\mu_2} - 1\right)n_2(t) + \mu_3\left(\frac{\delta \bar{n}_1}{\mu_3} - 1\right)n_3(t).\end{aligned}$$

Since $\gamma/\delta < \mu_2/\mu_3$, also $\gamma/\mu_2 < \delta/\mu_3$, we obtain

$$\dot{V}(t) \leq \mu_2\left(\frac{\delta \bar{n}_1}{\mu_3} - 1\right)n_2(t) + \mu_3\left(\frac{\delta \bar{n}_1}{\mu_3} - 1\right)n_3(t). \tag{1.19}$$

From condition (1.19), it follows that $\dot{V}(t) < 0$ whenever $\dfrac{\delta \bar{n}_1}{\mu_3} < 1$. Therefore $\dot{V}$ is negative definite along the trajectories of the solution to our system (1.2). Furthermore, it follows from (1.19) that

$$\dot{V}\big(n_1(t), n_2(t), n_3(t)\big) = 0, \text{ if and only if } n_2(t) = n_3(t) = 0.$$

Substituting $n_2(t) = n_3(t) = 0$ into our system (1.2), we get the linear differential equation

$$\dot{n_1}(t) = \Lambda - \mu_1 n_1(t).$$

Since $n_1(t) \leq \Lambda/\mu_1$, we conclude that $n_1(t)$ is an increasing function. Solving the above differential equation, we obtain its solution in the form

$$n_1(t) = e^{-\mu_1 t}\left(n_1(0) - \frac{\Lambda}{\mu_1}\right) + \frac{\Lambda}{\mu_1}$$

so that

$$\lim_{t\to+\infty} n_1(t) = \frac{\Lambda}{\mu_1}.$$

Therefore

$$\dot{V}\big(n_1(t), n_2(t), n_3(t)\big) = 0 \text{ if and only if } \big(n_1(t), n_2(t), n_3(t)\big) = E_0.$$

Thus $V$ is a Lyapunov function, and the largest compact invariant set in

$$\big\{\big(n_1(t), n_2(t), n_3(t)\big) \in \mathcal{D}\colon \dot{V} = 0\big\}$$

is the singleton $\{E_0\}$ since every solution of (1.2) in $\mathcal{D}$ converges towards $E_0$ as $t \to \infty$. Therefore $E_0$ is the unique accumulation point of the trajectories of the solution to our system (1.2), and by LaSalle's invariance principle [16], the equilibrium point $E_0$ is globally asymptotically stable whenever $\dfrac{\delta\Lambda}{\mu_1\mu_3} < 1$. □

## 1.5 Numerical results

In this section, we perform some numerical simulations in view of investigating both the impact of the time delay in the behavior of the solution and the asymptotic trend of the equilibria determined in the previous sections. To this end, we determine the solution to both systems with and without delay, namely systems (1.2) and (1.1), respectively, and compare the corresponding results. We have integrated the ODE systems (1.1) and (1.2) using the ode45 solver of the computing software *Matlab*®; see [17]. The results will be interpreted in light with the theoretical results of Sections 1.3 and 1.4.

We assume the following initial data:

$$\big(n_1^0, n_2^0, n_3^0\big) = (0.2, 0.2, 0.2)$$

and two different time delays, namely $\tau \in \{0.4, 0.8\}$. Concerning the input data, we assume the values for the parameters specified in the following two study cases:

**Case 1.1.** Convergence of the solution to the equilibrium state $E_0$. We consider

$$\Lambda = 0.119,\ \alpha = 0.21,\ \beta = 2,\ \gamma = 0.01,\ \rho = 0.025,\ \delta = 3,\ \mu_1 = 0.5,\ \mu_2 = 0.5,\ \mu_3 = 0.75.$$

In this case, we have

$$E_0 = (0.238, 0, 0)$$

and

$$\Lambda\gamma - \mu_1\mu_2 = -0.24881 < 0, \qquad \Lambda\delta - \mu_1\mu_3 = -0.018 < 0.$$

According to Theorem 1.3.2, the equilibrium state $E_0$ is locally asymptotically stable for any delay $\tau \geq 0$, and the solutions of systems (1.1) and (1.2) converge to $E_0$.

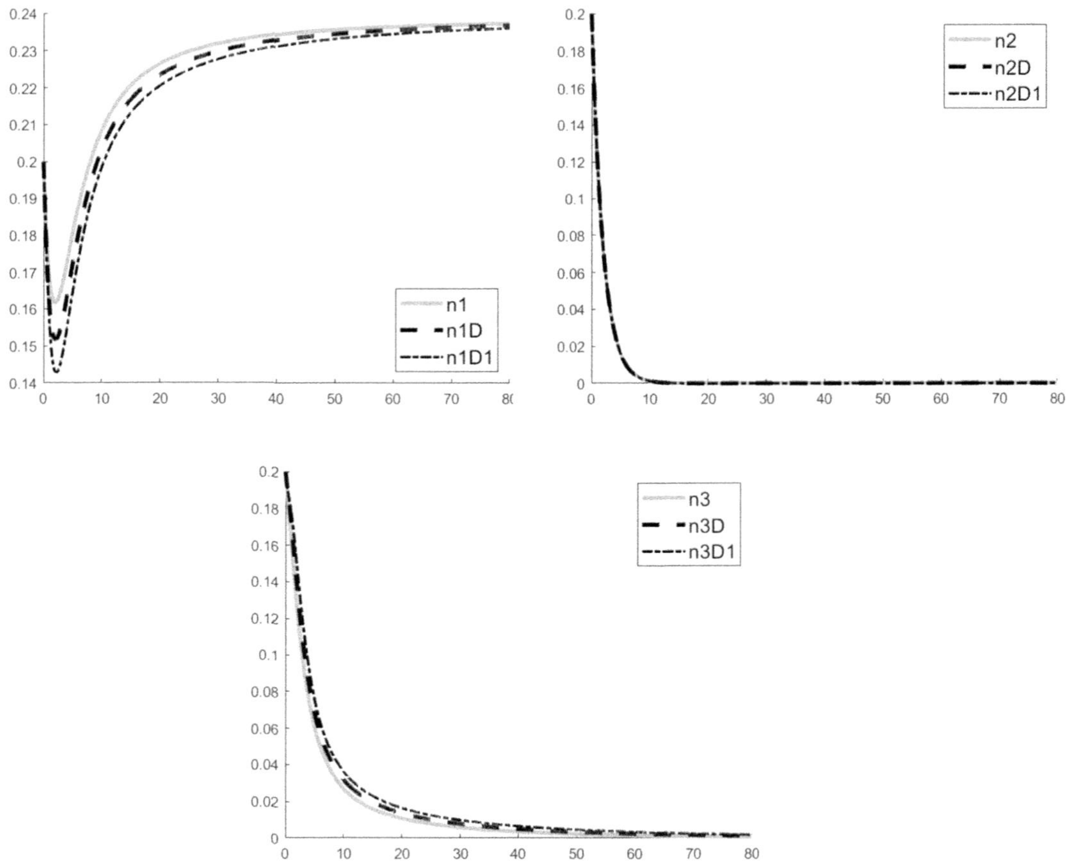


**FIGURE 1.1** Solution to system (1.1) without delay (full line) and to system (1.2) with delay $\tau = 0.4$ (D-dashed line); $\tau = 0.8$ (D1-dotted line). Dynamics of populations $n_1$, $n_2$, and $n_3$ showing the asymptotic trend to equilibrium $E_0$.

The results of our simulations are shown in Fig. 1.1. We can see that the effects of the delay are visible in the evolution of $n_1$ and $n_3$, whereas they are not detectable in the evolution of $n_2$. This is due to the fact that $n_1$ and $n_3$ depend directly on the delay $\tau$, and $n_2$ does not.

**Case 1.2.** Convergence of the solution to the equilibrium state $E_1$. We consider

$$\Lambda = 0.1,\ \alpha = 0.21,\ \beta = 0.3,\ \gamma = 0.21,\ \rho = 0.025,\ \delta = 0.35,\ \mu_1 = 0.1,\ \mu_2 = 0.2,\ \mu_3 = 0.21.$$

In this case, we have

$$\frac{\mu_3}{\delta} = 0.6, \qquad \frac{\Lambda\delta - \mu_1\mu_3}{\beta\mu_3} = 0.222,$$

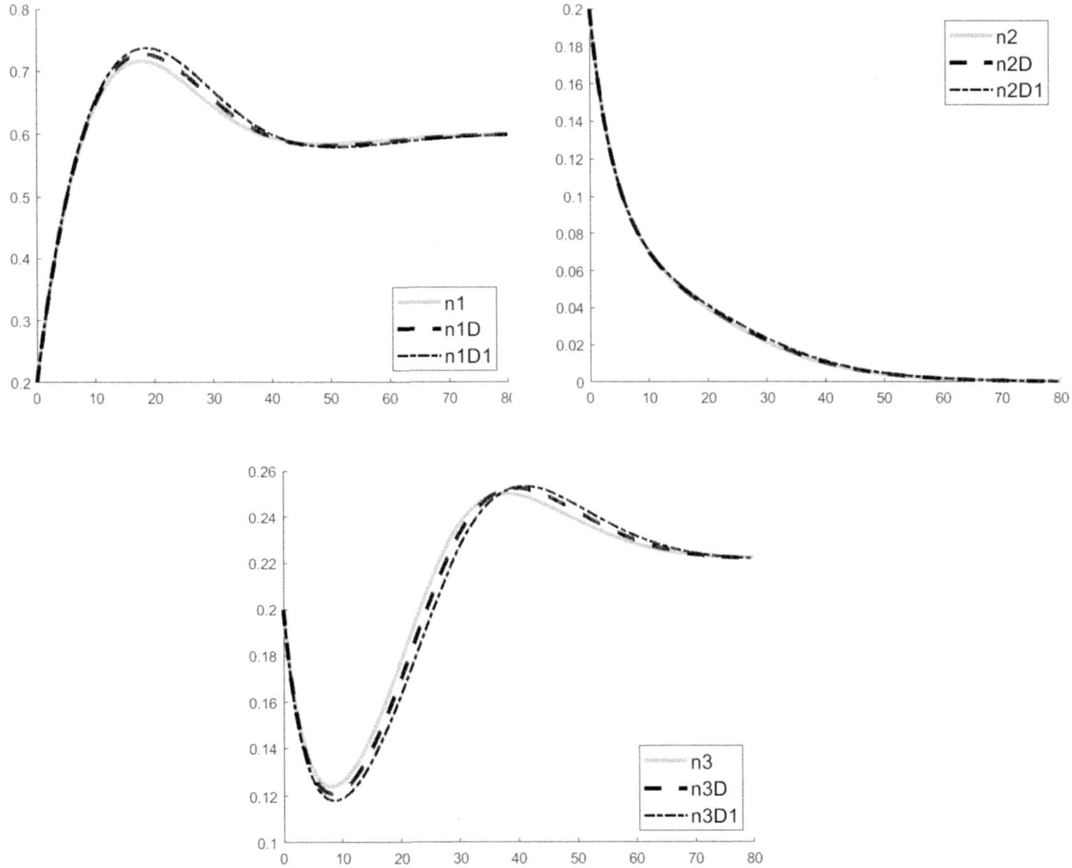


**FIGURE 1.2** Solution to system (1.1) without delay (full line) and to system (1.2) with delay $\tau = 0.4$ (D-dashed line); $\tau = 0.8$ (D1-dotted line). Dynamics of populations $n_1$, $n_2$, and $n_3$ showing the asymptotic trend to equilibrium $E_1$.

and the equilibrium state $E_1$ is given by

$$E_1 = (0.6, 0, 0.222).$$

Moreover, we have

$$\Lambda\delta - 2\mu_1\mu_3 = -0.07 < 0, \quad \Lambda - \frac{\beta\mu_3}{\rho\delta^2}(\gamma\mu_3 - \delta\mu_2) + \frac{\mu_1\mu_3}{\delta} = 5.37 > 0.$$

According to Theorem 1.3.3, the equilibrium state $E_1$ is locally asymptotically stable in both models with and without delay, and the solutions of systems (1.1) and (1.2) approach to $E_1$. The results are shown in Fig. 1.2. Again, the effects of the delay are visible in the evolution of $n_1$ and $n_3$, whereas they are almost undetectable in the evolution of $n_2$.

## 1.6 Summary and perspectives

In this work, we consider an existing macroscopic model for autoimmune diseases and introduce a discrete time delay in the dynamics. We determine the equilibrium states of the new model and study their local stability. Then we also investigate the global stability of a biologically meaningful disease-free equilibrium state. We perform some numerical simulations to estimate how the solution is influenced by the delay.

Even if the new model introduces the delay only in the terms describing the interactions between SAPCs and ISCs, the simulations show that the solution is affected by the delay, confirming the importance of studying the time-delayed generalization model.

Therefore the delayed model proposed in this work provides a rather comprehensive explanation for the active immune regulation, which can have a long time effect in controlling the development of autoimmunity. In particular, the delay describes the cumulative duration of the viral attachment and the interaction between the ISCs and SAPCs.

The work developed in this article did not exhaust the subject here initiated; the stability of the unique equilibrium state of co-existence of all populations cells was not studied and the global stability had been studied only for a disease-free equilibrium state. These challenging problems will be considered in a future work.

Another problem that we plan to address is the development and analysis of an extended version of the model considered here, possibly with more than one time delay. We plan to consider a different time delay for the interactions involving SRTCs. In this case, the delay can describe the lag that the self-reactive T cells take to activate their memory. More specifically, there are two types of self-reactive T cells, namely naïve cells and cells with memory. An important characteristic of the naïve self-reactive T cells response is the formation of immunological memory after initial antigen exposure, which helps the immune system learn with experience. After the initial interaction, naïve T cells undergo clonal expansion to become activated T cells with memory. See more details in [18]. However, according to [19] and references therein, memory T cells show a delayed recruitment to the infected site. The new idea is to consider this time delay in our system and investigate its effects on the local and global behavior of the solution.

## Acknowledgments

This work is part of L. Boudjellal first-year PhD program, under the support of CMAT PhD fellowship UMINHO/BID/2021/116.

It is also partially supported by CoSysM3–2022.03091.PTDC from national funds (OE) through FCT/MCTES (https://doi.org/10.54499/2022.03091.PTDC) and by the Portuguese FCT Projects UIDB/00013/2020, UIDP/00013/2020 of CMAT-UM, UIDB/04106/2020, UIDP/04106/2020 of CIDMA-UA.

The authors acknowledge the fruitful discussions with Professor Ana Jacinta Soares, from the University of Minho, Portugal.

# References

[1] G. Mondanelli, A. Iacono, A. Carvalho, et al., Amino acid metabolism as drug target in autoimmune diseases, Autoimmunity Reviews 18 (2019) 334–348.

[2] M. Delitala, U. Dianzani, T. Lorenzi, M. Melensi, A mathematical model for immune and autoimmune response mediated by T-cells, Computers & Mathematics with Applications 66 (2013) 1010–1023.

[3] M.P. Ramos, C. Ribeiro, A.J. Soares, A kinetic model of T cell autoreactivity in autoimmune diseases, Journal of Mathematical Biology 79 (2019) 2005–2031.

[4] J. Oliveira, M.P. Ramos, C. Ribeiro, A.J. Soares, Equilibrium of a kinetic model of oscillating patterns for chronic autoimmune diseases, Kinetic and Related Models 17 (6) (2024) 971–992, https://www.aimsciences.org/article/id/65e6e9703cb6602db4912013.

[5] M.F.P. Costa, M.P. Machado Ramos, C. Ribeiro, A.J. Soares, Optimal control model of immunotherapy for autoimmune diseases, Mathematical Methods in the Applied Sciences 44 (2021) 8883–8902.

[6] W. Zhang, L.M. Wahl, P. Yu, Modeling and analysis of recurrent autoimmune disease, SIAM Journal on Applied Mathematics 74 (2014) 1998–2025.

[7] R. Della Marca, M.P. Ramos, C. Ribeiro, A.J. Soares, Mathematical modelling of oscillating patterns for chronic autoimmune diseases, Mathematical Methods in the Applied Sciences 45 (2022) 7144–7161.

[8] H. Smith, An Introduction to Delay Differential Equations with Applications to the Life Sciences, Texts in Applied Mathematics, vol. 57, Springer, New York, 2011.

[9] A.P. Lemos Paião, Optimal Control and Mathematical Models in Epidemiology, Ph.D. Thesis, University of Aveiro, Portugal, 2019.

[10] K. Gu, V.L. Kharitonov, J. Chen, Stability of Time Delays Systems, Control Engineering Series, Birkhäuser, 2003.

[11] J. Dieudonné, Foundations of Modern Analysis, Academic Press, New York, 1960.

[12] B. Buonomo, M. Cerasuolo, The effect of time delay in plant-pathogen interactions with host demography, Mathematical Biosciences and Engineering 12 (2015) 473–490.

[13] R.V. Culshaw, S. Ruan, A delay-differential equation model of HIV infection of CD4+T-cells, Mathematical Biosciences 165 (2000) 27–39.

[14] F. Rodrigues, C.J. Silva, D.F.M. Torres, H. Maurer, Optimal control of a delayed HIV model, Discrete and Continuous Dynamical Systems. Series B 23 (1) (2018) 443–458.

[15] Cristiana J. Silva, Delfim F.M. Torres, Global stability for a HIV/AIDS model, Communications Faculty of Sciences University of Ankara. Series A1: Mathematics and Statistics 1 (67) (2018) 93–101.

[16] J. LaSalle, Some extensions of Liapunov's second method, I.R.E. Transactions on Circuit Theory 7 (4) (1960) 520–527.

[17] MATLAB, The MathWords website, https://www.mathworks.com/products/matlab.html. (Accessed October 2022).

[18] P. Devarajan, H. Chen, Autoimmune effector memory T cells: the bad and the good, Immunologic Research 57 (2013) 1–3.

[19] K. Steinbach, I. Vincenti, D. Merkler, Resident-memory T cells in tissue-restricted immune responses: for better or worse?, Frontiers in Immunology 9 (2018) 1–15.

2

# Alternative food for predators: elucidating their impact on a predation model with Allee effect on prey

Nicole Martínez-Jeraldo[a], Elizabeth Rozas-Torres[b], Eduardo González-Olivares[c], and Paulo C. Tintinago-Ruiz[d]

[a]Doctorado en Modelamiento Matemático Aplicado, Universidad Católica del Maule, Talca, Chile [b]Centro de Docencia de Ciencias Básicas para Ingeniería, Facultad de Ciencias de la Ingeniería, Universidad Austral de Chile, Valdivia, Chile [c]Pontificia Universidad Católica de Valparaíso, Valparaíso, Chile [d]Universidad del Quindío, Armenia, Colombia

## 2.1 Introduction

Due to its universal existence, the interaction between predators and their prey has been and will continue to be one of the dominant scopes in Theoretical Ecology [6]. Moreover, the understanding of this important relationship allows a better study and to analyze the behavior of food chains or trophic webs.

To respond to the criticism raised against the Lotka–Volterra model, Patrick Holt Leslie formulated in 1948 a new model [29], not conforming to the framework of the Lotka–Volterra model [40]. The equation for predators is an extension of the well-known logistic equation, assuming implicitly the competition among predators [30].

Leslie assumed that the predator would follow a standard logistic growth equation, but the carrying capacity would be determined by the prey density [26]. The Leslie–Gower model maintains the prey equation from the Volterra model [40], where the self-interference among the prey is assumed; besides, there exists a slanted predator isocline in the model [40].

Then, the Leslie predator-prey model is described by an autonomous bidimensional differential equation system, characterized by the following:

**i)** the *functional response* or *predator consumption rate* is linear [34,40], and

**ii)** the equation for the predators is a logistic-type growth function [1,40], where the habitual environmental carrying capacity for predators $K_y$ is proportional to prey abundance $x = x(t)$, that is, $K_y = K(x) = nx$ [34,40]. Because the two equations for the rate

Mathematical Modeling in Bioscience. https://doi.org/10.1016/B978-0-44-315445-4.00008-1

Copyright © 2025 Elsevier Inc. All rights are reserved, including those for text and data mining, AI training, and similar technologies.

growth of both populations are based in the logistic function, the model is also called as *logistic model* [40].

If $y = y(t)$ represents the predator population size in the logistic predation model, the quotient $\frac{y}{nx}$ is called the *Leslie–Gower term*, measuring the loss in the predator population due to the rarity (per capita $\frac{y}{x}$) of its favorite food [24].

In the case of severe shortage, some predators can switch over to other prey population, but its growth would be limited by the fact that its favorite food (represented by $x$) is not available in abundance [24].

This ability can be taken care of by adding a positive constant $c$ to the environmental carrying capacity for predators, being now described by $K(x) = nx + c$; in that case, it is said that the model is represented by a *Leslie–Gower scheme*, also known as *modified Leslie–Gower model* [24]: if $x = 0$, then $K(0) = c$; the latter indicates that the predators are generalist since they search for an alternative food, avoiding their extinction [24].

**The Allee effect**

A population is said to be under an Allee effect if the growth rate per capita is initially an increasing function, then decreases to zero at a higher density [15,38,39]. At large population sizes, this rate is negative as it happens in the logistic equation for all population sizes [5,22,31]. This is why for some populations, the growth rate is not always positive for small population density, and it may not be decreasing as suggested by the logistic model [32].

When a population is affected by an Allee effect, there exists a positive relationship between population density and the individual's fitness [8]. It is an ecological phenomenon that occurs at low population densities, where the per capita growth rate is a growing function of population abundance [27,38,39].

This phenomenon has received different names, such as *depensation* in Fisheries Science [11,31], or *negative competition effect, inverse density dependence* or *positive density dependence* in Population Dynamics [31].

To describe mathematically the Allee effect, we employ the most common form to express the growth rate of a population [4,11,13,22,27]; this is depicted by the cubic polynomial differential equation

$$\frac{dx}{dt} = r\left(1 - \frac{x}{K}\right)(x - m)\,x. \tag{2.1}$$

Although the parameter $m$ has the same measurement unit as $x$ the population size, when positive, Eq. (2.1) represents an Allee effect also when $m \leq 0$. That is, the Allee effect exists in a population for all values of the parameter $m$, when the per capita growth rate $\frac{1}{x}\frac{dx}{dt}$ is a growing function of population size [13,15,21]. This is fulfilled if and only if $-K < m \ll K$, with $K > 0$. We note if $m < -K$, the Allee effect disappears.

In Fig. 2.1, we show the marked difference between standard logistic growth and population growth incorporating Allee represented by Eq. (2.1).

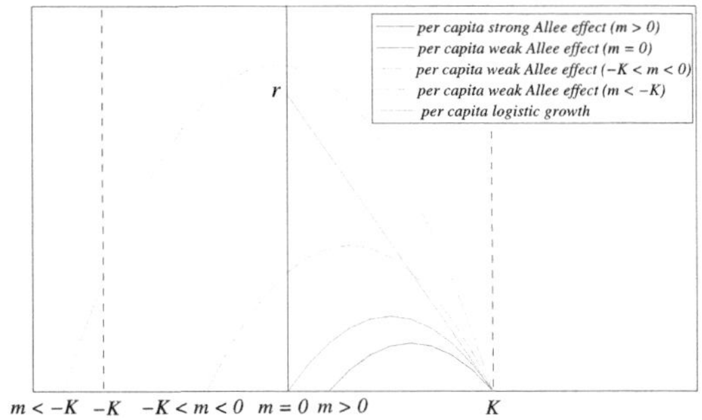


**FIGURE 2.1** Graph of the relation of the per capita growth rate $\frac{1}{x}\frac{dx}{dt}$ and the population size, comparing the corresponding to the logistic growth (in purpure [dark gray in print version]) and Eq. (2.1) for different values of the parameter $m$.

When $m > 0$, the population growth rate decreases if the population size is below the *threshold level* $m$ and the population goes to extinction. In this case, Eq. (2.1) describes a *strong Allee effect* [41,43,44].

If $m \leq 0$, it is said that the population is affected by a *weak Allee effect* [11,43]. In Fisheries Sciences, the same phenomena are called *critical* and *pure depensation,* respectively [11,31,43].

This effect can be due to a wide range of challenging biological phenomena, such as inbreeding depression, food exploitation, mate finding, and social dysfunction, predator avoidance, or defense (see Table 1 in [5] or Table 2.1 in [14]). The foregoing are important in some populations as it relates to their conservation and extinction problems; these are modeled by many algebraic forms [21,32], such as the following:

**i)** $\frac{dx}{dt} = r\left(1 - \frac{x}{K}\right)x - \frac{px}{x+b}$ proposed in [15,38];
**ii)** $\frac{dx}{dt} = r\left(1 - \frac{x}{K}\right)\left(x - \frac{m+b}{x+b}\right)x$ proposed in [7,8];
**iii)** $\frac{dx}{dt} = r\left(1 - \frac{x}{K}\right)\frac{px}{x+b} - Dx$ proposed in [42,46];
**iv)** $\frac{dx}{dt} = r\left(1 - \frac{x}{K}\right)x - bx\exp(-cx)$ proposed in [37].

Nonetheless, it is possible to prove that many of them are topologically equivalent [21]. Besides, different forms may produce changes in the dynamics of each model, particularly in the number of limit cycles surrounding a positive equilibrium point in predator-prey models [1,33].

Furthermore, two or more Allee effects produced by different causes can generate mechanisms that act simultaneously in a single population (see Table 2 on [5] or Table 2.2 on [14]). The combined influence of some of these phenomena is called *multiple Allee effect* [5].

### Limit cycles

We will focus our attention in describing the dynamics of the system, particularly the different bifurcations that it produces. A special interest is to establish the number of limit cycles originated by multiple Hopf bifurcations [10,16,28], also named Bautin bifurcation [18,19].

This is not an easy task, and it is related to solving the well-known Hilbert 16th problem (proposed by the German mathematician David Hilbert in 1900); it refers to the maximum number and relative position of limit cycles on bidimensional differential equation systems [12,18]. Yet, this question remains unsolved in the case of a polynomial differential equation system, which degree must be less than or equal to $p \in \mathbb{N}$ with $p \geq 2$ [18].

In particular, for predator-prey models, this problem also remains unanswered. This question starts with the seminal work by Kuo-Shung Cheng in 1981 [9], who was the first to prove the uniqueness of a limit cycle for a specific Gause-type predator-prey model [17], considering a Holling type II functional response [12]. So, one of our main goals involves describing the dynamical system and establishing the number of limit cycles exhibited by the proposed model. We will try to answer this problem using the method of calculations of the Lyapunov quantities [10], applied to a polynomial system, topologically equivalent to the original one.

Our analysis will have in mind the model in which the Allee effect is absent [24], where the prey growth rate is the logistic type and predators have an alternative food, for which the unique positive equilibrium point is globally asymptotically stable [24]. Some results obtained in [2] are also deepen.

We consider the determination of the number of limit cycles bifurcating from a weak focus; it is a good criterion for the classification of predator-prey models.

This chapter is organized as follows: In Section 2.2, we present the modified Leslie–Gower type model and a topologically equivalent system, which is obtained by means of a diffeomorphism; besides, the number of the equilibrium points is determined. In Section 2.3, the main dynamical properties of the model are established, particularly the nature of the equilibrium points and the number of the limit cycles. In Section 2.4, numerical simulations are presented to reinforce the results obtained in previous sections. Ecological consequences and a comparative study of the mathematical results are given in Section 2.5.

## 2.2 The model

The modified model Leslie–Gower [24], considering Allee effect, is given by the following bidimensional differential equations system of Kolmogorov type [17,20]:

$$X_\nu (x, y) : \begin{cases} \frac{dx}{dt} = \left(r \left(1 - \frac{x}{K}\right)(x - m) - qy\right) x \\ \frac{dy}{dt} = s\left(1 - \frac{y}{nx+c}\right) y \end{cases}, \tag{2.2}$$

where $x = x(t)$ and $y = y(t)$ indicate the prey and predator population sizes, respectively, for $t \geq 0$, measure as number of individuals, biomass, density for unit of area or volume, with $\nu = (r, K, q, s, n, c, m) \in \mathbb{R}_+^5 \times (\mathbb{R}_+ \cup \{0\}) \times ]-K, K[$, having the following meanings:

$r$ is the intrinsic prey growth rate or biotic potential;
$K$ is the prey environmental carrying capacity;
$q$ is the maximum number of prey that can be eaten by a predator in each time unit (satiation rate);
$s$ is the intrinsic predator growth rate;
$n$ is a measure of the food quality, indicating how the predators turn eaten prey into new predator births;
$c$ is the amount of alternative food available for predators.

This last parameter indicates that the predator is generalist, and that if an available prey does not exist, it has a source of alternative food. Clearly, if $c = 0$, the predator carrying capacity is $K(x) = nx$ and the Leslie–Gower model is not defined in $x = 0$.

The term $g(x) = -\frac{y^2}{K(x)}$ could be thought of as modeling intraspecific competition among the predators for the prey [26]. Other meaning has been given for the parameter $c > 0$, indicating that it measures the extent to which environment provides protection to prey [24].

System (2.2) or vector field $X_\nu(x, y)$ is defined in

$$\Omega = \left\{(x, y) \in \mathbb{R}^2 | \, x \geq 0, \; y \geq 0\right\} = \mathbb{R}_0^+ \times \mathbb{R}_0^+.$$

The equilibrium points of the system are $(0, 0)$, $(K, 0)$, $(0, c)$, $(m, 0)$ when $m > 0$, and $(x_e, y_e)$, satisfying the equation of the isoclines

$$y = nx + c \text{ and } y = \frac{r}{q}\left(1 - \frac{x}{K}\right)(x - m).$$

Clearly, $(x_e, y_e)$ can be a positive equilibrium point (or at interior of the first quadrant) or cannot exists there, depending on the sign of factor $1 - \frac{x}{K}$.

Non-autonomous versions of system (2.2) have been studied by incorporating delay, impulses or partial derivatives. In spite of these recent studies, the dynamics of system (2.2) has not been fully analyzed, and some existing results are not true or well understood.

### 2.2.1 Topological equivalence

To simplify the calculations, we follow the methodology used in [22,25], which is described in the following lemma:

**Lemma 2.1.** *System* (2.2) *is topologically equivalent to the Kolmogorov system [17,20]:*

$$Y_\eta(u, v): \begin{cases} \frac{dx}{dt} = ((1 - u)(u - M) - Qv)(u + C)u \\ \frac{dy}{dt} = S(u + C - v)v \end{cases} \tag{2.3}$$

*where* $\eta = (Q, C, S, M) \in \mathbb{R}_+^3 \times \,]-1, 1[$, *with* $Q = \frac{qn}{r}$, $C = \frac{c}{nK}$, $S = \frac{s}{rK}$ *and* $M = \frac{m}{K}$.

*Proof.* Using the change of variables given by $x = Ku$ and $y = nKv$, replacing in system (2.2) and after simplification, we obtain the new vector field

$$U_\nu(u,v):\begin{cases} K\frac{du}{dt} &= (r\ (1-u)(Ku-m)\ -qnKv)\ Ku \\ nK\frac{dv}{dt} &= s\left(1-\frac{nKv}{nKu+c}\right)nKv \end{cases}.$$

Factoring yields

$$U_\nu(u,v):\begin{cases} \frac{du}{dt} &= rK\left((1-u)\left(u-\frac{m}{K}\right)-\frac{qn}{r}v\right)u \\ \frac{dv}{dt} &= s\left(1-\frac{v}{u+\frac{c}{nK}}\right)v \end{cases}.$$

By means of the time rescaling given by $\tau = \frac{rK}{u+\frac{c}{nK}}t$, and using the chain rule, we get $\frac{du}{dt} = \frac{du}{d\tau}\frac{d\tau}{dt} = \frac{du}{d\tau}\frac{rK}{u+\frac{c}{nK}}$ and $\frac{dv}{dt} = \frac{dv}{d\tau}\frac{rK}{u+\frac{c}{nK}}$.

Then, we have the system

$$V_\nu(u,v):\begin{cases} \frac{rK}{u+\frac{c}{nK}}\frac{du}{d\tau} &= rK\left((1-u)\left(u-\frac{m}{K}\right)-\frac{qn}{r}v\right)u \\ \frac{rK}{u+\frac{c}{nK}}\frac{dv}{d\tau} &= s\left(1-\frac{v}{u+\frac{c}{nK}}\right)v \end{cases},$$

and we get

$$V_\nu(u,v):\begin{cases} \frac{du}{d\tau} &= \left((1-u)\left(u-\frac{m}{K}\right)-\frac{qn}{r}v\right)\left(u+\frac{c}{nK}\right)u \\ \frac{dv}{d\tau} &= \frac{s}{rK}\left(u+\frac{c}{nK}-v\right)v \end{cases}$$

Substituting the quotients by $Q$, $C$, $S$, and $M$, system (2.2) is obtained. □

**Remark 2.2.** We note that a diffeomorphism [10] has been constructed, with $\varphi : \tilde{\Omega}\times\mathbb{R} \longrightarrow \Omega\times\mathbb{R}$ such that

$$\varphi(u,v,\tau) = \left(Ku, nKv, \frac{u+\frac{c}{nK}}{rK}\tau\right) = (x,y,t),$$

where $\tilde{\Omega} = \left\{(u,v)\in\mathbb{R}^2 \mid u \geq 0,\ v \geq 0\right\}$.

As

$$\det D\varphi(u,v,\tau) = \det\begin{pmatrix} K & 0 & 0 \\ 0 & nK & 0 \\ \frac{1}{rK} & 0 & \frac{1}{rK}\left(u+\frac{c}{nK}\right) \end{pmatrix} = \frac{nK}{r}\left(u+\frac{c}{nK}\right) > 0.$$

Then, $\varphi$ is a diffeomorphism preserving time orientation; the vector field $X_\nu(x,y)$ in the new system of coordinates is topologically equivalent to the vector field $Y_\eta(u,v) = \varphi\circ X_\nu(x,y)$, which has the form $Y_\eta(u,v) = P(u,v)\frac{\partial}{\partial u} + Q(u,v)\frac{\partial}{\partial v}$ [16], being $P(u,v)$ and $Q(u,v)$ the right side of the equations in system (2.2). Vector field $Y_\eta$ is a continuous extension of system (2.1) at $\tilde{\Omega}$.

Our analysis will be made considering $M > 0$, $M = 0$ and $M < 0$.

## 2.2.2 Equilibrium points

The equilibrium points of system (2.2) or singularities of vector field $Y_\eta(u, v)$ are $(1, 0)$, $(0, 0)$, $(M, 0)$, when $M > 0$, $(0, C)$, when $C > 0$, and the points $(u_e, v_e)$ are in the intersection of isoclines curves $v = \frac{1}{Q}(1 - u)(u - M)$ and $v = u + C$.

Then, the abscissa $u_e$ of the positive equilibrium points is solution of the second degree equation:

$$p(u) = u^2 - (1 + M - Q)u + (CQ + M) = 0. \tag{2.4}$$

Considering the Descartes signs rule, and according to the sign of the coefficient $a_1 = 1 + M - Q$, Eq. (2.4) has two, one, or no positive roots.

The coefficient $a_0 = CQ + M > 0$ if and only if $M \geq 0$. But $a_0$ can be negative if $M < 0$ and $CQ < -M$.

**Case 2.1.** Supposing $M > 0$.

**1a)** Assuming $a_1 > 0$, then Eq. (2.4) has

**1a1.** two positive real roots if and only if $\Delta = (1 + M - Q)^2 - 4(CQ + M) > 0$, which are

$$u_1 = \tfrac{1}{2}\left((1 + M - Q) - \sqrt{\Delta}\right) \text{ and } u_2 = \tfrac{1}{2}\left((1 + M - Q) + \sqrt{\Delta}\right),$$

with $u_1 < u_2$.

**1a2.** one positive real roots if and only if $\Delta = 0$, which is given by

$$u^* = \frac{1}{2}(1 + M - Q).$$

**1a3.** no positive real roots if and only if $\Delta < 0$.

**1b)** no positive real roots if and only if $1 + M - Q \leq 0$.

Fig. 2.2 shows the case in which there exist six equilibrium points in (2.3).

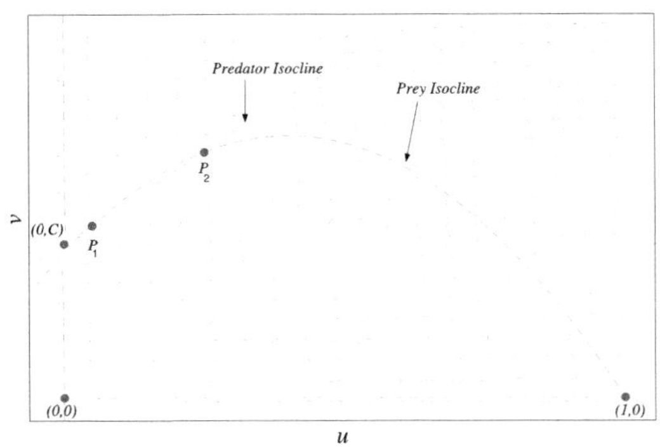


**FIGURE 2.2** For $1 - M - Q > 0$, when $M > 0$, the intersection of isoclines in system (2.3) can determine up six equilibrium points. When $M \leq 0$, there exist five equilibrium points.

**Case 2.2.** Supposing $M = 0$.

The equilibrium points of system (2.3) are $(1, 0)$, $(0, 0)$, $(0, C)$ (the point $(M, 0)$ coincides with $(0, 0)$) and the points $(u_e, v_e)$, with $u_e$ satisfying Eq. (2.4), which reduces to

$$p(u) = u^2 - (1 - Q)u + CQ = 0. \tag{2.5}$$

For this equation it has the following cases:

**2a)** Supposing $b_1 = 1 - Q > 0$, then Eq. (2.5) has

**2a1.** two positive real roots if and only if $\Delta_0 = (1 - Q)^2 - 4CQ > 0$, i.e. $C < \frac{(1-Q)^2}{4Q}$, which are

$$u_3 = \tfrac{1}{2}\left((1 - Q) - \sqrt{\Delta}\right) \text{ and } u_4 = \tfrac{1}{2}\left((1 - Q) + \sqrt{\Delta}\right),$$

with $u_3 < u_4$.

**2a2.** one positive real roots if and only if $\Delta_0 = 0$, which is given by

$$u^{**} = \frac{1}{2}(1 - Q).$$

**2a3.** no positive real roots if and only if $\Delta_0 < 0$.

**2b)** no positive real roots if and only if $b_1 = 1 - Q \leq 0$.

**Case 2.3.** Supposing $M < 0$. In this case $(M, 0)$ is in the second quadrant and new possibilities arise for the positive equilibria.

**3.1)** When $CQ + M > 0$, it has the same possibilities than Case 2.1, i.e., there are two, one, or no positive real roots.

**3.2)** When $CQ + M = 0$, Eq. (2.4) becomes

$$p(u) = \left(u^2 - (1 + M - Q)\right)u = 0.$$

Then,

**3.2i.** If $c_1 = 1 + M - Q > 0$, the unique positive real root is $u_5 = 1 + M - Q$, (the solution $u = 0$ is repeated).

**3.2ii.** If $1 + M - Q \leq 0$, there exists no positive real root.

**3.3)** When $CQ + M < 0$,

**3.3i.** If $1 + M - Q \neq 0$, let the sign of $1 + M - Q$, Eq. (2.4), has a unique positive real root

$$u_6 = \tfrac{1}{2}\left((1 + M - Q) + \sqrt{\Delta}\right).$$

**3.3ii.** If $1 + M - Q = 0$, Eq. (2.4) becomes

$$p(u) = u^2 + (CQ + M) = 0,$$

which has a unique positive real root, given by $u_7 = \sqrt{(CQ + M)}$, with $CQ + M < 0$. If $CQ + M \geq 0$, there no exists a positive real root.

All the above cases can be described by Table 2.1 with $\Delta = (1 + M - Q)^2 - 4(CQ + M)$ and $\nu(\mathbb{R}_+)$, the number of the positive real roots of system (2.3).

**Table 2.1** Number of positive real roots of Eq. (2.2).

| Case | $M$ | $1+M-Q$ | $CQ+M$ | $\Delta$ | $\nu(\mathbb{R}_+)$ |
|---|---|---|---|---|---|
| 1a1 | $+$ | $+$ | $+$ | $+$ | 2 |
| 1a2 | $+$ | $+$ | $+$ | 0 | 1 |
| 1a3 | $+$ | $+$ | $+$ | $-$ | 0 |
| 1b | $+$ | 0 or $-$ | $+$ | $-$ | 0 |
| 2a1 | 0 | $+$ | $+$ | $+$ | 2 |
| 2a2 | 0 | $+$ | $+$ | $+$ | 1 |
| 2a3 | 0 | $+$ | $+$ | $-$ | 0 |
| 2b | 0 | 0 or $-$ | $+$ | $+$ | 0 |
| 3.a1 | $-$ | $+$ | $+$ | $+$ | 2 |
| 3.a2 | $-$ | $+$ | $+$ | 0 | 1 |
| 3.a3 | $-$ | $+$ | $+$ | $-$ | 0 |
| 3.b | $-$ | 0 or $-$ | $+$ | $-$ | 0 |
| 3.2i | $-$ | $+$ | 0 | $+$ | 1 |
| 3.2ii | $-$ | 0 or $-$ | 0 | $-$ | 0 |
| 3.3i | $-$ | $+$ or $-$ | $-$ | $+$ | 1 |
| 3.3ii | $-$ | 0 | $-$ | $+$ | 1 |

To determine the local nature of the equilibrium points, we must have the Jacobian matrix of the system (2.3), which is given by

$$DY_\eta(u, v) = \begin{pmatrix} DY_\eta(u, v)_{11} & -Qu(u + C) \\ Sv & S(-2v + u + C) \end{pmatrix},$$

with

$$DY_\eta(u, v)_{11} = ((1 - u)(u - M) - Qv)(u + C) + ((1 - u)(u - M) - Qv)u \\ + u(u + C)(M - 2u + 1).$$

## 2.3 Main results

For system (2.3), we have the following general properties:

**Lemma 2.3** (Existence of a positive invariant region). *The set* $\tilde{\Gamma} = \{(u, v) \in \tilde{\Omega} / \ 0 \le u \le 1, v \ge 0\}$ *is a positive invariant region.*

*Proof.* Clearly the $u - axis$ and the $v - axis$ are invariant sets, because the system is a Kolmogorov type. If $u = 1$, we have

$$\frac{du}{d\tau} = -Qv(1 + C) < 0,$$

and whatever the sign of

$$\frac{dv}{d\tau} = S(1 + C - v)\, v,$$

the trajectories enter and remain in the region $\tilde{\Gamma}$. □

In system (2.2), the set $\Gamma = \{(x, y) \in \Omega : 0 \leq x \leq K,\ y \geq 0\}$ is a positive invariant region.

**Lemma 2.4** (Boundedness of the solutions). *The solutions are uniformly bounded.*

*Proof.* Using the Poincaré compactification [10,16]. Making a change of variables and a time rescaling given by the function $\theta : \tilde{\Omega} \times \mathbb{R} \longrightarrow \breve{\Omega} \times \mathbb{R}$ gives

$$\theta(X, Y, T) = \left( \frac{X}{Y}, \frac{1}{Y}, Y^3 T \right) = (u, v, \tau).$$

So, we have $\frac{dX}{d\tau} = \left( \frac{du}{d\tau} v - \frac{dv}{d\tau} u \right) \frac{1}{v^2}$ and $\frac{dY}{d\tau} \left( -\frac{dv}{d\tau} \right) \frac{1}{v^2}$.
Replacing in the system (2.3), we get

$$\hat{\Gamma}_\eta(X, Y) : \begin{cases} \frac{dx}{dt} = -\frac{X}{Y^3} \left( \begin{array}{l} X^3 - X^2Y - SY^2 + CMY^3 + CQY^2 + CSY^3 \\ -CXY + CX^2Y + MXY^2 - MX^2Y + SXY^2 + QXY \\ -CMXY^2 \end{array} \right) \\ \frac{dy}{dt} = -S(X + CY - 1). \end{cases}$$

Making a time rescaling given by $T = \frac{1}{Y^3}\tau$, it has $\frac{dT}{d\tau} = \frac{1}{Y^3}$, $\frac{dX}{d\tau} = \frac{dX}{dT}\frac{dT}{d\tau}$, and $\frac{dY}{d\tau} = \frac{dY}{dT}\frac{dT}{d\tau}$ and we obtain a new system topologically equivalent to the vector field $\hat{\Gamma}_\eta(X, Y)$, given by

$$\check{\Gamma}_\eta(X, Y) : \begin{cases} \frac{dx}{dt} = -X \left( \begin{array}{l} X^3 - X^2Y - SY^2 + CMY^3 + CQY^2 + CSY^3 \\ -CXY^2 + CX^2Y + MXY^2 - MX^2Y + SXY^2 + QXY \\ -CMXY^2 \end{array} \right) \\ \frac{dy}{dt} = -SY^3(X + CY - 1). \end{cases}$$

Evaluating the Jacobian matrix at (0, 0) gives

$$D\check{\Gamma}_\eta(0, 0) = \begin{pmatrix} 0 & 0 \\ 0 & 0 \end{pmatrix}.$$

To desingularize the origin of the vector field $\check{\Gamma}_\eta(X, Y)$, we apply the directional blowing up method [16], making the change of variables:

$$X = r,\ Y = r^2 p,\ \frac{dX}{dT} = \frac{dr}{dT},\ \frac{dp}{dT} = \left( \frac{dY}{dT} - \frac{dr}{dT} 2rp \right) \frac{1}{r^2}.$$

Then, the vector field $\check{\Gamma}_\eta(X,Y)$ is transformed in the following system:

$$\bar{F}_\eta(r,p):\begin{cases}\frac{dr}{dT} = -r^4\begin{pmatrix} Qp - rp - Srp^2 - Cr^2p^2 + Mr^2p^2 + Sr^2p^2 \\ +Crp - Mrp + CQrp^2 - CMr^2p^2 + CMr^3p^3 \\ +CSr^3p^3 + 1 \end{pmatrix} \\ \frac{dp}{dT} = r^3p\begin{pmatrix} 2Qp - 2rp - Srp^2 - 2Cr^2p^2 + 2Mr^2p^2 + Sr^2p^2 \\ +2Crp - 2Mrp + 2CQrp^2 - 2CMr^2p^2 + 2CMr^3p^3 \\ +CSr^3p^3 + 2 \end{pmatrix}\end{cases}$$

To desingularize the origin of the vector field $\bar{F}_\eta(r,p)$, we make a new time rescaling given by

$$\lambda = r^3T, \text{ with } \frac{dr}{dT} = \frac{dr}{d\lambda}\frac{d\lambda}{dT} \text{ and } \frac{dp}{dT} = \frac{dp}{d\lambda}\frac{d\lambda}{dT},$$

obtaining a vector field topologically equivalent to $\bar{F}_\eta(r,p)$ given by

$$\tilde{F}_\eta(r,p):\begin{cases}\frac{dr}{d\lambda} = -r\begin{pmatrix} Qp - rp - Srp^2 - Cr^2p^2 + Mr^2p^2 + Sr^2p^2 \\ +Crp - Mrp + CQrp^2 - CMr^2p^2 + CMr^3p^3 \\ +CSr^3p^3 + 1 \end{pmatrix} \\ \frac{dp}{d\lambda} = p\begin{pmatrix} 2Qp - 2rp - Srp^2 - 2Cr^2p^2 + 2Mr^2p^2 + Sr^2p^2 \\ +2Crp - 2Mrp + 2CQrp^2 - 2CMr^2p^2 + 2CMr^3p^3 \\ +CSr^3p^3 + 2 \end{pmatrix}\end{cases}.$$

Evaluating the Jacobian matrix at $(0,0)$ of the vector field $\tilde{F}_\eta(r,p)$, it has

$$D\tilde{F}_\eta(0,0) = \begin{pmatrix} -1 & 0 \\ 0 & 2 \end{pmatrix}.$$

Therefore $\det D\tilde{F}_\eta(0,0) = -2 < 0$.

Then the point $(0,0)$ is the hyperbolic saddle point of the vector field $\tilde{F}_\eta(r,p)$. Thus the equilibrium $(0,0)$ is the non-hyperbolic saddle of the vector field $\bar{F}_\eta(r,p)$, of the vector field $\check{\Gamma}_\eta(X,Y)$, and the vector field $\hat{\Gamma}_\eta(X,Y)$.

Therefore the point $(0,\infty)$ is the non-hyperbolic saddle point of the compactified vector field $Y_\eta(u,v)$.

Then, the solutions are bounded. □

## 2.3.1 Model with strong Allee effect

In the following, we assume $0 < M < 1$, remembering that $C > 0$.

#### 2.3.1.1 Nature of equilibrium points over the axis

**Lemma 2.5.** *The singularities* $(0,0)$ *and* $(1,0)$ *are hyperbolic saddle points for all* $\eta = (Q,C,S,M) \in \mathbb{R}^3_+ \times ]0,1[$.

*Proof.* Evaluating the Jacobian matrix in each point, we have

1.

$$DY_\eta(1,0) = \begin{pmatrix} -(1+C)(1-M) & -Q(1+C) \\ 0 & S(1+C) \end{pmatrix},$$

and $\det DY_\eta(1,0) = -S(1+C)^2(1-M) < 0$.
Therefore applying the trace and the determinant theorem, the equilibrium $(1,0)$ is a saddle point.

2.

$$DY_\eta(0,0) = \begin{pmatrix} -CM & 0 \\ 0 & SC \end{pmatrix},$$

and $\det DY_\eta(0,0) = -SC^2M < 0$.

Then, applying the trace and the determinant theorem, the equilibrium $(0,0)$ is a saddle point. □

**Lemma 2.6.** *The equilibrium point $(M,0)$ is a hyperbolic repeller for all parameter values.*

*Proof.* It is immediate since the Jacobian matrix evaluated in $(M,0)$ is

$$DY_\eta(M,0) = \begin{pmatrix} M(1-M)(C+M) & -QM(C+M) \\ 0 & S(C+M) \end{pmatrix}.$$

Then, $\det DY_\eta(M,0) = M(1-M)(C+M)^2$ and

$$\mathrm{tr} DY_\eta(M,0) = (M(1-M)+S)(C+M) > 0.$$

So, applying the trace and the determinant theorem, the thesis holds. □

**Lemma 2.7.** *The equilibrium point $(0,C)$ is a hyperbolic attractor point for all parameter values.*

*Proof.* The Jacobian matrix evaluated in the point $(0,C)$ is

$$DY_\eta(0,C) = \begin{pmatrix} -C(M+QC) & 0 \\ SC & -SC \end{pmatrix}.$$

Then,

$$\det DY_\eta(0,C) = SC^2(M+QC) > 0$$

and

$$\mathrm{tr} DY_\eta(0,C) = -C(M+QC) - SC < 0.$$

Therefore the point $(0,C)$ is an attractor point. □

#### *2.3.1.2 Nature of positive equilibria*

We recall that if $a_1 = 1 + M - Q > 0$, then the roots of Eq. (2.3) depend on the sign of

$$\Delta = (1 + M - Q)^2 - 4(CQ + M) > 0.$$

Suppose $\Delta > 0$, then there are two real solutions:

$$u_1 = \frac{1}{2}\left(a_1 - \sqrt{\Delta}\right) \text{ and } u_2 = \frac{1}{2}\left(a_1 + \sqrt{\Delta}\right),$$

with $u_1 < u_2$.

Evaluating the Jacobian matrix in the positive equilibria, we have

$$DY_\eta(u, u + C) = \begin{pmatrix} u(u + C)(M - 2u + 1) & -Qu(u + C) \\ S(u + C) & -S(u + C) \end{pmatrix}.$$

Then,

$$\det DY_\eta(u, u + C) = -S(u + C)^2(M - 2u + 1 - Q),$$

whose sign depends on the factor

$$T = M - 2u + 1 - Q.$$

In addition, the trace is

$$\mathrm{tr} DY_\eta(u, u + C) = (u + C)(u(M - 2u + 1) - S),$$

whose sign depends on the factor

$$T_1 = u(M - 2u + 1) - S.$$

To determine whether the hyperbolic equilibrium points are nodes or foci, we define the quantity

$$\begin{aligned} P &= \left(\mathrm{tr} DY_\eta(u, u + C)\right)^2 - 4\det DY_\eta(u, u + C) \\ &= ((u + C)(u(M - 2u + 1) - S))^2 - 4\left(-S(u + C)^2(M - 2u + 1 - Q)\right). \end{aligned}$$

**Theorem 2.8** (Nature of the first positive equilibrium point). *The point $P_1 = (u_1, u_1 + C)$ is a hyperbolic saddle point.*

*Proof.* Evaluating the factor $T$ at $u_1$, we have

$$T(u_1) = M - 2u_1 + 1 - Q = \sqrt{\Delta} > 0.$$

Then, $\det DY_\eta(u, u + C) < 0$.

By the trace and the determinant theorem, the point $P_1 = (u_1, u_1 + C)$ is a hyperbolic saddle point. □

**Remark 2.9** (Existence of a separatrix curve). By the above lemma, the point $(u_1, u_1 + C)$ is a hyperbolic saddle point. Let $W_+^s(P_1)$ and $W_+^u(P_1)$ be the upper stable manifold and the right unstable manifold of the saddle point $P_1$, respectively.

The stable manifold $W_+^s(P_1)$ determines a separatrix curve $\bar{\Sigma}$, which divides the behavior of the trajectories in the phase plane; any solution above the manifold $\bar{\Sigma} = W_+^s(P_1)$ have the equilibrium $(0, C)$ as its *ω-limit.* Those trajectories with initial conditions below the separatrix curve can have different *ω-limit.*

Before establishing the nature of point $P_2 = (u_2, u_2 + C)$, we must show some properties of the system related to the stable and unstable manifolds of the equilibrium $P_1$.

**Theorem 2.10** (Existence of a homoclinic curve). *There is a subset of parameter values for which the two manifolds coincide, forming a homoclinic loop, i.e.,* $W_-^s(u_1, v_1) \cap W_+^u(u_1, v_1) \neq \phi$.

*Proof.* Using the theorem of existence and uniqueness of solutions [10] and the geometry of the stable and unstable manifolds of the saddle point $P_1 = (u_1, u_1 + C)$, the following hold: The right unstable manifold $W_+^u(P_1)$ of the saddle point $P_1$ cannot cross the line to the right of the line $u = 1$, which is the border of the invariance region. Therefore this solution must change its direction and return to the left.

In turn, the *α-limit* of the upper stable variety $W_+^s(P_1)$ can be the following:

- **i)** the point $(\infty, 0)$ of the domain compact of the vector field $Y_\eta(u, v)$,
- **ii)** the point $P_2$, when it is repeller, or
- **iii)** an unstable limit cycle surrounding the equilibrium point $P_2$, when this is a local attractor.

Then, there are conditions in the parameter space, for which $W_+^s(P_1) \cap W_+^u(P_1) \neq \phi$.

So, there is a homoclinic curve, created by the saddle point $P_1$ and surrounding the point $P_2$. □

**Remark 2.11.** When the homoclinic orbit is broken, a non-infinitesimal limit cycle is generated. To determine the stability of this limit cycle we adopt the criterion based on the calculation of the absolute value of the quotient between the product of the negative eigenvalues and the product of the positive eigenvalues of the saddle points based on the Andronov–Leontovich theorem [4].

We denote by $\lambda^-(u_1, v_1)$ and $\lambda^+(u_1, v_1)$ the negative and positive eigenvalues of the saddle point $(u_1, v_1)$, respectively.

Then, we will calculate $R$, the absolute value of the ratio between the negative and positive eigenvalues of the Jacobian matrix evaluated in a saddle point $(P_1)$. If $R = 1$, the limit cycle is neutrally stable [4]. Thus in this case, $R$ depends on the sign of the difference $R_d = |\lambda^-| - \lambda^+$.

It is easy to see that for $P_1 = (u_1, u_1 + C)$, it has

$$\det DY_\eta(u_1, u_1 + C) = -S(u_1 + C)^2(M - 2u_1 + 1 - Q),$$

$$\operatorname{tr}DY_\eta(u_1, u_1 + C) = (u_1 + C)(u_1(M - 2u_1 + 1) - S).$$

Then, the eigenvalues of the Jacobian matrix evaluated on $P_1$ are

$$\lambda^- = \frac{1}{2}(C + u_1)\left(\operatorname{tr}DY_\eta(u_1, u_1 + C) - \sqrt{\Delta_1}\right),$$
$$\lambda^+ = \frac{1}{2}(C + u_1)\left(\operatorname{tr}DY_\eta(u_1, u_1 + C) + \sqrt{\Delta_1)}\right),$$

with $\lambda^- < 0 < \lambda^+$ and

$$\Delta_1 = \left(\operatorname{tr}DY_\eta(u_1, u_1 + C)\right)^2 - 4\det DY_\eta(u_1, u_1 + C).$$

Then the following result is obtained:

**Theorem 2.12** (Stability of the non-infinitesimal limit cycle). *When the homoclinic orbit is broken, a non-infinitesimal limit cycle is generated [4], which is*

**a)** *stable if and only if* $S < u_1(M - 2u_1 + 1)$.
**b)** *unstable if and only if* $S > u_1(M - 2u_1 + 1)$.
**c)** *neutrally stable if and only if* $S = u_1(M - 2u_1 + 1)$.

*Proof.* It has that $R_d = |\lambda^-| - \lambda^+ = (C + u_1)\operatorname{tr}DY_\eta(u_1, u_1 + C)$.
The sign of $\operatorname{tr}DY_\eta(u_1, u_1 + C)$ depends on the sign of the factor

$$T(u_1, M, S) = u_1(M - 2u_1 + 1) - S).$$

Considering $R = 1$ or $R_d = 0$, we have

$$S = u_1(M - 2u_1 + 1)$$

and the non-infinitesimal limit cycle is neutrally stable.

The other cases a and b are obtained when $R > 1$ or $R_d < 0$ and $R < 1$ or $R_d > 0$, respectively. □

**Remark 2.13.** The nature of the second equilibrium point $P_2 = (u_2, u_2 + C)$ depends on the relative position of $W_+^u(1, 0)$, the unstable manifold of the saddle point $(1, 0)$ and $W_+^s(P_1)$, the stable manifold of the saddle point $(P_1)$.

**Theorem 2.14** (Nature of the second positive equilibrium point). *Let's consider the points* $(u^*, v^u) \in W_+^u(1, 0)$ *and* $(u^*, v^s) \in W_+^s(P_1)$ *with* $0 < M < u_1 < u^* < 1$.

1. *Suppose* $v^s > v^u$, *then the point* $P_2 = (u_2, u_2 + C)$ *is*
   **i)** *a hyperbolic attractor if and only if* $u_2(M - 2u_2 + 1) < S$.
   *Moreover,* $P_2$ *is*

   **ia)** *an attractor node if and only if* $P > 0$.
   **ib)** *an attractor focus if and only if* $P < 0$.

**ii)** *a hyperbolic repeller if and only if* $u_2(M-2u_2+1)>S$.
*Moreover,* $P_2$ *is*

**iia)** *a repeller node if and only if* $P>0$.
**iib)** *a repeller focus if and only if* $P>0$. *In this case, the point* $P_2$ *is surrounded by a stable limit cycle.*

**iii)** *a weak focus, if and only if,* $u_2(M-2u_2+1)=S$.

**2.** *Suppose that* $v^s<v^u$, *then, the equilibrium point* $P_2=(u_2,u_2+C)$ *is*

**2.1** *a hyperbolic attractor surrounded by an unstable limit cycle.*
**2.2** *a hyperbolic repeller node or focus and the trajectories of the system* (2.2) *have the point* $(0,C)$ *as their* $\omega$-limit, *which is an almost globally stable equilibrium [35].*

*Proof.* 1. Suppose $v^s>v^u$.

Then, $W_+^s(P_1)$ the upper stable manifold of $P_1$ is above $W_+^u(1,0)$ the unstable manifold of the saddle point $(1,0)$.

Thus the $\omega$-*limit* of $W_+^s(P_1)$ must be the following:

**i)** the point $P_1=(u_1,u_1+C)$, originating a heteroclinic curve,
**ii)** the point $P_2=(u_2,u_2+C)$, when this is an attractor node, or
**iii)** an attractor limit cycle.

Now, evaluating the factor $T$ at the point $P_2$, we have

$$T(u_2)=M-2u_2+1-Q=-\sqrt{\Delta}<0.$$

Hence $\det DY_\eta(u,u+C)>0$.

Therefore the nature of $P_2$ depends on the sign of the trace.

Then, when $T_1(u_2)=u_2(M-2u_2+1)-S$ is positive, negative, or zero, the thesis holds.

If the trace is positive, $P_2$ is a hyperbolic repeller; according to the existence and uniqueness theorem and the Poincaré–Bendixson theorem, there must be a limit cycle surrounding the equilibrium point $P_2$.

2. Suppose $v^s<v^u$.

Therefore $W_+^s(P_1)$, the upper stable manifold of $P_1$ is below $W_+^u(1,0)$, the unstable manifold of the saddle point $(1,0)$. Then,

**a)** If $u_2(M-2u_2+1)<S$, the equilibrium point $P_2$ is a hyperbolic attractor. Thus the $\omega$-*limit* of $W_+^u(P_1)$ is
**a1)** the point $P_2$, or else,
**a2)** an attractor limit cycle surrounding an unstable limit cycle and the equilibrium point $P_2$.

**b)** If $u_2(M-2u_2+1)>S$, the equilibrium point $P_2$ is a hyperbolic repeller (focus or node). Then, the solutions with $\alpha$-*limit* in the neighborhoods of the point $P_2$ have as $\omega$-*limit*
**b1)** an attractor (or stable) limit cycle, when $H<0$, or else,
**b2)** the point $(0,C)$, when $H>0$ or $H<0$, which is almost globally stable equilibrium [35]. □

**Corollary 2.15** (Transversality condition). *There is a Hopf bifurcation at the equilibrium point* $P_2 = (u_2, u_2 + C)$ *for the bifurcation value* $u_2 (M - 2u_2 + 1) = S$.

*Proof.* The proof follows from the theorem described above, when

$$T = (M - 2u_2 + 1 - Q) \text{ is negative, and}$$
$$T_1 = u_2 (M - 2u_2 + 1) - S \text{ changes sign.}$$

In addition, the transversality condition [10,18,36] is verified since

$$\frac{\partial}{\partial S}\left(\text{tr}DY_\eta (u, u + C)\right) = -(u_2 + C) < 0.$$

□

**Theorem 2.16** (Existence of two limit cycles). *The equilibrium point* $P_2 = (H, H + C)$ *is at least a two-order weak focus, when* $S = H(M - 2H + 1)$ *and the second Lyapunov quantity is zero, which depends of the polynomial* $F(H, C, M, W)$ *defined below.*

*Proof.* Remembering that the Jacobian matrix evaluated at the point $P_2 = (H, H + C)$ is

$$DY_\eta (H, H + C) = \begin{pmatrix} H(H + C)(M - 2H + 1) & -QH(H + C) \\ S(H + C) & -S(H + C) \end{pmatrix}.$$

Then, when $S = H(M - 2H + 1) > 0$, the equilibrium $P_2 = (H, H + C)$ is a weak focus.

We are going to calculate the weakness of the focus $(H, H + C)$ by means of the method of calculating the Lyapunov quantities [28], making a transfer to the point $(0, 0)$ of a new coordinates system.

Let $u = U + H$ and $v = V + H + C$, with $\frac{du}{d\tau} = \frac{dU}{d\tau}$ and $\frac{dv}{d\tau} = \frac{dV}{d\tau}$.

Substituting in system (2.2), we have

$$Z_\eta (U, V) : \begin{cases} \frac{dU}{d\tau} = \begin{pmatrix} (1 - (U + H))(U + H - M) \\ -Q(V + H + C) \end{pmatrix}(U + H + C)(U + H) \\ \frac{dV}{d\tau} = S(U + H + C - (V + H + C))(V + H + C) \end{cases}.$$

The Jacobian matrix of the translated vector field $Z_\eta (U, V)$ evaluated at $(0, 0)$ is the same that $DY_\eta (H, H + C)$.

Therefore

$$\det DZ_\eta (0, 0) = HS(C + H)^2 (2H - M + Q - 1),$$
$$\text{tr}DZ_\eta (0, 0) = \left(H - S - 2H^2 + HM\right)(C + H).$$

From Eq. (2.3), we have that

$$H^2 - (1 + M - Q)H + (CQ + M) = 0,$$

i.e., $Q = \frac{-(H^2 + (-M - 1)H + M)}{(C + H)}$.

So,

$$\det DZ_\eta(0,0) = W^2 = H^2(C+H)(M-2H+1)\left(H^2+2CH-(C+M+CM)\right).$$

The calculation of the Lyapunov quantities requires a normal form of the vector field $Z_\eta(U,V)$, for which we must make a suitable change of variables. We will use the Jordan matrix [3] associated to vector field, which is

$$J = \begin{pmatrix} 0 & -W \\ W & 0 \end{pmatrix},$$

where $W = \sqrt{\det DZ_\eta(0,0)}$.

The first Lyapunov quantity is $\eta_1 = \mathrm{tr} DZ_\eta(0,0) = 0$.

The basis change matrix is

$$M = \begin{pmatrix} S(C+H) & -W \\ S(C+H) & 0 \end{pmatrix}.$$

Its inverse matrix is $M^{-1} = \begin{pmatrix} 0 & \frac{1}{S(C+H)} \\ -\frac{1}{W} & \frac{1}{W} \end{pmatrix}$.

Considering the change of variables given by

$$\begin{pmatrix} U \\ V \end{pmatrix} = \begin{pmatrix} S(C+H) & -W \\ S(C+H) & 0 \end{pmatrix}\begin{pmatrix} x \\ y \end{pmatrix},$$

or

$$\begin{pmatrix} x \\ y \end{pmatrix} = \begin{pmatrix} 0 & \frac{1}{S(C+H)} \\ -\frac{1}{W} & \frac{1}{W} \end{pmatrix}\begin{pmatrix} U \\ V \end{pmatrix},$$

it has the new system

$$\bar{Z}_\eta(U,V): \begin{cases} \frac{dx}{d\tau} = \frac{1}{S(C+H)}\frac{dV}{d\tau} \\ \frac{dy}{d\tau} = -\frac{1}{W}\frac{dU}{d\tau} + \frac{1}{W}\frac{dV}{d\tau} \end{cases}.$$

Replacing $x$ and $y$ in the vector field $Z_\eta(U,V)$, after a long algebraic calculations and by means of the time rescaling given by $\tau = WT$, we obtain the normal form:

$$\tilde{Z}_\nu(x,y): \begin{cases} \frac{dx}{dT} = -y - Sxy \\ \frac{dy}{dT} = \begin{pmatrix} x + b20x^2 + b11xy \\ \left(-C-2H+3CH-CM-2HM+5H^2\right)y^2 \\ +\frac{S^3(C+H)^2\left(-C-M+5CH-CM+C^2+3H^2\right)}{W^2}x^3 \\ -b21x^2y - b12xy^2 + \frac{S^4(C+H)^4}{W^2}x^4 - \frac{4S^3(C+H)^3}{W}x^3y \\ 6S^2(C+H)^2x^2y^2 - 4SW(C+H)xy^3 + W^2y^4 \end{pmatrix} \end{cases}$$

where,

$$b20 = \frac{S^2 (C+H) \left(7CH^2 + 3C^2H - C^2M - 2CH - CM - 2HM - C^2 + 3H^3 - 2CHM\right)}{W^2},$$

$$b11 = \frac{S^2 (C+H) \left(7CH^2 + 3C^2H - C^2M - 2CH - CM - 2HM - C^2 + 3H^3 - 2CHM\right)}{W},$$

$$b21 = \frac{S^2 y (C+H) \left(-3C - H - 2M + 15CH - 3CM - HM + 3C^2 + 10H^2\right)}{W},$$

$$b12 = S\left((3C + 2H + 1)M - 3C^2 + 3(1-5H)C + H(2-11H)\right).$$

Bearing in mind that

$$S = H(M - 2H + 1), \text{ and}$$
$$W^2 = H^2 (C+H)(M-2H+1)\left(H^2 + 2CH - (C + M + CM)\right),$$

and using the Mathematica package [45], the second Lyapunov quantity is obtained, being given by

$$\eta_2 (H, C, M, W) = \frac{F(H, C, M, W)}{8\left(-C + 2CH + H^2 - M - CM\right)^2 W}.$$

As the denominator is positive its sign depends on the $F(H, C, M, W)$, which is a polynomial function given by

$$F(H, C, M, W) = \left( \begin{array}{c} f_0(H, C, W) + f_1(H, C, M, W) + f_2(H, C, M, W) \\ + f_3(H, C, M, W) + f_4(H, C, M, W) + f_5(H, C, M, W) \end{array} \right)$$

with

$$\begin{aligned} f_0(H, C, M, W) &= f_0(H, C, W), \\ f_1(H, C, M, W) &= M f_1(H, C, W), \\ f_2(H, C, M, W) &= M^2 f_2(H, C, W), \\ f_3(H, C, M, W) &= M^3 f_3(H, C, W), \\ f_4(H, C, M, W) &= M^4 f_{41}(H, C, W), \\ f_5(H, C, M, W) &= M^5 f_5(H, C, W), \end{aligned}$$

where,

$$f_0(H, C, W) = (1-H)H \left( \begin{array}{c} 3H^2 (2H-1)^3 C^4 + f_{03}(H, C) C^3 + \\ f_{02}(H, C) C^2 + f_{01}(H, C) C + f_{00}(H, C, W) \end{array} \right),$$

with

$$f_{03}(H, C) = H^2 \left(-13H + 28H^2 + 1\right)(2H-1)^2,$$

$$f_{02}(H,C) = 2(3H-1)W + \left(158H^7 - 203H^6 + 90H^5 - 16H^4 + H^3 - 6H + 2\right),$$
$$f_{01}(H,C) = H\left((13H-4)W + \left(86H^7 - 85H^6 + 25H^5 - 2H^4 - 13H + 4\right)\right),$$
$$f_{00}(H,C,W) = H^2\left((6H-1)W + \left(16H^7 - 10H^6 + H^5 - 6H + 1\right)\right).$$

$$f_1(H,C,W) = H^2(2H-1)\left(-C + 2CH + H^2\right)(C+H)\,f_{11}(H,C) + f_{12}(H,C,W),$$

where

$$f_{11}(H,C) = 6(2H-1)C^3 + 6(6H-1)(2H-1)C^2 + f_{111}(H)C + f_{110}(H),$$
$$f_{111}(H) = \left(2 - 9H - 35H^2 + 84H^3\right),$$
$$f_{110}(H) = H\left(2 - 20H + 17H^2 + 19H^3\right),$$
$$f_{12}(H,C,W) = W(H-1)\left(7H^3 + (17C+1)H^2 + \left(8C^2 - 3C\right)H - 2C^2\right).$$

$$f_2(H,C,W) = H^2\left(\begin{array}{c} 60H(1-H)C^5 + f_{24}(H,C,W)C^4 + f_{23}(H,C,W)C^3 \\ + f_{22}(H,C,W)C^2 + f_{21}(H,C,W)C + f_{20}(H,C,W) \end{array}\right),$$

with

$$f_{24}(H,C,W) = \left(471H^2 - 450H^3 - 141H - 15\right),$$
$$f_{23}(H,C,W) = \left(515H^3 - 845H^4 + 159H^2 - 143H + 9\right),$$
$$f_{22}(H,C,W) = \left(595H^3 - 149H^4 - 614H^5 - 203H^2 + H + 20\right),$$
$$f_{21}(H,C,W) = \left(332H^4 - 334H^5 - 173H^6 - 15H^3 - 34H^2 + 5H + 4\right),$$
$$f_{20}(H,C,W) = -W(H-1)\left(C + 2H + 4CH + 2C^2 + H^2\right).$$

$$f_3(H,C,W) = H^2(C+H)\left(f_{34}(H)C^4 + f_{33}(H)C^3 + f_{32}(H)C^2 + f_{31}(H)C + f_{30}(H)\right),$$

with

$$f_{34}(H) = 12(2H-1),$$
$$f_{33}(H) = \left(362H^2 - 36H + 3\right),$$
$$f_{32}(H) = \left(164H^3 + 189H^2 - 200H + 36\right),$$
$$f_{31}(H) = \left(35H^4 + 202H^3 - 12H^2 - 78H + 18\right),$$
$$f_{30}(H) = \left(H^5 + 30H^4 + 48H^3 - 25H^2 - 5H + 2\right).$$

$$f_4(H,C,W) = -H^2(C+1)(C+H)\,f_4(H,C),$$

with

$$f_4(H,C) = 3C^3 + 3(13H-4)C^2 + \left(24H^2 + 38H - 17\right)C + \left(2H^3 + 18H^2 + 5H - 4\right).$$

$$f_5(H,C,W) = H^2(C+1)^2(3C+H+2)(C+H).$$

Numerically, it can show that the polynomial $F(H,C,M,W)$ changes its sign. For instance, choosing $M=0$, it has $f_0(H,C,W)$. Picking out $W=1$, it has

$$\begin{aligned} f_{02}(H) &= H^3(2H-1)\left(14H - 62H^2 + 79H^3 - 1\right),\\ f_{01}(H) &= H^5(2H-1)\left(-21H + 43H^2 + 2\right),\\ f_{00}(H) &= H^2\left((6H-1) + \left(16H^7 - 10H^6 + H^5 - 6H + 1\right)\right). \end{aligned}$$

Then,

$$f_0(H,C) = (1-H)H^3(2H-1)(C+H)(3C+8H-1)\left(-C+2CH+H^2\right)^2.$$

Evaluating this polynomial for $H=0.4$ and $H=0.6$, we get

$$f_0(0.4,C) = -0.007\,68\,(C+0.4)(0.2C-0.16)^2(3C+2.2),$$

which is negative for all value of $C>0$, and

$$f_0(0.6,C) = 0.017\,28\,(3C+3.8)(C+0.6)(0.2C+0.36)^2,$$

which is positive for all value of $C>0$.

Therefore there exists a value $0.4 < H^* < 0.6$, such as $f_0(H,C)=0$.

Then, the function $F(H,C,M,W)$ change of sign, and, at least, a second limit cycle exists. □

Now, considering $a_1 = 1+M-Q>0$ and $\Delta = (1+M-Q)^2 - 4(CQ+M) = 0$, we have

**Theorem 2.17** (Bogdanov–Takens bifurcation). *Let $u^* = \frac{1}{2}a_1$; then, the point $P_* = (u^*, u^*+C)$, collapse of $P_1$ and $P_2$, is*

**a)** *a saddle-node attractor if and only if $S > u^*(M - 2u^* + 1)$,*
**b)** *a saddle-node repeller if and only if $S < u^*(M - 2u^* + 1)$,*
**c)** *a codimension two cusp point if and only if $S = u^*(M - 2u^* + 1)$.*

*Proof.* It is clear that $\det DY_\eta(u^*, u^*+C) = 0$, when the points $P_1$ and $P_2$ collapse.

Before this coincidence, the points $P_1$ and $P_2$ are saddle point and node attractor, respectively.

Then, $P_*$ is a saddle-node attractor [28].

As $T_1(u^*) = u^*(M - 2u^* + 1) - S$, we get the different cases considering the sign of $T_1$. □

## 2.3.2 Model with special weak Allee effect

Now, we consider the special case when $M = 0$. The system (2.3) becomes

$$\Gamma_\eta(u, v): \begin{cases} \frac{du}{d\tau} = ((1-u)u - Qv)(u+C)u, \\ \frac{dv}{d\tau} = S(u + C - v)v \end{cases} \tag{2.6}$$

where $\eta = (Q, S, C) \in \mathbb{R}^3$.

The equilibrium points of the system (2.6) are $(0,0)$, $(1,0)$, $(0,C)$ and the interior points, obtained by intersecting the isoclines; the existence of these solutions depend on the discriminant of the second-degree equation

$$u^2 - (1-Q) + QC = 0. \tag{2.7}$$

Let $\Delta_2 = (1-Q)^2 - 4QC$. Eq. (2.7) has

1. Two positive real roots if and only if $1 - Q > 0$ and $\Delta_2 > 0$ so that

$$u_3 = \frac{1 - Q - \sqrt{(1-Q)^2 - 4QC}}{2} \text{ and } u_4 = \frac{1 - Q + \sqrt{(1-Q)^2 - 4QC}}{2}.$$

   In this case, we obtain two equilibrium points: $(u_3, v_3)$ and $(u_4, v_4)$, where $v_i = u_{i+2} + C$ with $i = 1, 2$.
2. A positive real root if and only if $1 - Q > 0$ and $\Delta_2 = 0$, obtaining that the equilibrium point is $(u^*, v^*) = \left(\frac{1-Q}{2}, u^* + C\right)$.
3. It has no positive real roots if $1 - Q > 0$ and $\Delta_2 < 0$, or else $1 - Q \leq 0$.

   The Jacobian matrix associated with the system (2.6) is

$$D\Gamma_\eta(u, v) = \begin{pmatrix} D\Gamma(u,v)_{11} & -Qu(C+u) \\ Sv & S(u + C - 2v) \end{pmatrix},$$

where $D\Gamma(u, v)_{11} = -4u^3 + 3(1-C)u^2 + 2(C - Qv)u - CQv$.

The system (2.6) has the following properties:

**Lemma 2.18** (Nature of equilibrium over the axes). *For all values of the parameters:*

**i)** *The equilibrium* $(1,0)$ *is a hyperbolic saddle point.*
**ii)** *The point* $(0,C)$ *is a hyperbolic attractor node.*
**iii)** *The equilibrium* $(0,0)$ *is a repeller saddle-node, a non-hyperbolic point.*

*Proof.* i) and ii) are immediate.

iii) As in the system (2.6), the equilibria $(M, 0)$ and $(0, 0)$ coincide. Then, the non-hyperbolic equilibrium point $(0,0)$ becomes on a saddle-node attractor. □

The following theorems are based on the similar theorems presented in the case $M > 0$, and their proofs are similar, and they will not be given here:

**Theorem 2.19** (Nature of the equilibrium $P_3$). *The equilibrium point* $(u_3, v_3)$ *is a saddle point for all parameter values.*

Let $W_+^s(P_3)$ and $W_+^u(P_3)$ be the upper stable manifold and the right unstable manifold of saddle point $P_3$, respectively, and $W_+^u(1,0)$ the unstable manifold of the saddle point $(1,0)$.

**Theorem 2.20** (Nature of the equilibrium $P_4$). *Consider the points* $(u^*, v^u) \in W_+^u(1,0)$ *and* $(u^*, v^s) \in W_+^s(P_3)$ *with* $0 < u_3 < u^* < 1$.

1. *Suppose* $v^s > v^u$, *then the point* $P_4 = (u_4, u_4 + C)$ *is*
   - **i)** *a hyperbolic attractor if and only if* $S > u_4(1 - 2u_4)$.
     *Moreover,* $P_4$ *is*
     - **ia)** *an attractor node if and only if* $P > 0$.
     - **ib)** *an attractor focus if and only if* $P < 0$.
   - **ii)** *a hyperbolic repeller if and only if* $S < u_4(1 - 2u_4)$.
     *Moreover,* $P_4$ *is*
     - **iia)** *a repeller node if and only if* $P > 0$.
     - **iib)** *a repeller focus if and only if* $P > 0$. *In this case, the point* $P_4$ *is surrounded by a stable limit cycle.*
   - **iii)** *a weak focus if and only if* $S = u_4(1 - 2u_4)$.
2. *Suppose that* $v^s < v^u$, *then the equilibrium point* $P_4 = (u_4, u_4 + C)$ *is*
   - **2.1** *an attractor surrounded by an unstable limit cycle.*
   - **2.2** *a hyperbolic repeller node or focus, and the orbits of system* (2.6) *have the point* $(0, C)$ *as their* $\omega$-limit, *being almost globally stable equilibrium [35].*

**Theorem 2.21** (Multiple limit cycles). *The equilibrium* $P_2 = (H, H + C)$ *is at least a two-order weak focus when* $S = H(M - 2H + 1)$. *The second Lyapunov quantity de is zero, which depends on the polynomial* $F(H, C, M, W)$ *defined in the proof of Theorem 2.16.*

*Proof.* The proof follow from Theorem 2.16, considering the factor $f_0(H, C, W)$, picking out $W = 1$, i.e.,

$$f_0(H, C) = (1 - H) H^3 (2H - 1)(C + H)(3C + 8H - 1)\left(-C + 2CH + H^2\right)^2$$

and evaluating this polynomial for $H = 0.4$ and $H = 0.6$ □

Now, considering $a_1 = 1 - Q > 0$ and $\Delta_{22} = 0$, we have

**Theorem 2.22** (Bogdanov–Takens bifurcation [28]). *Let* $u^{**} = \frac{1}{2}a_1$; *then,* $\det D\Gamma_\eta(u^{**}, u^{**} + C) = 0$, *and the point* $P_{**} = (u^{**}, u^{**} + C)$, *collapse of* $P_3$ *and* $P_4$, *is*

**a)** *a saddle-node attractor if and only if* $S > \frac{CQ(Q+1)}{2C-Q+1}$,
**b)** *a saddle-node repeller if and only if* $S < \frac{CQ(Q+1)}{2C-Q+1}$,
**c)** *a codimension two cusp point if and only if* $S = \frac{CQ(Q+1)}{2C-Q+1}$.

*Proof.* It follows immediately from Theorem 2.17. □

### 2.3.3 Model with weak Allee effect

Assuming $-1 < M < 0$, the system (2.3) has up five equilibrium points; they are $(0,0)$, $(1,0)$, $(0,C)$, and the points in the intersection of the isoclines curves: $v = \frac{1}{Q}(1-u)(u-M)$ and $v = u + C$.

The point $(M,0)$ is an equilibrium that lies in the second quadrant; it has no ecological interest, but it can have influence in the dynamics of system (2.3).

Fig. 2.3 shows the possible intersections among the isoclines, when $M < 0$.

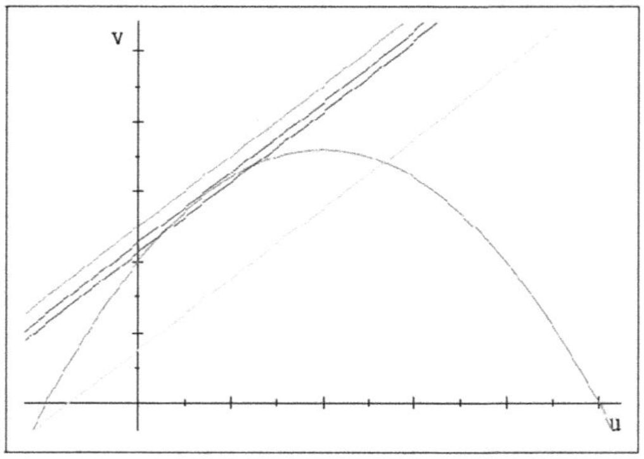


**FIGURE 2.3** Graph of the isoclines: the curve $v = \frac{1}{Q}(1-u)(u-M)$ is fixed, meanwhile the straight $v = u + C$ is dependent on the values of the parameter $C$.

There are four different cases for the intersection of the nullclines, as shown in Fig. 2.3, or else, if the solutions of Eq. (2.4) are considered, as was seen earlier for the case $M > 0$.

**Theorem 2.23** (Nature of equilibrium over the axes).

**i)** *The equilibrium* $(1,0)$ *is a hyperbolic saddle point.*
**ii)** *The equilibrium* $(0,0)$ *is a hyperbolic repeller.*
**iii)** *The point* $(0,C)$ *is*
  **iiia)** *an attractor point if and only if* $M + QC > 0$.
  **iiib)** *a hyperbolic saddle point if and only if* $M + QC < 0$.
  **iiic)** *a non-hyperbolic saddle-node if and only if* $M + QC = 0$.

*Proof.* i) and ii) are immediate.

ii) As the Jacobian matrix evaluated in the point $(0,C)$ is

$$DY_\eta(0,C) = \begin{pmatrix} -C(M+QC) & 0 \\ SC & -SC \end{pmatrix}.$$

Then, $\det DY_\eta(0,C) = SC^2(M+QC)$ depends on the sign of the factor

$$T_2 = M + QC.$$

iiia) If $M+QC<0$, the nature of the point $(0,C)$ depends on

$$\mathrm{tr}DY_\eta(0,C) = -C(M+QC) - SC,$$

which is negative.

Thus the point $(0,C)$ is an attractor point.

iiib) If $M+QC<0$, $\det DY_\eta(0,C)$ is negative, and the point $(0,C)$ is a hyperbolic saddle equilibrium.

iiic) If $M+QC=0$, $\det DY_\eta(0,C)$ is null, then the point $(0,C)$ is a non-hyperbolic equilibrium. Furthermore, as $\mathrm{tr}DY_\eta(0,C)$ is negative, then the point $(0,C)$ is a saddle-node. □

**Remark 2.24.** The result obtained by the precedent theorem indicates that the relationship between the parameter $C$, the intercept of the second (predator) isocline with the vertical axis, and the quotient between the parameters $M$ and $Q$ has significant consequences on the nature of the equilibrium point $(0,C)$.

To study the nature of the positive equilibria, we remember that the abscissa $u_e$ of the positive equilibrium points is solution of Eq. (2.4):

**1)** When $CQ+M>0$, and $a_1 = 1+M-Q>0$, there exist the same possibilities as in Case 2.1, revised above; i.e., there are two, one, or no positive real roots, depending on the sign of $\Delta$.

**1a)** Then Eq. (2.4) has

**1a.1)** two positive real roots if and only if $\Delta = (1+M-Q)^2 - 4(CQ+M) > 0$, which are

$$u_1 = \tfrac{1}{2}\left(1+M-Q-\sqrt{\Delta}\right) \text{ and } u_2 = \tfrac{1}{2}\left(1+M-Q+\sqrt{\Delta}\right),$$

with $0<u_1<u_2<1$.

**1a.2)** one positive real root if and only if $\Delta = 0$, which is given by

$$u^* = \frac{1}{2}(1+M-Q).$$

**1a.3)** no positive real roots if and only if $\Delta < 0$.

**1b)** no positive real roots if and only if $a_1 = 1+M-Q \le 0$.

**2)** When $CQ+M=0$, Eq. (2.4) becomes

$$p_1(u) = \left(u^2 - (1+M-Q)\right)u = 0.$$

Then,

**2i.** If $a_1 = 1 + M - Q > 0$, the unique positive real root is $u_5 = 1 + M - Q$; the solution $u = 0$ is repeated.

**2ii.** If $1 + M - Q \leq 0$, there exists no positive real root.

**3)** When $CQ + M < 0$,

**3i.** If $1 + M - Q \neq 0$, any let the sign of $1 + M - Q$, Eq. (2.4) has a unique positive real root given by

$$u_6 = \frac{1}{2}\left(1 + M - Q + \sqrt{\Delta}\right).$$

**3ii.** If $1 + M - Q = 0$, Eq. (2.4) becomes

$$p_2(u) = u^2 + (CQ + M) = 0.$$

If $CQ + M < 0$, the equation $p_2(u) = 0$ has a unique positive real root, given by $u_6 = \sqrt{-(CQ + M)}$.

If $CQ + M \geq 0$, there exists no positive real root.

When $a_1 = 1 + M - Q > 0$, $CQ + M > 0$, and $\Delta > 0$, it has the same cases obtained for $M > 0$. The nature of the equilibrium point is similar to that case.

By this, we analyze only the cases appearing as a unique equilibrium point, (different to the point $(u^*, u^* + C)$, when $\Delta = 0$), that this, when in (2.4)

**a)** $CQ + M < 0$, and $1 + M - Q < 0$, or

**b)** $CQ + M < 0$, and $1 + M - Q > 0$.

In such situations $\Delta > 0$; the unique solution of Eq. (2.4) is given by

$$u_2 = \frac{1}{2}\left(1 + M - Q + \sqrt{\Delta}\right),$$

with $u_1 < 0 < u_2 < 1$.

For both conditions, the equilibrium $(0, C)$ is a hyperbolic saddle point. Then, the stable manifold of this saddle point $\bar{\Sigma} = W_+^s(0, C)$, the straight line $u = 1$, and the coordinates axes determine a compact region $\bar{\Lambda}$.

Moreover, due to the nature of the equilibrium points $(0, 0)$, $(1, 0)$, and the Poincaré–Bendixson theorem, which is applicable in the region $\bar{\Lambda}$, the *$\omega$-limit* of the solutions of system (2.3) can be

**i)** a stable positive equilibrium point, or

**ii)** a stable limit cycle, or

**iii)** a stable positive equilibrium point and a stable limit cycle.

**Theorem 2.25** (Nature of the unique positive equilibrium point). *The equilibrium* $(u_2, u_2 + C)$ *is*

**a)** *an attractor if and only if* $S > u_2\left(Q + \sqrt{\Delta}\right)$,

**b)** *a repeller if and only if* $S < u_2\left(Q + \sqrt{\Delta}\right)$,

**c)** *a weak focus if and only if* $S = u_2\left(Q + \sqrt{\Delta}\right)$.

*Proof.* For the Jacobian matrix evaluated in the positive equilibria $(u_2, u_2 + C)$, with $u_2 = \frac{1}{2}\left((1 + M - Q) + \sqrt{\Delta}\right)$, it has

$$\det DY_\eta\,(u_2, u_2 + C) = -S\,(u_2 + C)^2\,(M - 2u_2 + 1 - Q)\,,$$

whose sign depends on the factor,

$$T = M - 2u_2 + 1 - Q, \text{ or}$$
$$T = M - \left((1 + M - Q) + \sqrt{\Delta}\right) + 1 - Q = -\sqrt{\Delta}.$$

Thus $\det DY_\eta\,(u_2, u_2 + C) > 0$.

Hence, the nature of the equilibrium depends of the trace, which is

$$\mathrm{tr} DY_\eta\,(u_2, u_2 + C) = (u_2 + C)\,(u_2\,(M - 2u_2 + 1) - S)\,,$$

whose sign depends on the factor

$$T_1 = u_2\,(M - 2u_2 + 1) - S,$$

which can be written as

$$T_1 = u_2\left(M - (1 + M - Q) + \sqrt{\Delta} + 1\right) - S = u_2\left(Q + \sqrt{\Delta}\right) - S.$$

Then, the thesis is obtained considering the sign of the factor $T_1$. □

**Corollary 2.26** (Existence of a limit cycle). *When* $S < u_2\left(Q + \sqrt{\Delta}\right)$, *there exists a infinitesimal limit cycle generated by Hopf bifurcation.*

*Proof.* The proof follows from the theorem described above, given change of sign. Moreover, the transversality condition [10,18,36] is verified since

$$\frac{\partial}{\partial S}\left(\mathrm{tr} DY_\eta\,(u, u + C)\right) = -\,(u_2 + C) < 0.$$

□

**Theorem 2.27** (Weakness of the fine focus). *The equilibrium* $(u_2, u_2 + C)$ *is at least a two-order weak focus if and only if* $S = H\,(M - 2H + 1)$.

*Proof.* Newly considering the last factor obtained in Theorem 2.16.

$$f_0\,(H, C, W)$$

and picking out $W = 1$, the following is obtained:

$$f_0\,(H, C) = (1 - H)\,H^3\,(2H - 1)\,(C + H)\,(3C + 8H - 1)\left(-C + 2CH + H^2\right)^2.$$

Evaluating this polynomial for $H = 0.4$ and $H = 0.6$ follows that $f_0(H, C, W)$ changes of sign □

**Remark 2.28.** The cases 2i and 3ii, where there is also a unique positive equilibrium point $(u_5, u_5 + C)$ and $(u_6, u_6 + C)$, with $u_5 = 1 + M - Q$ and $u_6 = \sqrt{-(CQ + M)}$, have a behavior similar to that studied in this subsection. In this case, there no exists the Bogdanov–Takens bifurcation.

### 2.3.4 Model without Allee effect

Suppose now that $M < -1$, i.e., the Allee effect is absent (see Fig. 2.1).

In Eq. (2.4), it has that $a_1 = 1 + M - Q < 0$.

Then, it fulfills that

**i)** If $CQ + M < 0$, there exists a unique positive real root, given by

$$u_2 = \tfrac{1}{2}\left(1 + M - Q + \sqrt{\Delta}\right).$$

**ii)** If $CQ + M \geq 0$, there is no positive real root.

Furthermore, there exists the compact region $\bar{\Lambda}$, determined by the stable manifold $\bar{\Sigma} = W_+^s(0, C)$ of the saddle point $(0, C)$, the straight line $u = 1$, and the coordinates axes. So, the Poincaré–Bendixson theorem applies in $\bar{\Lambda}$.

**Theorem 2.29** (Nature of equilibria over the horizontal axis). *For all* $\eta = (Q, C, S, M) \in \mathbb{R}_+^3 \times ]\infty, -1[$, *the singularity*

**I)** $(0, 0)$ *is a hyperbolic repeller,*

**II)** $(1, 0)$ *is a hyperbolic saddle point.*

*Proof.* Evaluating the Jacobian matrix in each equilibrium point, it has

**1.**

$$DY_\eta(0, 0) = \begin{pmatrix} -CM & 0 \\ 0 & SC \end{pmatrix},$$

$\det DY_\eta(0, 0) = -SC^2M > 0$ and $\mathrm{tr}DY_\eta(0, 0) = -C(M - S) > 0$.

Then, by the trace and the determinant theorem, the equilibrium $(0, 0)$ a hyperbolic repeller.

**2.**

$$DY_\eta(1, 0) = \begin{pmatrix} -(1 + C)(1 - M) & -Q(1 + C) \\ 0 & S(1 + C) \end{pmatrix},$$

and $\det DY_\eta(1, 0) = -S(1 + C)^2(1 - M) < 0$.

Therefore applying the trace and the determinant theorem, the equilibrium $(1, 0)$ is a saddle point. □

**Theorem 2.30** (Nature of equilibrium over the vertical axis). *The equilibrium point* $(0, C)$ *is*

**i)** *a saddle point if* $CQ + M < 0$,
**ii)** *an attractor if* $CQ + M > 0$,
**iii)** *an attractor-node if* $CQ + M = 0$.

*Proof.* The Jacobian matrix evaluated in the point $(0, C)$ is

$$DY_\eta (0, C) = \begin{pmatrix} -C(CQ + M) & 0 \\ SC & -SC \end{pmatrix}.$$

Then,

$$\det DY_\eta (0, C) = SC^2 (CQ + M),$$

whose sign depends on the factor $CQ + M$. Thus the equilibrium $(0, C)$ is

**i)** a saddle point if $CQ + M < 0$.
**ii)** an attractor if $CQ + M > 0$ since

$$\mathrm{tr} DY_\eta (0, C) = -(CQ + M + S)C < 0.$$

**iii)** an attractor-node if $CQ + M = 0$ since $\det DY_\eta (0, C) = 0$, and $\mathrm{tr} DY_\eta (0, C) = -SC < 0$. □

**Theorem 2.31** (Nature of the unique positive equilibrium point). *If* $CQ + M < 0$, *the point* $P_7 = (u_2, u_2 + C)$ *is globally asymptotically stable.*

*Proof.* Evaluating the Jacobian matrix in the positive equilibrium $(u_2, u_2 + C)$, we have that

$$\det DY_\eta (u_2, u_2 + C) = -S(u_2 + C)^2 (1 + M - Q - 2u_2) > 0$$

since the factor $T = 1 + M - Q - 2u_2 < 0$.

Thus its nature depends of the trace,

$$\mathrm{tr} DY_\eta (u_2, u_2 + C) = (u_2 + C)(u_2 (M - 2u_2 + 1) - S),$$

which depends on the sign of

$$T_1 = u_2 (M - 2u_2 + 1) - S,$$

which is negative, in this case.

Then, $\mathrm{tr}DY_\eta(u_2, u_2 + C) < 0$ and the unique positive equilibrium point $(u_2, u_2 + C)$ is a hyperbolic attractor.

Therefore as the equilibrium $(0, C)$ is a hyperbolic saddle, $(0, 0)$ is a hyperbolic repeller and $(1, 0)$ is a hyperbolic saddle point, by Poincaré–Bendixson theorem, the equilibrium $(u_2, u_2 + C)$ is globally asymptotically stable. □

To reinforce the result set forth in the above theorem, we show the inexistence of periodic orbits in the model.

**Theorem 2.32** (Non-existence of cycles). *When $M < -1$ in system (2.3), there is no limit cycles.*

*Proof.* Applying the Bendixson–Dulac criterion.

Considering the function $g(u, v) = \frac{1}{S(u+C)uv}$, we calculate the sum

$$\frac{d}{du}(P(u, v)\, g(u, v)) + \frac{d}{dv}(Q(u, v)\, g(u, v))$$

with

$$P(u, v) = ((1 - u)(u - M) - Qv)(u + C)u, \text{ and } Q(u, v) = S(u + C - v)v.$$

It has

$$\begin{aligned}
&\frac{d}{du}\left(((1 - u)(u - M) - Qv)\frac{1}{Sv}\right) + \frac{d}{dv}\left((u + C - v)\frac{1}{(u + C)u}\right) \\
&= \frac{1}{Sv}(M - 2u + 1) - \frac{1}{u(C + u)} < 0.
\end{aligned}$$

Therefore there is no limit cycles in system (2.3) when $M < -1$. □

**Remark 2.33.** We observe that in this case, the behavior of system (2.3) is similar to the behavior of the original Leslie–Gower-type [29] model, in which the Allee effect is not considered. Uf $CQ + M \geq 0$, the equilibrium $(0, C)$, is globally asymptotically stable in absence of positive equilibrium points.

## 2.4 Numerical simulations

Now, we show some simulations regarding the behavior of system (2.3) and the nature of the equilibrium points using the Pplane package.

1. In Figs. 2.4 to 2.8, we show the distinct situation of the model when $M > 0$.
2. In Figs. 2.9 to 2.12, the parameter $M < 0$.
   **2.1** In Figs. 2.9 to 2.11, there exist two positive equilibrium points.
   **2.2** In Figs. 2.12 to 2.15, there is one positive equilibrium point.

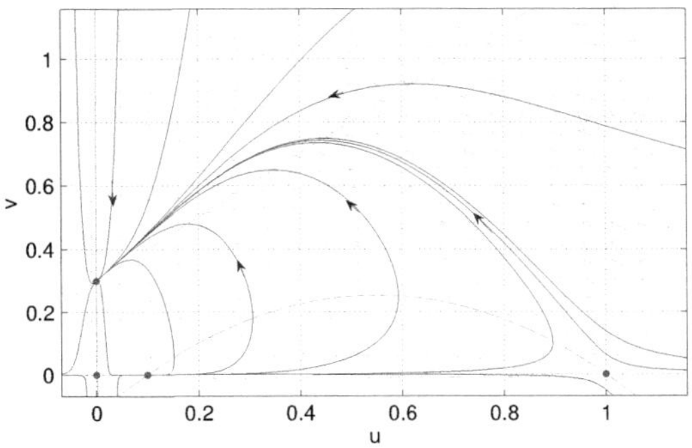


**FIGURE 2.4** For $M = 0.1$, $C = 0.3$, $Q = 0.8$, and $S = 1$, it has that $1 + M - Q \leq 0$; there are no interior equilibrium points, and $(0, C)$ is a global attractor node.

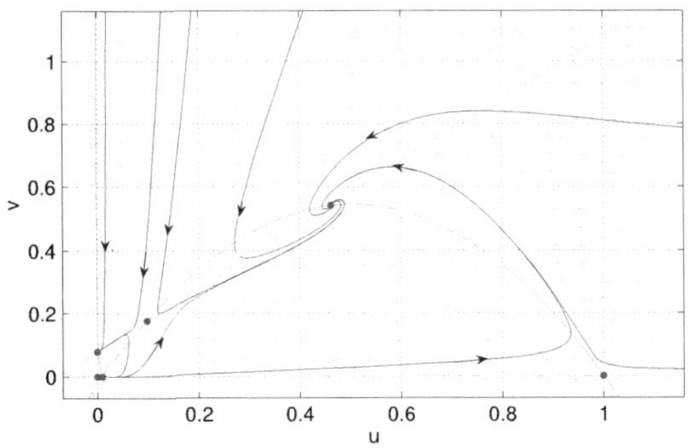


**FIGURE 2.5** For $M = 0.01$, $C = 0.0785$, $Q = 0.45$, and $S = 0.317$ there exist two positive equilibrium points. The equilibrium $P_2$ is an attractor focus, $P_1$ is a saddle point, and $(0, C)$ is an attractor node,.

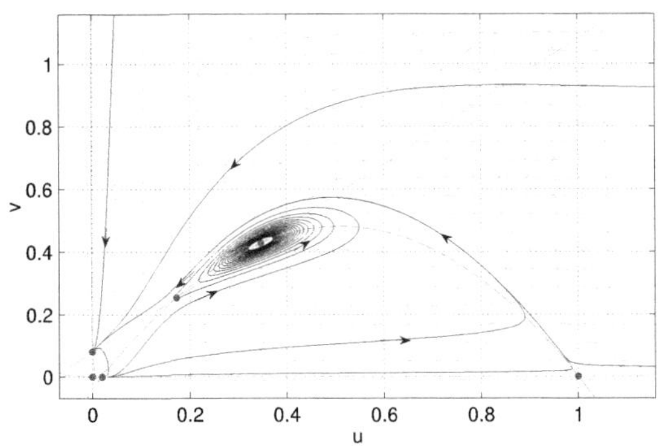


**FIGURE 2.6** For $M = 0.02$, $C = 0.08$, $Q = 0.5$, and $S = 0.114$, there are two interior equilibrium points. The equilibrium point $P_2$ is a local attractor focus, $P_1$ is a saddle point, and $(0, C)$ is an attractor node,.

## 2.5 Conclusions

Mathematical models are a fundamental tool to understand and analyze the dynamics of food chains. In population ecology, a classic and significant theme is the interaction be-

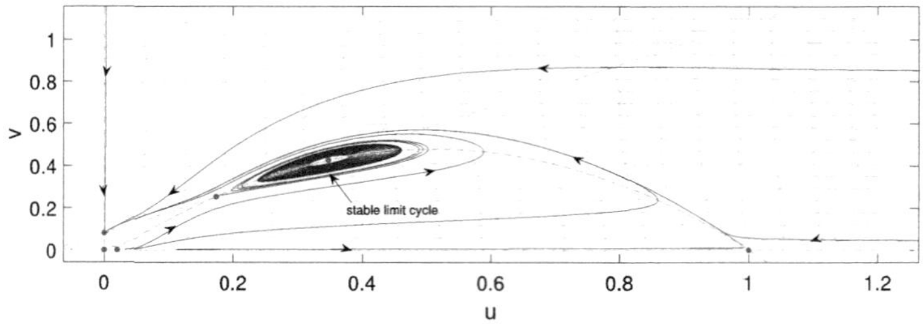


**FIGURE 2.7** For $M = 0.02$, $C = 0.08$, $Q = 0.5$, and $S = 0.111$ there are two interior equilibrium points. The equilibrium $P_2$ is repeller surrounded by a stable limit cycle; $P_1$ is a saddle point, and $(0, C)$ is an attractor node.

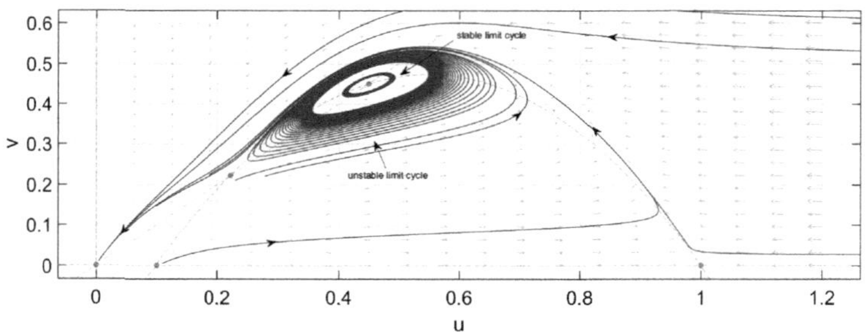


**FIGURE 2.8** For $M = 0.1$, $C = 0.0001$, $Q = 0.42778$, and $S = 0.0891875$, there are two interior equilibrium points. The point $P_2$ is a repeller surrounded by two limit cycles: the innermost stable and the outermost unstable, and $(0, C)$ is an attractor node.

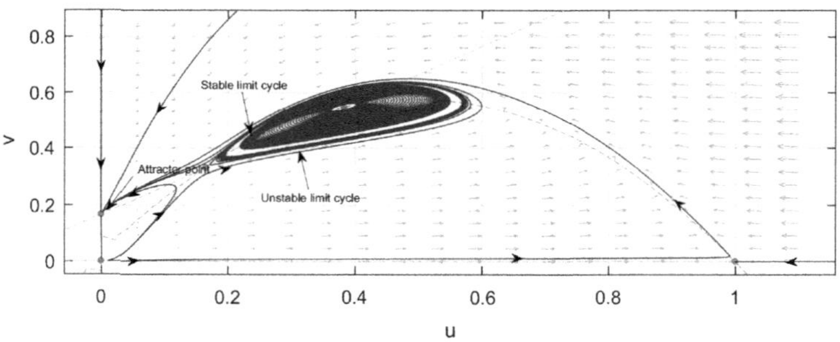


**FIGURE 2.9** For $M = -0.01$, $C = 0.165$, $Q = 0.442785$, and $S = 0.084075$, there are two interior equilibrium points. The point $P_2$ is a repeller surrounded by two limit cycles: the innermost stable and the outermost unstable; moreover $(0, C)$ is an attractor node.

tween predators and their prey, which are an important members of those chains. For this reason, in recent years, Leslie-type predation models have received increased attention to investigate dynamic behaviors between species.

In this work we have shown the main properties of a Leslie–Gower type predation model, assuming that the predators have an alternative food in the event that their favorite prey is scarce. Moreover, it has been assumed that the prey population is affected by the Allee effect [5,7,8,31].

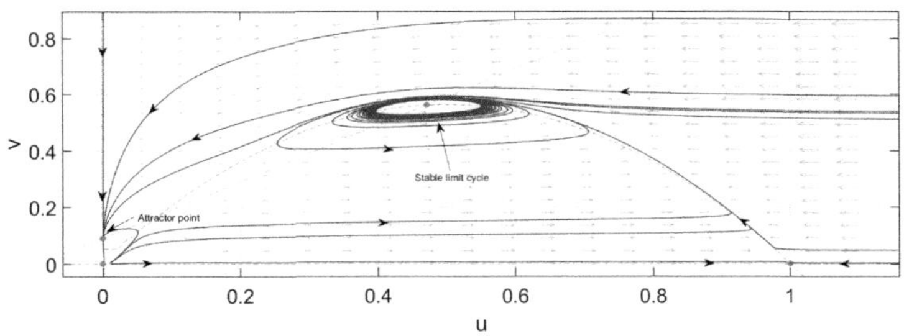


**FIGURE 2.10** For $M = -0.001$, $C = 0.08$, $Q = 0.445$, and $S = 0.02$ there are two interior equilibrium points. The equilibrium $P_2$ is repeller surrounded by a stable limit cycle, $P_1$ is saddle point, and $(0, C)$ is an attractor node.

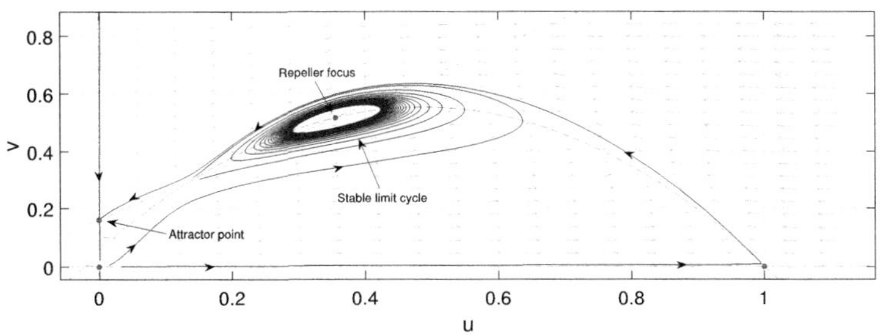


**FIGURE 2.11** For $M = -0.0255$, $C = 0.1605$, $Q = 0.47575$, and $S = 0.09275$ are two positive equilibrium points. The equilibrium $P_2$ is repeller surrounded by a stable limit cycle, $P_1$ is saddle point, and $(0, C)$ is an attractor node.

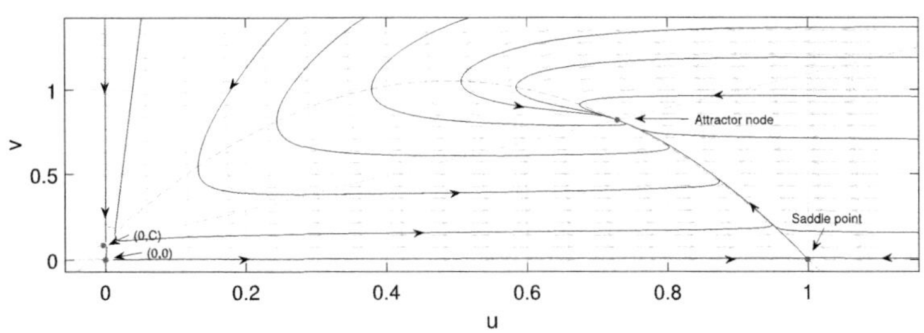


**FIGURE 2.12** For $M = -0.025$, $C = 0.09$, $Q = 0.25$, and $S = 0.035$ there is one positive equilibrium point. The equilibrium $P_2$ is an attractor node, and $(0, C)$ is a saddle point.

By means of a diffeomorphism, we obtained a topologically equivalent system to the original one, dependent only on four parameters. We show that in this model (and therefore in the original) there are different dynamic situations according to the relationships obtained between the parameters.

Among the most important results, the existence of a separator curve $\bar{\Sigma}$ has been demonstrated, dividing the behavior of the solutions or trajectories of the system in the phase plane. This separatrix curve exists when there are two positive equilibrium points within the first quadrant. It is generated by the upper stable manifold $W^s_+(u_{e1}, u_{e1} + C)$ of the saddle point $P_1$. This implies that for the same set of parameters—two very close so-

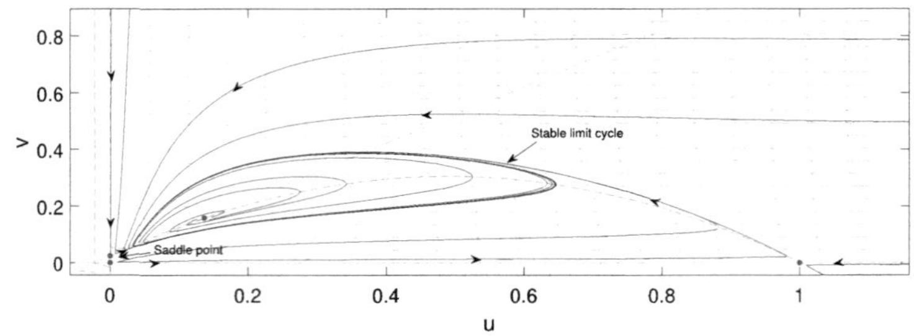


**FIGURE 2.13** For $M=-0.02$, $C=0.0225$, $Q=0.85$ and $S=0.07$ there is one positive equilibrium point. The equilibrium $P_2$ is repeller surrounded by a stable limit cycle, and $(0, C)$ is a saddle node.

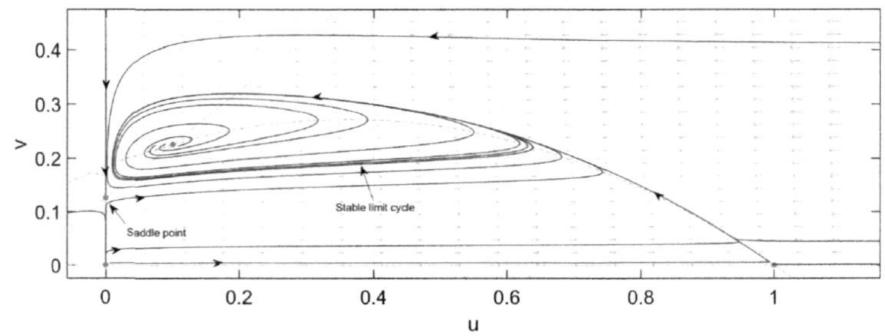


**FIGURE 2.14** For $M=-0.275$, $C=0.125$, $Q=1.5$, and $S=0.025$ there exists one positive equilibrium point. The equilibrium $P_2$ is repeller surrounded by a stable limit cycle, and $(0, C)$ is a saddle point.

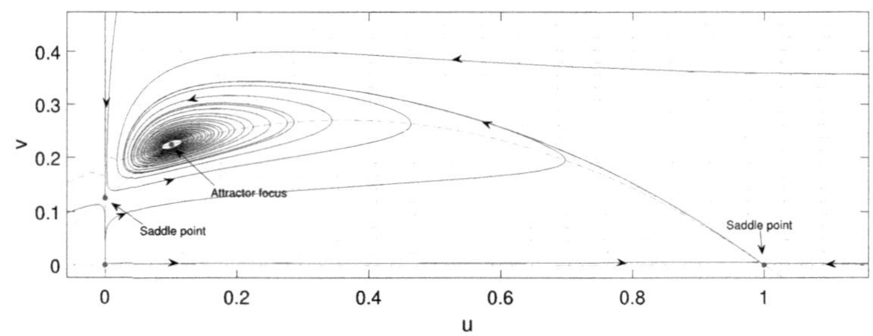


**FIGURE 2.15** For $M=-0.275$, $C=0.125$, $Q=1.5$, and $S=0.055$ there is one interior equilibrium point. The equilibrium $P_2$ is an attractor focus, and $(0, C)$ is a saddle point.

lutions but on a different side of that separatrix $\bar{\Sigma}$—would have different and far $\omega$-*limits*, which means that the solutions are highly sensitive to the initial conditions.

The studied model has a rich and interesting dynamic, just as other modified Leslie–Gower type models, which stands out when compared with the Leslie–Gower model, where the growth rate of the prey is of the logistic type and the predators also have an alternative food, but in which the Allee effect is absent. In this simple model, there is a single positive equilibrium point that is globally stable asymptotic (gas), which is proved by constructing adequate Lyapunov functions [24].

Besides, there are significant differences between the model (2.3) studied here, and the model that does not consider alternative food for predators, when $C = 0$, studied in [24]. One of them is that in the case of $M = 0$, our model has two limit cycles, whereas in the previous model there is only a single limit cycle.

Another difference is that in the model with alternative food for predators, there is the new equilibrium point $(0, C)$ on the vertical axis, which can be attractor in the case where $M > 0$, and can be a chair point if $M < 0$; it fulfills the condition $M + QC \leq 0$.

We have established the number of limit cycles, surrounding a positive equilibrium point of the center type focus, originated by multiple Hopf bifurcations [10,18,19,23]. We also believe that the determination of the number of limit cycles that branch from a weak focus [10,28,36] is a good criterion for the classification of predator-prey models.

Our results also confirm the fact that if the predator population is generalist, the extinction of each species can be avoided for certain initial conditions [22,25]. Furthermore, the analyzed model satisfies the recommendations suggested in [6] for being credible, simple, and useful.

Some results obtained in [2] are also extended, showing the existence of i) a homoclinic curve, ii) two infinitesimal limit cycles surrounding a positive equilibrium point, and iii) a non-infinitesimal limit cycle, generated when the homoclinic is broken; its stability is determined by the absolute value of the quotient between the negative and the positive eigenvalues [4].

# References


[1] P. Aguirre, E. González-Olivares, E. Sáez, Three limit cycles in a Leslie-Gower predator-prey model with additive Allee effect, SIAM Journal on Applied Mathematics 69 (5) (2009) 1244–1269.

[2] C. Arancibia-Ibarra, The basins of attraction in a modified May-Holling-Tanner predator-prey model with Allee effect, Nonlinear Analysis 185 (2019) 15–28.

[3] D.K. Arrowsmith, C.M. Place, Dynamical System. Differential Equations, Maps and Chaotic Behaviour, Chapman and Hall, 1992.

[4] A.D. Bazykin, Nonlinear Dynamics of Interacting Populations, World Scientific Publishing Co. Pte. Ltd., 1998.

[5] L. Berec, E. Angulo, F. Courchamp, Multiple Allee effects and population management, Trends in Ecology & Evolution 22 (2007) 185–191.

[6] A.A. Berryman, A.P. Gutierrez, R. Arditi, Credible, parsimonious and useful predator-prey models - a reply to Abrams, Gleeson and Sarnelle, Ecology 76 (1995) 1980–1985.

[7] D.S. Boukal, L. Berec, Single-species models and the Allee effect: extinction boundaries, sex ratios and mate encounters, Journal of Theoretical Biology 218 (2002) 375–394.

[8] D.S. Boukal, M.W. Sabelis, L. Berec, How predator functional responses and Allee effects in prey affect the paradox of enrichment and population collapses, Theoretical Population Biology 72 (2007) 136–147.

[9] K-S. Cheng, Uniqueness of a limit cycle for a predator-prey system, SIAM Journal on Mathematical Analysis 12 (1981) 541–548.

[10] C. Chicone, Ordinary Differential Equations with Applications, 2nd edition, Texts in Applied Mathematics, vol. 34, Springer, 2006.

[11] C.W. Clark, Mathematical Bioeconomics: The Optimal Management of Renewable Resources, 2nd edition, John Wiley and Sons, 1990.

[12] C.S. Coleman, Hilbert's 16th. problem: how many cycles?, in: M. Braun, C.S. Coleman, D. Drew (Eds.), Differential Equations Model, Springer Verlag, 1983, pp. 279–297.

[13] F. Courchamp, T. Clutton-Brock, B. Grenfell, Inverse density dependence and the Allee effect, Trends in Ecology & Evolution 14 (10) (1999) 405–410.

[14] F. Courchamp, L. Berec, J. Gascoigne, Allee Effects in Ecology and Conservation, Oxford University Press, 2007.

[15] B. Dennis, Allee effects: population growth, critical density, and the chance of extinction, Natural Resource Modeling 3 (4) (1989) 481–538.

[16] F. Dumortier, J. Llibre, J.C. Artés, Qualitative Theory of Planar Differential Systems, Springer, 2006.

[17] H.I. Freedman, Deterministic Mathematical Models in Population Ecology, Marcel Dekker, 1980.

[18] V.A. Gaiko, Global Bifurcation Theory and Hilbert's Sixteenth Problem, Mathematics and Its Applications, vol. 559, Kluwer Academic Publishers, 2003.

[19] V.A. Gaiko, C. Vuik, Global dynamics in the Leslie-Gower model with the Allee effect, International Journal of Bifurcation and Chaos 28 (12) (2018) 1850151.

[20] B-S. Goh, Management and Analysis of Biological Populations, Elsevier Scientific Publishing Company, 1980.

[21] E. González-Olivares, B. González-Yañez, J. Mena-Lorca, R. Ramos-Jiliberto, Modelling the Allee effect: are the different mathematical forms proposed equivalents?, in: R. Mondaini (Ed.), Proceedings of the 2006 International Symposium on Mathematical and Computational Biology, E-Papers Serviços Editoriais Ltda., Rio de Janeiro, 2007, pp. 53–71.

[22] E. González-Olivares, J. Mena-Lorca, A. Rojas-Palma, J.D. Flores, Dynamical complexities in the Leslie-Gower predator-prey model as consequences of the Allee effect on prey, Applied Mathematical Modelling 35 (2011) 366–381.

[23] E. González-Olivares, A. Rojas-Palma, B. González-Yañez, Multiple limit cycles in a Leslie-Gower type predator-prey model considering weak Allee effect on prey, Nonlinear Analysis: Modelling and Control 22 (3) (2017) 347–365.

[24] E. González-Olivares, C. Arancibia-Ibarra, A. Rojas-Palma, B. González-Yañez, Bifurcations and multistability on the May-Holling-Tanner predation model considering alternative food for the predators, Mathematical Biosciences and Engineering 16 (5) (2019) 4274–4298.

[25] B. González-Yañez, E. González-Olivares, J. Mena-Lorca, Multistability on a Leslie-Gower type predator-prey model with nonmonotonic functional response, in: R. Mondaini, R. Dilao (Eds.), BIOMAT 2006 - International Symposium on Mathematical and Computational Biology, World Scientific Co. Pte. Ltd., 2007, pp. 359–384.

[26] G.W. Harrison, Global stability of predator-prey interactions, Journal of Mathematical Biology 8 (1979) 159–171.

[27] M. Kot, Elementary Mathematical Ecology, Cambridge University Press, 2001.

[28] Y.A. Kuznetsov, Elements of Applied Bifurcation Theory, 3rd edition, Springer, 2004.

[29] P.H. Leslie, Some further notes on the use of matrices in population mathematics, Biometrika 35 (1948) 213–245.

[30] P.H. Leslie, J.C. Gower, The properties of a stochastic model for the predator-prey type of interaction between two species, Biometrika 47 (1960) 219–234.

[31] M. Liermann, R. Hilborn, Depensation: evidence, models and implications, Fish and Fisheries 2 (2001) 33–58.

[32] N. Martínez-Jeraldo, P. Aguirre, Allee effect acting on the prey species in a Leslie-Gower predation model, Nonlinear Analysis: Real World Applications 45 (2019) 895–917.

[33] N. Martínez-Jeraldo, E. Rozas-Torres, E. González-Olivares, A Leslie-Gower-type predation model considering generalist predators and the Allee effect on prey, Selecciones Matemáticas 8 (1) (2021) 147–160 (in Spanish).

[34] R.M. May, Stability and Complexity in Model Ecosystems, 2nd ed., Princeton University Press, 2001.

[35] P. Monzón, Almost global attraction in planar systems, Systems & Control Letters 54 (2005) 753–758.

[36] L. Perko, Differential Equations and Dynamical Systems, 3rd edition, Springer-Verlag, 2001.

[37] J.R. Philip, Sociality and sparce populations, Ecology 38 (1957) 107–111.

[38] P.A. Stephens, W.J. Sutherland, Consequences of the Allee effect for behaviour, ecology and conservation, Trends in Ecology & Evolution 14 (1999) 401–405.

[39] P.A. Stephens, W.J. Sutherland, R.P. Freckleton, What is the Allee effect?, Oikos 87 (1999) 185–190.

[40] P. Turchin, Complex Population Dynamics. A Theoretical/Empirical Synthesis, Monographs in Population Biology, vol. 35, Princeton University Press, 2003.

[41] G.A.K. van Voorn, L. Hemerik, M.P. Boer, B.W. Kooi, Heteroclinic orbits indicate overexploitation in predator-prey systems with a strong Allee, Mathematical Biosciences 209 (2007) 451–469.
[42] G. Wang, X-G. Liang, F-Z. Wang, The competitive dynamics of populations subject to an Allee effect, Ecological Modelling 124 (1999) 183–192.
[43] M-H. Wang, M. Kot, Speeds of invasion in a model with strong or weak Allee effects, Mathematical Biosciences 171 (2001) 83–97.
[44] J. Wang, J. Shi, J. Wei, Predator-prey system with strong Allee effect in prey, Journal of Mathematical Biology 62 (2011) 291–331.
[45] S. Wolfram, Mathematica: A System for Doing Mathematics by Computer, 2nd edition, Wolfram Research, Addison Wesley, 1991.
[46] S-R. Zhou, Y-F. Liu, G. Wang, The stability of predator-prey systems subject to the Allee effect, Theoretical Populations Biology 67 (2005) 23–31.

3

# Modeling and analysis of an eco-epidemiological model with Caputo–Fabrizio derivative

**Muhammad Altaf Khan**[a] **and Behzad Ghanbari**[b]

[a] *Institute of Ground Water Studies, Faculty of Natural and Agricultural Sciences, University of the Free State, Bloemfontein, South Africa* [b] *Department of Engineering Science, Kermanshah University of Technology, Kermanshah, Iran*

## 3.1 Introduction

The mathematical modeling of species and its interaction with each other is an interesting research area. In the field of Ecology and Biology, the modeling of prey-predator and their interaction is gaining much interest from researchers, and hence many valuable research has been published [1,2]. The mathematical modeling tool has been employed to investigate the dynamics mentioned above by modifying and developed other models [3,4]. Mathematical models in deterministic form can be classified as ordinary, or partial differential equations, fractional differential equations, and/or stochastic differential equations; see, for example, the population dynamics with Allee effect is considered in [5]. The ecological study in partial differential equation is discussed in [3]. The interaction among species has been discussed in [2], and the prey-predator model in discrete sense discussed in [6]. The prey-predator model has a vital role in my ecological study of the interaction of the species and to determine the possible relations of existence of these in nature.

The mathematical model constructed in fractional derivative was designed to explore the behaviors of species or individuals in more broad sense, given the limitations of other models. So, the model with fractional order are more dynamical than the integer order, and researchers can benefit from the rich dynamics of the model. In Mathematical Biology, fractional calculus concept has been utilized by many researchers, and still the authors are using it for their models. The field of fractional calculus gains much attention when the new operators are developed, such as the Caputo–Fabrizio and the Atangana–Baleanu derivative. These operators are extensively used for the modeling purposes of the real-world problems; note, for example, that the authors use the concept of fractional operator in Atangana–Baleanu sense to study the coronavirus model [7]. The application of fractional derivative to investigate the dynamics of malaria model with fractional optimal control problem has been studied in [8]. The dengue fever model and its analysis in fractional derivative is explored in [9]. The authors used the fractional derivative in studying

**Mathematical Modeling in Bioscience. https://doi.org/10.1016/B978-0-44-315445-4.00009-3**

Copyright © 2025 Elsevier Inc. All rights are reserved, including those for text and data mining, AI training, and similar technologies.

the coronavirus model using genetic algorithm-based control strategy in [10]. The dynamics of cancer model using the fractional derivative with the motivation of immune response has been studied in [11]. It is obvious that the fractional-order models capture many dynamics properties, which cannot be obtained using ordinary order models. Due to many advantages of the fractional-order models, we use it in the present work to explore the dynamical analysis of an eco-epidemiological model in Caputo–Fabrizio derivative.

The present investigations consider a eco-epidemiological model published in [12] with three components that describes the prey-predator interaction. We extend this model into fractional-order using the concept of Caputo–Fabrizio derivative and investigate its dynamical analysis and present various numerical results that were not explored yet. We write the model in integer order, Section 3.2, and then using the definition of fractional derivative, we apply the Caputo–Fabrizio derivative. We present in Section 3.3 some related mathematical results of the model. In Section 3.4, we explain the numerical scheme for the fractional model in details, and in Section 3.4.1, we use many important parameters of the model and present the graphical results. We finally write the main findings of the model in the conclusion, Section 3.5.

## 3.2 Mathematical model

We consider a 3-dimensional model of prey-predator system. The model has three compartments: the susceptible prey, infected prey, and the predator population—$S(t)$, $I(t)$ and $P(t)$, respectively. The model considered in [12] is given by

$$\begin{aligned}
\frac{dS(t)}{dt} &= \left[r - bS(t) - cl(t) - \frac{\beta l(t)}{a+S(t)} - \frac{\alpha_1 P(t)}{e+S(t)}\right] S(t),\\
\frac{dl(t)}{dt} &= \left[\frac{\beta S(t)}{a+S(t)} - \frac{\alpha_2 P(t)}{d+1(t)} - \mu\right] I(t),\\
\frac{dP(t)}{dt} &= \left[\left(\frac{c_1\alpha_1 S(t)}{e+S(t)} + \frac{c_2\alpha_2 I(t)}{d+l(t)}\right)\frac{P(t)}{\tau+P(t)} - m\right] P(t).
\end{aligned} \tag{3.1}$$

In above model, the $r$ is the growth rate of the susceptible prey. The infected prey dies with the death rate $\mu$, whereas the infected prey dies naturally with the rate $m$. $c_1$ and $c_2$, respectively, present the conversion efficacy of predator on susceptible and infected prey. The intra-class competition among the susceptible is shown by $b$, whereas the parameter $c$ is for the intra-class impact on infected prey. The parameters $\alpha_1$ and $\alpha_2$, respectively, show the attacking rate of predator on the susceptible, and infected prey. $d, e, a$ are the half saturation constants of predator on infected, susceptible, and the disease transmission, respectively. Respectively, $\beta$ and $\tau$ denote the infection rate of contact and the Allee parameter. All the parameters are non-negative, and the model has the non-negative initial condition: $S(0) \geq 0$, $I(0) \geq 0$, $P(0) \geq 0$. Next, we present the fractional version of the model (3.1).

### 3.2.1 Fractional model

We follow the definition given by for the formulation of fractional model.

**Definition 3.1.** [13] Consider $w \in M^1(b_1, b_2)$, $b_2 > b_1$ and $\theta \in [0, 1]$. Then we can define the Caputo–Fabrizio derivative given by

$$ {}_0^{CF}D_t^{\theta} w(t) = \frac{C(\theta)}{1-\theta} \int_a^t w'(\kappa) \exp\left[-\frac{\theta}{1-\theta}(t-\kappa)\right] \mathrm{d}\kappa, \tag{3.2} $$

where $C(\theta)$ is the normalization function with $C(0) = C(1) = 1$.

Using the above Definition (3.2), we write the fractional model given by

$$ \begin{aligned} {}_0^{CF}D_t^{\theta} \mathrm{S}(t) &= \left[r - b\,\mathrm{S}(t) - \mathrm{cl}(t) - \frac{\beta \mathrm{l}(t)}{a + \mathrm{S}(t)} - \frac{\alpha_1 \mathrm{P}(t)}{e + \mathrm{S}(t)}\right] \mathrm{S}(t), \\ {}_0^{CF}D_t^{\theta} \mathrm{l}(t) &= \left[\frac{\beta \mathrm{S}(t)}{a + \mathrm{S}(t)} - \frac{\alpha_2 \mathrm{P}(t)}{d + 1(t)} - \mu\right] \mathrm{I}(t), \\ {}_0^{CF}D_t^{\theta} \mathrm{P}(t) &= \left[\left(\frac{c_1\alpha_1\,\mathrm{S}(t)}{e + \mathrm{S}(t)} + \frac{c_2\alpha_2 l(t)}{d + l(t)}\right) \frac{\mathrm{P}(t)}{\tau + \mathrm{P}(t)} - m\right] \mathrm{P}(t). \end{aligned} \tag{3.3} $$

Next, we present the model possible equilibrium points and its stability. To have these equilibrium points, we refer the reader to follow [12].

## 3.3 Existence and uniqueness

We give the existence and uniqueness results for the fractional model (3.3) below using [14]:

$$ \begin{aligned} S(t) - S(0) &= {}_0^{CF}I_t^{\theta}\left\{\left[r - bS(t) - cl(t) - \frac{\beta l(t)}{a + S(t)} - \frac{\alpha_1 P(t)}{e + S(t)}\right] S(t)\right\}, \\ I(t) - I(0) &= {}_0^{CF}I_t^{\theta}\left\{\left[\frac{\beta S(t)}{a + S(t)} - \frac{\alpha_2 P(t)}{d + 1(t)} - \mu\right] I(t)\right\}, \\ P(t) - P(0) &= {}_0^{CF}I_t^{\theta}\left\{\left[\left(\frac{c_1\alpha_1 S(t)}{e + S(t)} + \frac{c_2\alpha_2 I(t)}{d + l(t)}\right) \frac{P(t)}{\tau + P(t)} - m\right] P(t)\right\}. \end{aligned} \tag{3.4} $$

Furthermore, we use [14] and write the following:

$$ \begin{aligned} S(t) - S(0) = &\frac{2(1-\theta)}{C(\theta)(2-\theta)}\left\{\left[r - bS(t) - cl(t) - \frac{\beta l(t)}{a + S(t)} - \frac{\alpha_1 P(t)}{e + S(t)}\right] S(t)\right\} \\ &+ \frac{2\theta}{C(\theta)(2-\theta)} \int_0^t \left\{\left[r - bS(t) - cl(t) - \frac{\beta l(t)}{a + S(t)} - \frac{\alpha_1 P(t)}{e + S(t)}\right] S(t)\right\} \mathrm{d}\kappa \\ E(t) - E(0) = &\frac{2(1-\theta)}{C(\theta)(2-\theta)}\left\{\left[\frac{\beta S(t)}{a + S(t)} - \frac{\alpha_2 P(t)}{d + 1(t)} - \mu\right] I(t)\right\} \end{aligned} $$

$$
\begin{aligned}
&+\frac{2\theta}{C(\theta)(2-\theta)}\int_0^t\left\{\left[\frac{\beta S(t)}{a+S(t)}-\frac{\alpha_2 P(t)}{d+1(t)}-\mu\right]I(t)\right\}\mathrm{d}\kappa\\
A(t)-A(0)=&\frac{2(1-\theta)}{C(\theta)(2-\theta)}\left\{\left[\left(\frac{c_1\alpha_1 S(t)}{e+S(t)}+\frac{c_2\alpha_2 I(t)}{d+l(t)}\right)\frac{P(t)}{\tau+P(t)}-m\right]P(t)\right\}\\
&+\frac{2\theta}{C(\theta)(2-\theta)}\int_0^t\left\{\left[\left(\frac{c_1\alpha_1 S(t)}{e+S(t)}+\frac{c_2\alpha_2 I(t)}{d+l(t)}\right)\frac{P(t)}{\tau+P(t)}-m\right]P(t)\right\}\mathrm{d}\kappa. \qquad (3.5)
\end{aligned}
$$

Let

$$
\begin{aligned}
N_1(t,S)&=\left\{\left[r-bS(t)-cl(t)-\frac{\beta l(t)}{a+S(t)}-\frac{\alpha_1 P(t)}{e+S(t)}\right]S(t)\right\},\\
N_2(t,I)&=\left\{\left[\frac{\beta S(t)}{a+S(t)}-\frac{\alpha_2 P(t)}{d+1(t)}-\mu\right]I(t)\right\},\\
N_3(t,P)&=\left\{\left[\left(\frac{c_1\alpha_1 S(t)}{e+S(t)}+\frac{c_2\alpha_2 I(t)}{d+l(t)}\right)\frac{P(t)}{\tau+P(t)}-m\right]P(t)\right\}. \qquad (3.6)
\end{aligned}
$$

**Theorem 3.1.** *The kernels given by $N_i$ for $i, 2, 3$ satisfy the Lipschitz condition and contraction if $J_1=\left\|b+c\overline{I}+\beta\overline{l}+\alpha_1\overline{P}\right\|$.*

*Proof.* Let us start with $N_1$ and consider $S$ and $S_1$ as functions, and we have the following:

$$
\begin{aligned}
\|N_1(t,S)-N_1(t,S_1)\|=&\left\|-b\left(S(t)-S(t_1)\right)-cI(t)-\frac{\beta l(t)}{a+(S(t)-S(t_1))}-\frac{\alpha_1 P(t)}{e+(S(t)-S(t_1))}\right\|\\
&\times\|(S(t)-S(t_1))\|,\\
\leq&\left\|-b\left(S(t)-S(t_1)\right)-cI(t)-\beta l(t)-\alpha_1 P(t)\right\|\|(S(t)-S(t_1))\|,\\
\leq&\left[b\|(S(t)-S(t_1))\|+\|cI(t)+\beta l(t)+\alpha_1 P(t)\|\left\|\right]\left(S(t)-S(t_1)\right)\right\|,\\
\leq& b\|(S(t)-S(t_1))\|\|(S(t)-S(t_1))\|+\|cI(t)+\beta l(t)+\alpha_1 P(t)\|\\
&\times\|(S(t)-S(t_1))\|,\\
\leq& b\|(S(t)-S(t_1))\|+\|cI(t)+\beta l(t)+\alpha_1 P(t)\|\|(S(t)-S(t_1))\|\\
\leq&\|b+cI(t)+\beta l(t)+\alpha_1 P(t)\|\|(S(t)-S(t_1))\|\\
\leq& J_1\|(S(t)-S(t_1))\|, \qquad (3.7)
\end{aligned}
$$

where $J_1=\left\|b+c\overline{I}+\beta\overline{l}+\alpha_1\overline{P}\right\|$, $\|I(t)\|\leq\bar{I}$ and $\|P(t)\|\leq\bar{P}$ are bounded and gives

$$
\|N_1(t,S)-N_1(t,S_1)\|=J_1\|(S(t)-S(t_1))\|. \qquad (3.8)
$$

Thus, we obtained the expression for $N_1$ and with the additional condition $0\leq J_1<1$ contraction is implied. So the remaining cases can be achieved as the above rule, given by

$$
\|N_2(t,I)-N_2(t,I_1)\|\leq J_2\|I(t)-I(t_1)\|,\ \|N_3(t,P)-N_3(t,P_1)\|\leq J_3\|P(t)-P(t_1)\|. \qquad (3.9)
$$

We write Eq. (3.5) with the kernels above:

$$\begin{aligned}
S(t) &= S(0) + \frac{2(1-\theta)}{(2-\theta)C(\theta)} N_1(t, S) + \frac{2\theta}{(2-\theta)C(\theta)} \int_0^t (N_1(\kappa, S))\,\mathrm{d}\kappa, \\
I(t) &= I(0) + \frac{2(1-\theta)}{(2-\theta)C(\theta)} N_2(t, I) + \frac{2\theta}{(2-\theta)C(\theta)} \int_0^t (N_2(\kappa, I))\,\mathrm{d}\kappa, \\
P(t) &= P(0) + \frac{2(1-\theta)}{C(\theta)(2-\theta)} N_3(t, P) + \frac{2\theta}{(2-\theta)C(\theta)} \int_0^t (N_3(\kappa, P))\,\mathrm{d}\kappa.
\end{aligned} \tag{3.10}$$

Furthermore, we have the formula

$$S_n(t) = \frac{2(1-\theta)}{(2-\theta)C(\theta)} N_1(t, S_{n-1}) + \frac{2\theta}{(2-\theta)C(\theta)} \int_0^l (N_1(\kappa, S_{n-1}))\,\mathrm{d}\kappa, \tag{3.11}$$

$$I_n(t) = \frac{2(1-\theta)}{(2-\theta)C(\theta)} N_2(t, I_{n-1}) + \frac{2\theta}{(2-\theta)C(\theta)} \int_0^t (N_2(\kappa, I_{n-1}))\,\mathrm{d}\kappa, \tag{3.12}$$

$$A_n(t) = \frac{2(1-\theta)}{(2-\theta)C(\theta)} N_3(t, P_{n-1}) + \frac{2\theta}{(2-\theta)C(\theta)} \int_0^t (N_3(\kappa, P_{n-1}))\,\mathrm{d}\kappa, \tag{3.13}$$

and the initial conditions are $(S(0), I(0), P(0)) = (S_0 I_0, P_0)$. We have then further the equation, given by

$$\begin{aligned}
\phi_n(t) &= S_n(t) - S_{n-1}(t) \\
&= \frac{2(1-\theta)}{(2-\theta)C(\theta)} (N_1(t, S_{n-1}) - N_1(t, S_{n-2})) \\
&\quad + \frac{2\theta}{(2-\theta)C(\theta)} \int_0^t (N_1(\kappa, S_{n-1}) - N_1(\kappa, S_{n-2}))\,\mathrm{d}\kappa, \\
\psi_n(t) &= I_n(t) - I_{n-1}(t) \\
&= \frac{2(1-\theta)}{(2-\theta)C(\theta)} (N_2(t, I_{n-1}) - N_2(t, I_{n-2})) \\
&\quad + \frac{2\theta}{(2-\theta)C(\theta)} \int_0^t (N_2(\kappa, E_{n-1}) - N_2(\kappa, E_{n-2}))\,\mathrm{d}\kappa, \\
\xi_n(t) &= P_n(t) - P_{n-1}(t) \\
&= \frac{2(1-\theta)}{(2-\theta)C(\theta)} (N_3(t, P_{n-1}) - N_3(t, P_{n-2})) \\
&\quad + \frac{2\theta}{(2-\theta)C(\theta)} \int_0^t (N_3(\kappa, P_{n-1}) - N_3(\kappa, P_{n-2}))\,\mathrm{d}\kappa.
\end{aligned} \tag{3.14}$$

It is necessary to mention that

$$S_n(t) = \sum_{i=1}^n \phi_i(t), \quad I_n(t) = \sum_{i=1}^n \psi_i(t), \quad P_n(t) = \sum_{i=1}^n \xi_i(t). \tag{3.15}$$

We have the detailed calculation, given by

$$\|\phi_n(t)\| = \|S_n(t) - S_{n-1}(t)\|$$

$$= \| \frac{2(1-\theta)}{(2-\theta)C(\theta)} \left(N_1(t, S_{n-1}) - N_1(t, S_{n-2})\right) + \frac{2\theta}{C(\theta)(2-\theta)} \int_0^t \left(N_1(\kappa, S_{n-1}) - N_1(\kappa, S_{n-2})\right) \mathrm{d}\kappa \|. \tag{3.16}$$

Applying on Eq. (3.16) the triangular inequality, we have

$$\|S_n(t) - S_{n-1}(t)\| \leq \frac{2(1-\theta)}{(2-\theta)C(\theta)} \|(N_1(t, S_{n-1}) - N_1(t, S_{n-2}))\| + \frac{2\theta}{C(\theta)(2-\theta)} \left\| \int_0^t \left(N_1(\kappa, S_{n-1}) - N_1(\kappa, S_{n-2})\right) \mathrm{d}\kappa \right\|. \tag{3.17}$$

We have on the fulfilment of Lipschitz condition

$$\|S_n(t) - S_{n-1}(t)\| \leq \frac{2(1-\theta)}{(2-\theta)C(\theta)} J_1 \|S_{n-1} - S_{n-2}\| + \frac{2\theta}{(2-\theta)C(\theta)} J_1 \int_0^t \|S_{n-1} - S_{n-2}\| \mathrm{d}\kappa. \tag{3.18}$$

Then, we have

$$\|\phi_n(t)\| \leq \frac{2(1-\theta)}{(2-\theta)C(\theta)} J_1 \|\phi_{n-1}(t)\| + \frac{2\theta}{(2-\theta)C(\theta)} J_1 \int_0^t \|\phi_{n-1}(\kappa)\| \mathrm{d}\kappa. \tag{3.19}$$

Similarly, we get the following results:

$$\begin{aligned}
\|\psi_n(t)\| &\leq \frac{2(1-\theta)}{(2-\theta)C(\theta)} J_2 \|\psi_{n-1}(t)\| + \frac{2\theta}{C(\theta)(2-\theta)} J_2 \int_0^t \|\psi_{n-1}(\kappa)\| \mathrm{d}\kappa, \\
\|\xi_n(t)\| &\leq \frac{2(1-\theta)}{(2-\theta)C(\theta)} J_3 \|\xi_{n-1}(t)\| + \frac{2\vartheta}{C(\theta)(2-\theta)} J_3 \int_0^t \|\xi_{n-1}(\kappa)\| \mathrm{d}\kappa.
\end{aligned} \tag{3.20}$$

□

Based on the above theorem, we present the following result:

**Theorem 3.2.** *The solution for the model (3.3) exists if there exists some* $t_0$ *such that*

$$\frac{2(1-\theta)}{(2-\theta)C(\theta)} J_1 + \frac{2\theta}{(2-\theta)C(\theta)} J_1 t_0 < 1.$$

*Proof.* Here, the functions $S(t)$, $I(t)$, and $P(t)$ hold the Lipschitz condition for the kernels and bounded. So, based on (3.19) and (3.20), we have the relation below using the recursive method:

$$\begin{aligned}
\|\phi_n(t)\| &\leq \|S_n(0)\| \left[ \left( \frac{2(1-\theta)}{C(\theta)(2-\theta)} J_1 \right) + \left( \frac{2\theta}{C(\theta)(2-\theta)} J_1 t \right) \right]^n, \\
\|\psi_n(t)\| &\leq \|I_n(0)\| \left[ \left( \frac{2(1-\theta)}{C(\theta)(2-\theta)} J_2 \right) + \left( \frac{2\theta}{C(\theta)(2-\theta)} J_2 t \right) \right]^n,
\end{aligned}$$

$$\|\xi_n(t)\| \le \|P_n(0)\| \left[ \left( \frac{2(1-\theta)}{C(\theta)(2-\theta)} J_3 \right) + \left( \frac{2\theta}{C(\theta)(2-\theta)} J_3 t \right) \right]^n . \tag{3.21}$$

So, the existence result is proved. Furthermore, to prove that the function is the solution of (3.3), we set

$$\begin{aligned} &S(t) - S(0) = S_n(t) - D_n^1(t), I(t) - I(0) = D_n^2(t) - Y_n(t), \\ &P(t) - P(0) = P_n(t) - D_n^3(t). \end{aligned} \tag{3.22}$$

Therefore we have

$$\begin{aligned} \|B_n(t)\| &= \left\| \frac{2(1-\theta)}{C(\theta)(2-\theta)} (U(t,S) - U(t,S_{n-1})) + \frac{2\theta}{C(\theta)(2-\theta)} \int_0^t (U(\kappa,S) - U(\kappa,S_{n-1}))\,\mathrm{d}\kappa \right\| \\ &\le \frac{2(1-\theta)}{(2-\theta)C(\theta)} \|(U(t,S) - U(t,S_{n-1}))\| + \frac{2\theta}{(2-\theta)C(\theta)} \int_0^t \|(U(\kappa,S) - U(\kappa,S_{n-1}))\|\,\mathrm{d}\kappa \\ &\le \frac{2(1-\theta)}{(2-\theta)C(\theta)} J_1 \|S - S_{n-1}\| + \frac{2\theta}{(2-\theta)C(\theta)} J_1 \|S - S_{n-1}\| t \end{aligned} \tag{3.23}$$

We get the following in the same manner:

$$\|B_n(t)\| \le \left( \frac{2(1-\theta)}{2C(\theta) - \theta C(\theta)} + \frac{2\theta}{2C(\theta) - \theta C(\theta)} t \right)^{n+1} J_1^{n+1} a. \tag{3.24}$$

Then, at $t_0$, we have

$$\left\| D_n^1(t) \right\| \le \left( \frac{2(1-\theta)}{C(\theta)(2-\theta)} + \frac{2\theta}{C(\theta)(2-\theta)} t_0 \right)^{n+1} J_1^{n+1} a. \tag{3.25}$$

When applying the limit on Eq. (3.25), when $n$ tends to $\infty$, we have $\left\| D_n^1(t) \right\|$ tending to 0. Similarly, it also holds for $\| D_n^2(t) |$, $\| D_n^3(t) \|$. For the uniqueness of solutions of system (3.3), we consider other solutions for (3.3), $S_1$ (t), $I_1(t)$, and $P_1(t)$, then

$$S(t) - S_1(t) = \frac{2(1-\theta)}{(2-\theta)C(\theta)} (N_1(t,S) - N_1(t,S_1)) + \frac{2\theta}{(2-\theta)C(\theta)} \int_0^t (N_1(\kappa,S) - N_1(\kappa,S_1))\,\mathrm{d}\kappa. \tag{3.26}$$

We have the following by applying norm:

$$\begin{aligned} \|S(t) - S_1(t)\| \le & \frac{2(1-\theta)}{(2-\theta)C(\theta)} \|N_1(t,S) - N_1(t,S_1)\| \\ & + \frac{2\theta}{(2-\theta)C(\theta)} \int_0^t \|N_1(\kappa,S) - N_1(\kappa,S_1)\|\,\mathrm{d}\kappa. \end{aligned} \tag{3.27}$$

Using the Lipschitz condition of the kernel leads to

$$\|S(t) - S_1(t)\| \le \frac{2(1-\theta)}{(2-\theta)C(\theta)} J_1 \|S(t) - S_1(t)\| + \frac{2\theta}{(2-\theta)C(\theta)} J_1 t \|S(t) - S_1(t)\| , \tag{3.28}$$

which gives

$$\|S(t)-S_1(t)\|\left(1-\frac{2(1-\theta)}{2C(\theta)-\theta C(\theta)}J_1-\frac{2\theta}{2C(\theta)-\theta C(\theta)}J_1 t\right)\leq 0. \tag{3.29}$$

□

**Theorem 3.3.** *There exists a unique solution for the system (3.3) if*

$$\left(1-\frac{2(1-\theta)}{2C(\theta)-\theta C(\theta)}J_1-\frac{2\theta}{2C(\theta)-\theta C(\theta)}J_1 t\right)>0. \tag{3.30}$$

*Proof.* If the condition (34) holds, then

$$\|S(t)-S_1(t)\|\left(1-\frac{2(1-\theta)}{2C(\theta)-\theta C(\theta)}J_1-\frac{2\theta}{2C(\theta)-\theta C(\theta)}J_1 t\right)\leq 0$$

leads to

$$\|S(t)-S_1(t)\|=0.$$

So, we have

$$S(t)=S_1(t).$$

Similarly, we have

$$I(t)=I_1(t),\ P(t)=P_1(t),$$

which proved the uniqueness for the model (3.3). □

## 3.4 Numerical scheme

We consider in detail the numerical scheme for the solution of an eco-epidemiological model considered in (3.3). This scheme is developed in [15] and used extensively for modeling many fractional models in a fractional Caputo–Fabrizio sense. We first give of the scheme in general, and then make it particular for the model.

$${}_0^{CF}D_t^{\theta}w(t)=g(t,w(t)), \tag{3.31}$$

or we can write it in the form given by

$$\frac{C(\theta)}{1-\theta}\int_0^t y'(\kappa)\exp\left[-\frac{\theta}{1-\theta}(t-\kappa)\right]\mathrm{d}\kappa=g(t,w(t)).$$

Using the concept of calculus theorem, we write the above equation in the form given by

$$w(t)-w(0)=\frac{1-\theta}{C(\theta)}g(t,w(t))+\frac{\theta}{C(\theta)}\int_0^t g(\kappa,w(\kappa))\mathrm{d}\kappa \tag{3.32}$$

so that

$$w(t_{n+1}) - w(0) = \frac{1-\theta}{C(\theta)} g(t_n, w(t_n)) + \frac{\theta}{C(\theta)} \int_0^{t_{n+1}} g(t, w(t))\mathrm{d}t, \tag{3.33}$$

and

$$w(t_n) - w(0) = \frac{1-\theta}{C(\theta)} g(t_{n-1}, w(t_{n-1})) + \frac{\theta}{C(\theta)} \int_0^{t_n} g(t, w(t))\mathrm{d}t. \tag{3.34}$$

Eqs. (3.33) and (3.34) gives

$$w(t_{n+1}) - w(t_n) = \frac{1-\theta}{C(\theta)} \{g(t_n, w_n) - g(t_{n-1}, w_{n-1})\} + \frac{\theta}{C(\theta)} \int_{t_n}^{t_{n+1}} g(t, w(t))\mathrm{d}t, \tag{3.35}$$

where

$$\begin{aligned} \int_{t_n}^{t_{n+1}} g(t, w(t))\mathrm{d}t &= \int_{t_n}^{t_{n+1}} \left\{ \frac{g(t_n, w_n)}{h}(t - t_{n-1}) - \frac{g(t_{n-1}, w_{n-1})}{h}(t - t_n) \right\} \mathrm{d}t \\ &= \frac{3h}{2} g(t_n, w_n) - \frac{h}{2} g(t_{n-1}, w_{n-1}). \end{aligned} \tag{3.36}$$

Thus

$$\begin{aligned} w(t_{n+1}) - w(t_n) =& \frac{1-\theta}{C(\theta)} [g(t_n, w_n) - g(t_{n-1}, w_{n-1})] + \frac{3\theta h}{2C(\theta)} g(t_n, w_n) \\ &- \frac{\theta h}{2C(\theta)} g(t_{n-1}, w_{n-1}), \end{aligned} \tag{3.37}$$

which implies that

$$w(t_{n+1}) - w(t_n) = \left( \frac{1-\theta}{C(\theta)} + \frac{3\theta h}{2C(\theta)} \right) g(t_n, w_n) - \left( \frac{1-\theta}{C(\theta)} + \frac{\theta h}{2C(\theta)} \right) g(t_{n-1}, w_{n-1}). \tag{3.38}$$

Hence,

$$w_{n+1} = w_n + \left( \frac{1-\theta}{C(\theta)} + \frac{3\theta h}{2C(\theta)} \right) g(t_n, w_n) - \left( \frac{1-\theta}{C(\theta)} + \frac{\theta h}{2C(\theta)} \right) g(t_{n-1}, w_{n-1}). \tag{3.39}$$

The numerical scheme then for the eco-epidemiological model can be written as

$$\begin{aligned} S_{n+1} = S_n &+ \left( \frac{1-\theta}{C(\theta)} + \frac{3\theta h}{2C(\theta)} \right) g(t_n, S_n, I(n), P(n)) \\ &- \left( \frac{1-\theta}{C(\theta)} + \frac{\theta h}{2C(\theta)} \right) g(t_{n-1}, S_{n-1}, I_{n-1}, P_{n-1}), \\ I_{n+1} = I_n &+ \left( \frac{1-\theta}{C(\theta)} + \frac{3\theta h}{2C(\theta)} \right) g(t_n, S_n, I(n), P(n)) \\ &- \left( \frac{1-\theta}{C(\theta)} + \frac{\theta h}{2C(\theta)} \right) g(t_{n-1}, S_{n-1}, I_{n-1}, P_{n-1}), \end{aligned}$$

$$P_{n+1} = P_n + \left(\frac{1-\theta}{C(\theta)} + \frac{3\theta h}{2C(\theta)}\right) g\left(t_n, S_n, I(n), P(n)\right)$$
$$- \left(\frac{1-\theta}{C(\theta)} + \frac{\theta h}{2C(\theta)}\right) g\left(t_{n-1}, S_{n-1}, I_{n-1}, P_{n-1}\right). \tag{3.40}$$

### 3.4.1 Numerical results

We present the numerical solution of the model (3.3) using the scheme shown above (3.40). The graphical results shown in Fig. 3.1 to 3.2 are obtained using the value of the parameters given by $\beta = 0.95$, $a = 0.3636$, $b = 1$, $\alpha_1 = 0.01$, $e = 15$, $c = 0.01$, $r = 1$, $\alpha_2 = 0.05$, $d = 0.5$, $\mu = 0.4$, $c_1 = 2$, $\kappa = 0$, $c_2 = 1$, and $m = 0.01$. We use the initial conditions $(S(0), I(0), P(0)) = (0.7, 0.1, 0.1)$, $\beta = 0.7$, and keep the rest of parameters value the same, as shown Fig. 3.1. Fig. 3.1 has subgraphs (a–f), in which $(a - b)$ describes the dynamical behavior of the model with different values of fractional order $\theta = 1, 0.96, 0.92, 0.88, 0.84, 0.8$, whereas (d–f) denote the time phase portraits of the model with different $\theta$ value. We plotted Fig. 3.2, which describes 3D phase trajectories of the system using different fractional order. We have Fig. 3.3 and 3.4 when choosing $\beta = 0.95$ and initial conditions $(S(0), I(0), P(0)) = (0.2, 0.05, 0.05)$ with different fractional-order values. Fig. 3.4 is the 3D phase trajectories of the model (3.3). We fixed the fractional-order $\theta = 0.96$, fixed and varied $\beta$, and presented the graphical results in Fig. 3.5 and 3.6. Fig. 3.5 shows the dynamics of the model for different $\beta$'s, whereas their corresponding phase portraits are shown in Fig. 3.6. Fig. 3.7 and Fig. 3.8 show the effect of parameter $b$ value on the model when keeping fixed the value of $\theta$. Fig. 3.7 displays the dynamics of the model for $b$'s different values, whereas its corresponding phase portraits are shown in Fig. 3.8.

## 3.5 Conclusion

We presented an eco-epidemiological model in fractional derivative in the Caputo–Fabrizio sense. The epidemiological model studied in this paper has give significant results in terms of fractional derivative. Various graphical results for the fractional-order parameters were given. Many possible different portraits in different planes were achieved, which shows that the model has many rich dynamical properties. We provided in detail the existence and uniqueness results. We presented a novel numerical scheme for the solution of the fractional model, and then finally obtained a scheme for the proposed model. We solved the model using the proposed derived scheme and obtained various simulation results. The model has rich dynamics, which utilized and presented the graphical results. By applying the fractional-order model and varying some key parameters, we obtained valuable graphical results that provide important insights into the system's behavior. The behavior of the model which showed oscillations are identified and presented graphically. We obtained various phase portraits for the fractional-order values. The graphical results obtained in the present paper reveals the significance of the fractional-order derivative proposed in this work. The previous literature related to such model did not obtain such

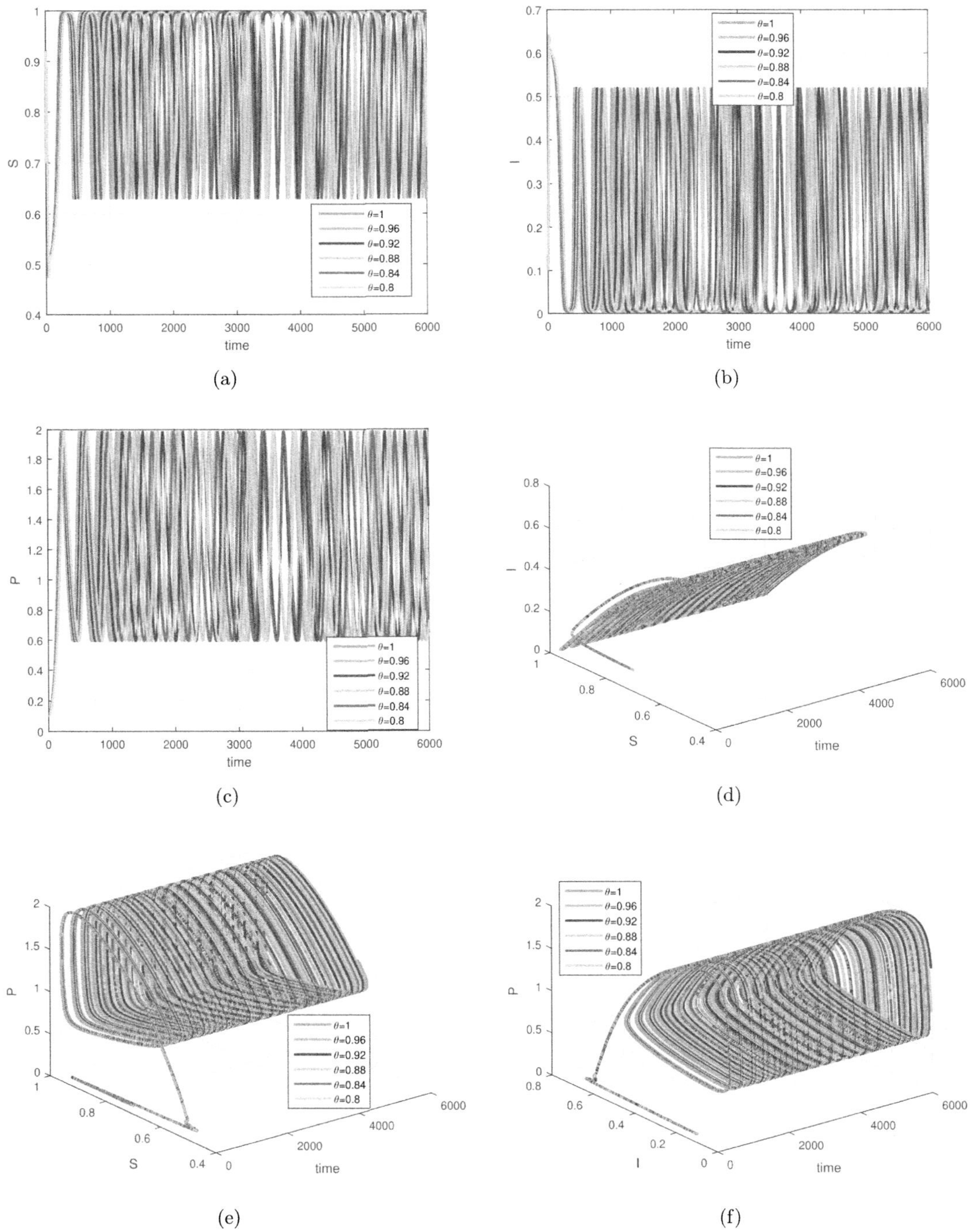


**FIGURE 3.1** Numerical simulation of the model when $\beta = 0.7$, and $\theta = 1, 0.96, 0.92, 0.88, 0.84, 0.8$.

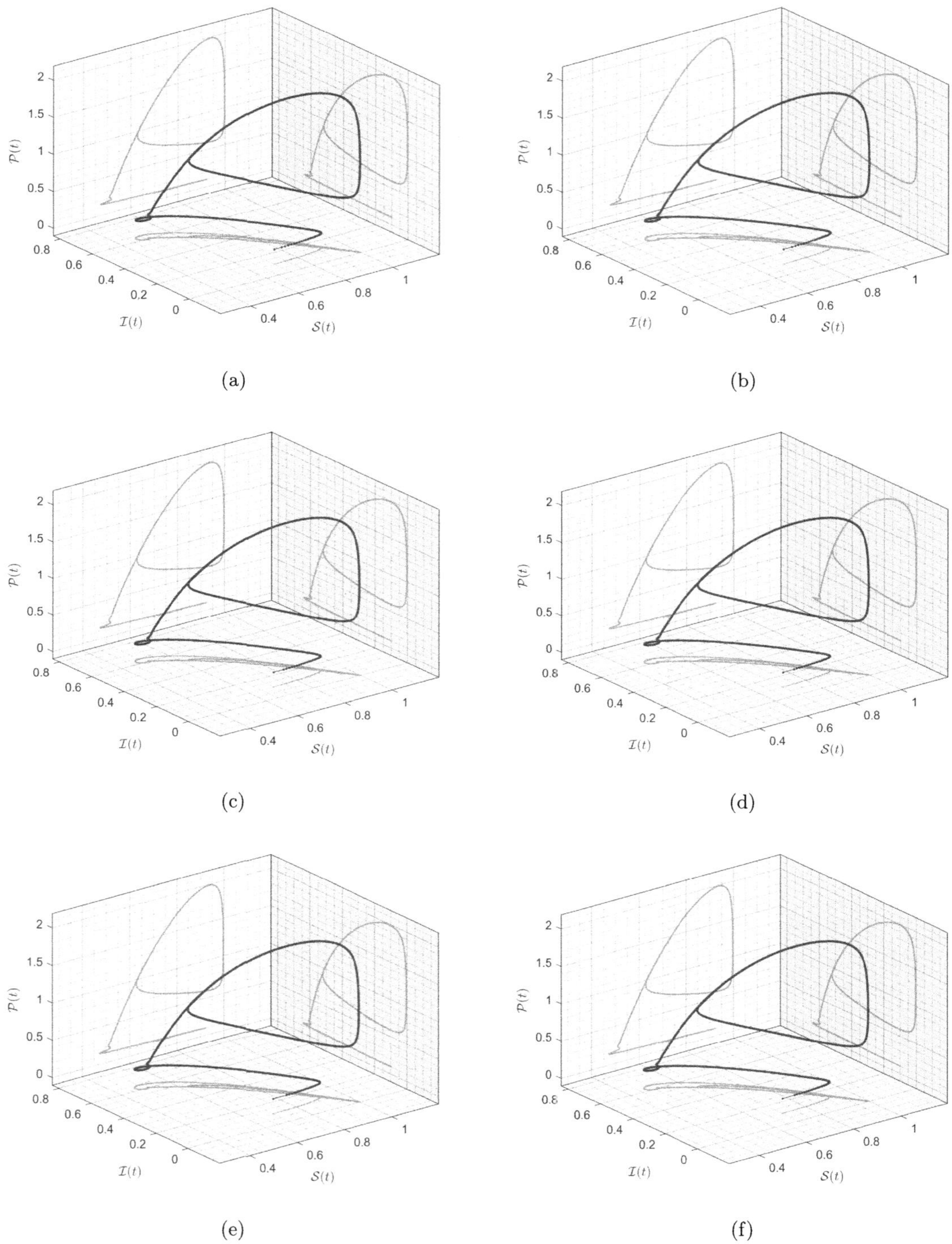


**FIGURE 3.2** 3D phase trajectories of the model (3.3) when $\beta = 0.7$, $(S(0), I(0), P(0)) = (0.7, 0.1, 0.1)$ and different $\theta$'s.

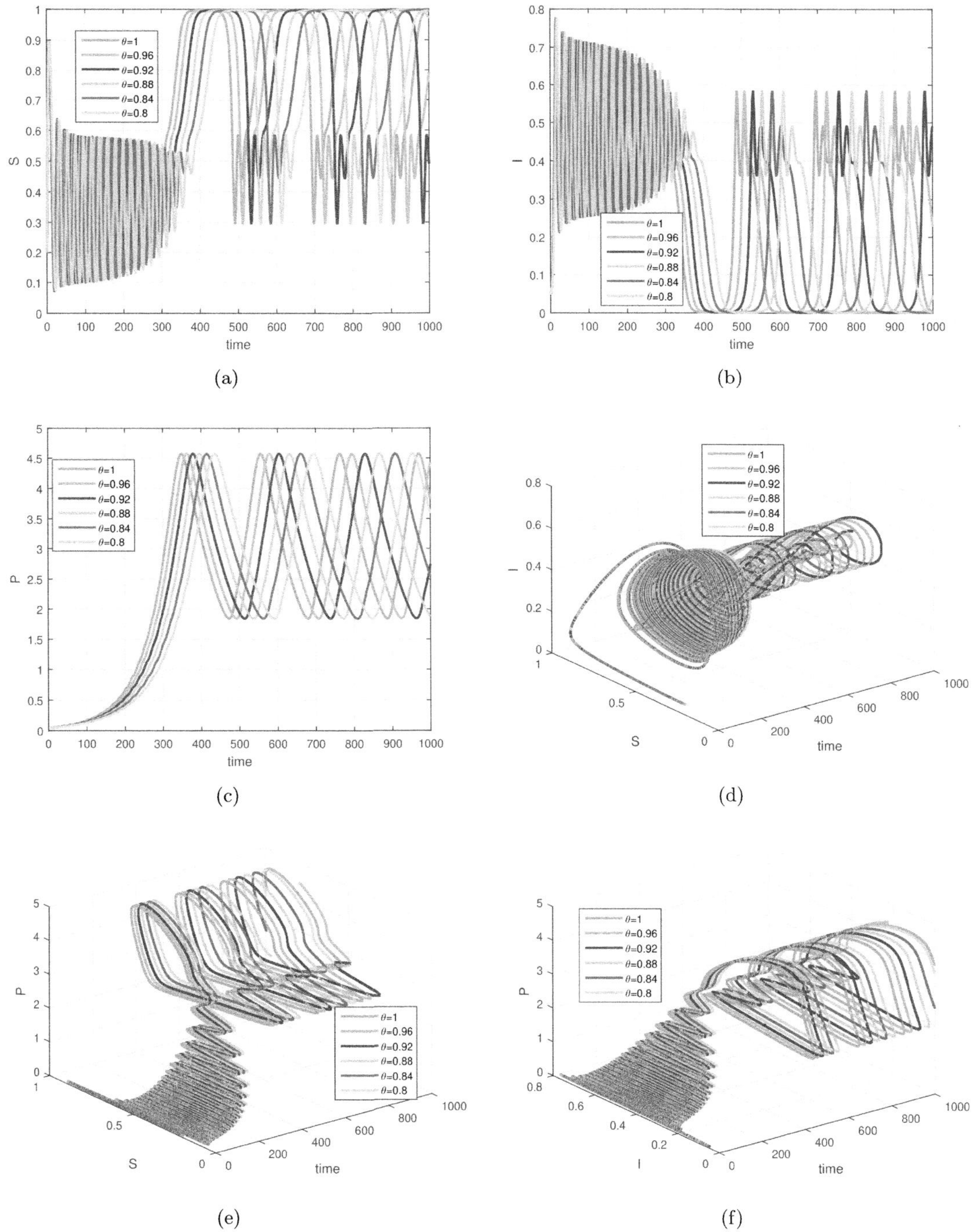

**FIGURE 3.3** When $\beta = 0.95$ and different $\theta$'s.

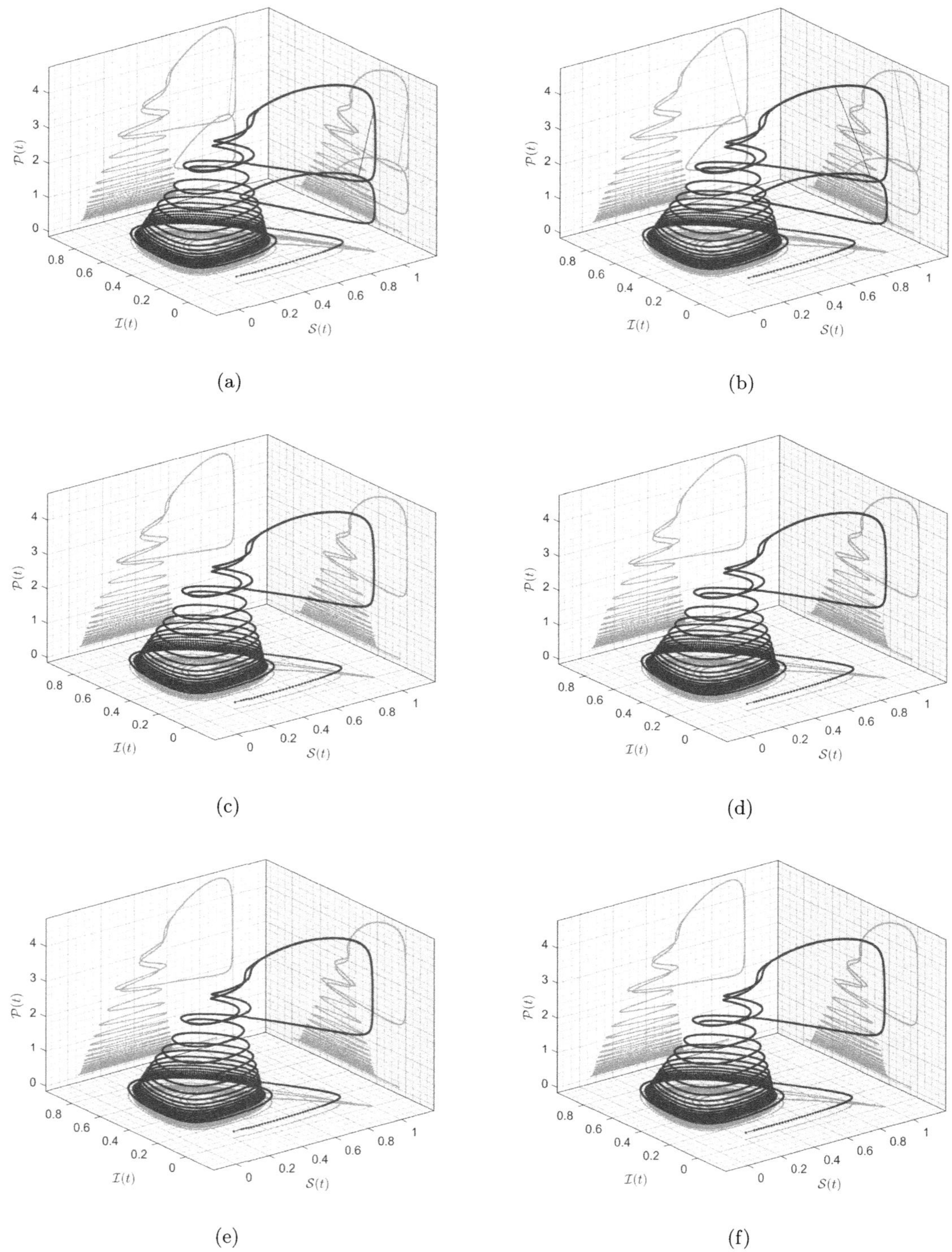


**FIGURE 3.4** 3D phase trajectories when $\beta = 0.95$ and different $\theta$'s.

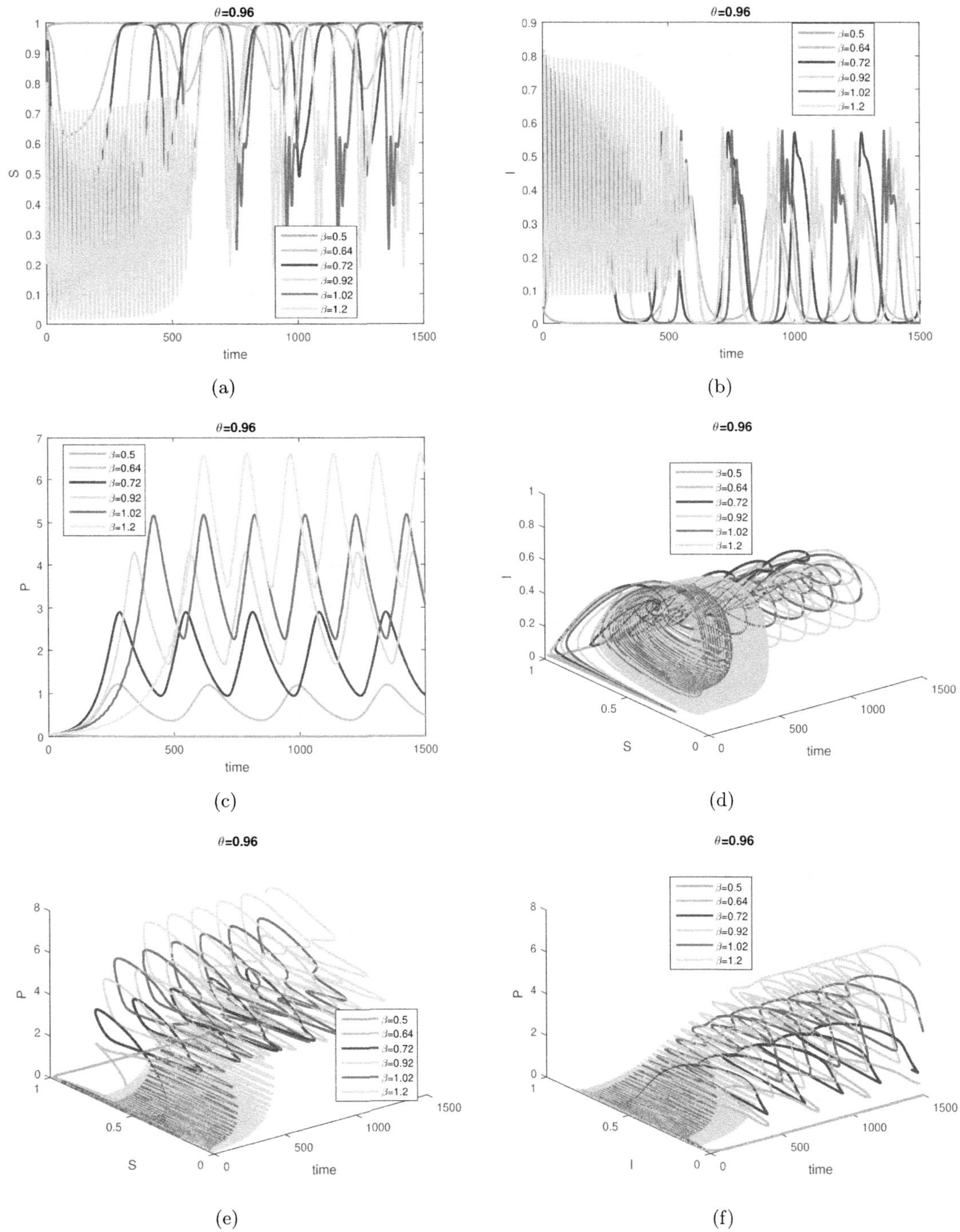


**FIGURE 3.5** When $\beta = 0.5, 0.64, 0.78, 0.92, 1.02, 1.2$, $\theta = 0.96$, and $(S(0), I(0), P(0)) = (0.2, 0.05, 0.05)$.

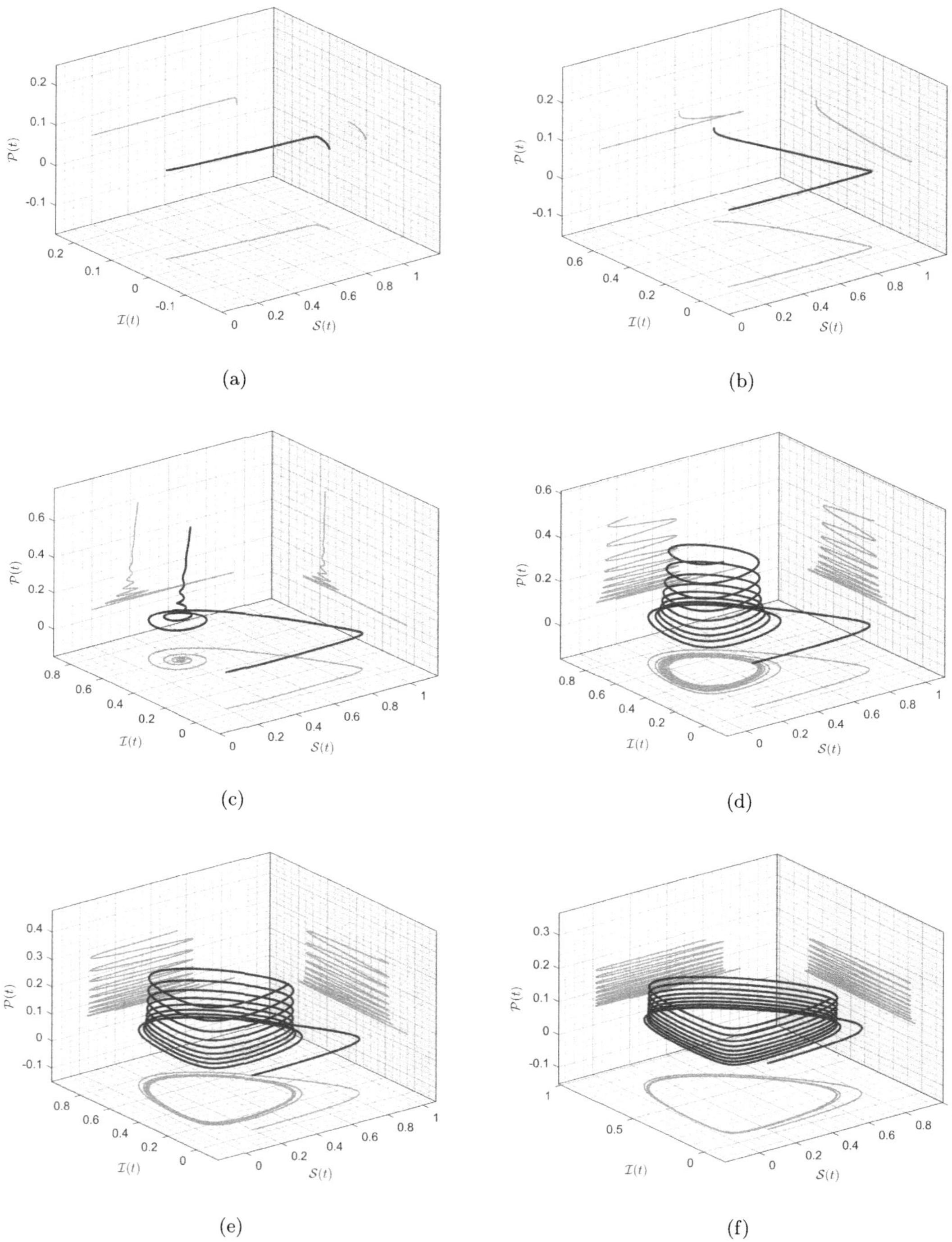


**FIGURE 3.6** 3D phase trajectories when $\beta = 0.5, 0.64, 0.78, 0.92, 1.02, 1.2$, $\theta = 0.96$, and $(S(0), I(0), P(0)) = (0.2, 0.05, 0.05)$.

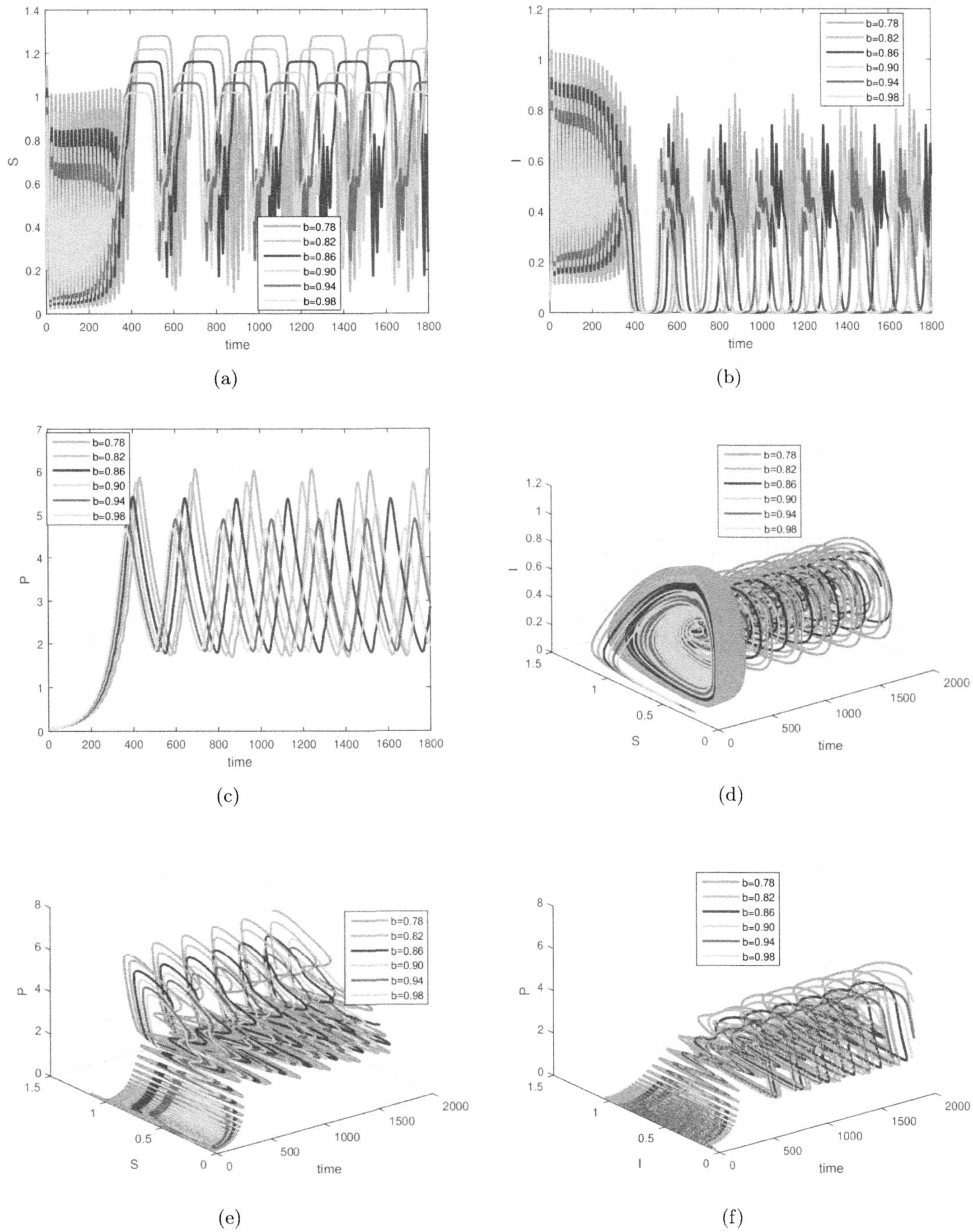


**FIGURE 3.7** Simulations of the model (3.3) when $\theta = 0.96$ and different $b$'s.

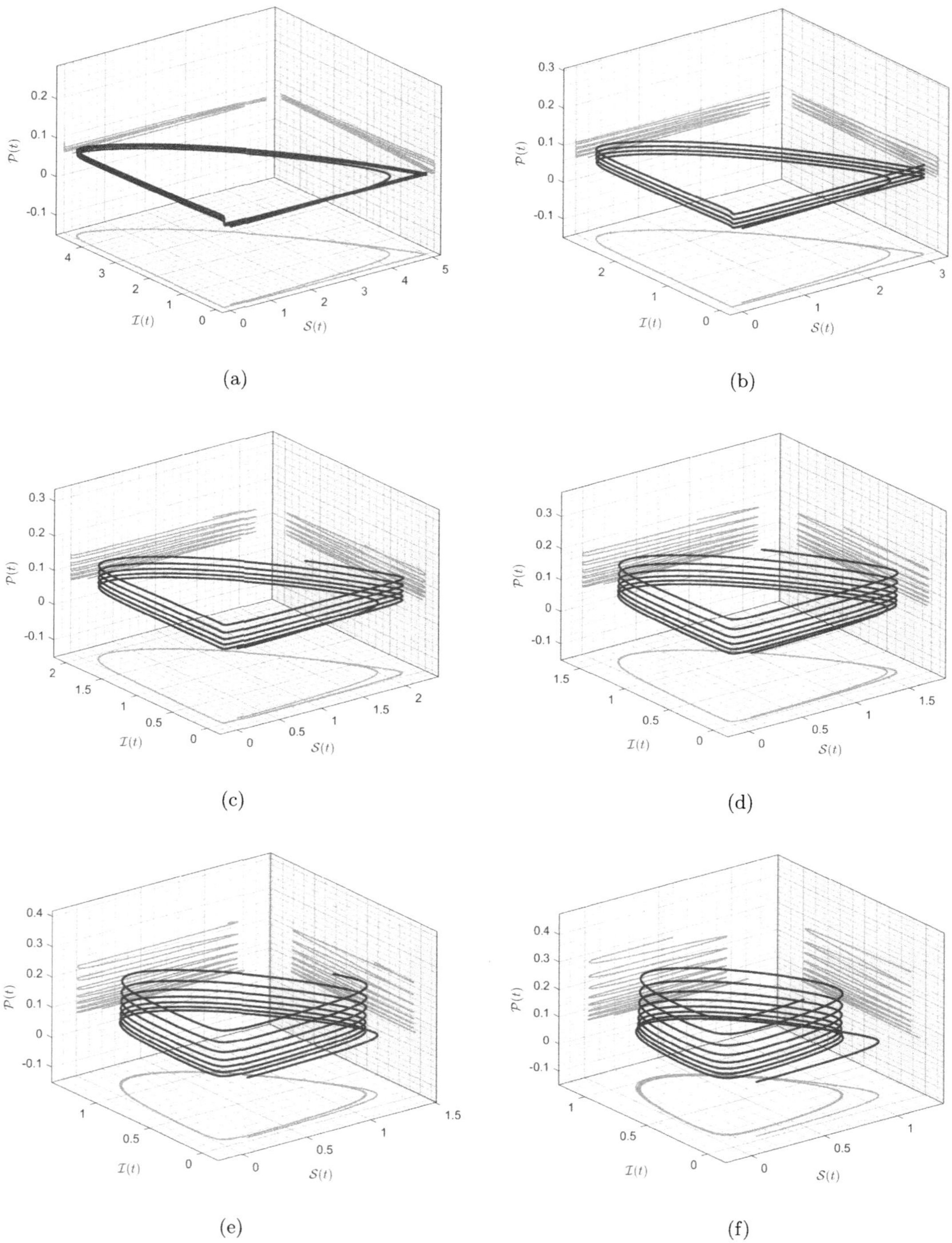


**FIGURE 3.8** 3D phase trajectories of the model (3.3) when $\theta = 0.96$ and different $b$'s.

useful dynamical results compared to this fractional study. Overall, the fractional model provides good foundational understanding for the dynamical analysis of the model.

## References


[1] Xinzhu Meng, Rui Liu, Tonghua Zhang, Adaptive dynamics for a non-autonomous Lotka–Volterra model with size-selective disturbance, Nonlinear Analysis: Real World Applications 16 (2014) 202–213.

[2] Nanako Shigesada, Kohkichi Kawasaki, Ei Teramoto, Spatial segregation of interacting species, Journal of Theoretical Biology 79 (1) (1979) 83–99.

[3] Elizabeth E. Holmes, Mark A. Lewis, J.E. Banks, R.R. Veit, Partial differential equations in ecology: spatial interactions and population dynamics, Ecology 75 (1) (1994) 17–29.

[4] J.D. Murray, Mathematical Biology II: Spatial Models and Biomedical Applications, vol. 3, Springer-Verlag, 2001.

[5] Mikko Kuussaari, Ilik Saccheri, Mark Camara, Ilkka Hanski, Allee effect and population dynamics in the Glanville fritillary butterfly, Oikos (1998) 384–392.

[6] Wan-Xiong Wang, Yan-Bo Zhang, Chang-zhong Liu, Analysis of a discrete-time predator–prey system with Allee effect, Ecological Complexity 8 (1) (2011) 81–85.

[7] Muhammad Altaf Khan, Abdon Atangana, Modeling the dynamics of novel coronavirus (2019-ncov) with fractional derivative, Alexandria Engineering Journal 59 (4) (2020) 2379–2389.

[8] N.H. Sweilam, S.M. AL-Mekhlafi, A.O. Albalawi, Optimal control for a fractional order malaria transmission dynamics mathematical model, Alexandria Engineering Journal 59 (3) (2020) 1677–1692.

[9] Hamed Al-Sulami, Moustafa El-Shahed, Juan J. Nieto, Wafa Shammakh, On fractional order Dengue epidemic model, Mathematical Problems in Engineering (2014) 2014.

[10] M. Higazy, Maryam Ahmed Alyami, New Caputo-Fabrizio fractional order seiasqeqhr model for covid-19 epidemic transmission with genetic algorithm based control strategy, Alexandria Engineering Journal 59 (6) (2020) 4719–4736.

[11] Esmehan Uçar, Necati Özdemir, A fractional model of cancer-immune system with Caputo and Caputo–Fabrizio derivatives, The European Physical Journal Plus 136 (1) (2021) 1–17.

[12] Md Saifuddin, Santanu Biswas, Sudip Samanta, Susmita Sarkar, Joydev Chattopadhyay, Complex dynamics of an eco-epidemiological model with different competition coefficients and weak Allee in the predator, Chaos, Solitons and Fractals 91 (2016) 270–285.

[13] Michele Caputo, Mauro Fabrizio, A new definition of fractional derivative without singular kernel, Progr. Fract. Differ. Appl 1 (2) (2015) 1–13.

[14] Jorge Losada, Juan J. Nieto, Properties of a new fractional derivative without singular kernel, Progress in Fractional Differentiation and Applications 1 (2) (2015) 87–92.

[15] Abdon Atangana, Kolade M. Owolabi, New numerical approach for fractional differential equations, Mathematical Modelling of Natural Phenomena 13 (1) (2018) 3.

# 4

# The Topp–Leone-exponentiated half logistic-generalized-G family of distributions with applications

B. Oluyede[a], M. Gabanakgosi[a], D. Wanduku[b], T. Moakofi[a], and F. Chipepa[a]

[a]*Department of Mathematics and Statistical Sciences, Botswana International University of Science and Technology, Palapye, Botswana* [b]*Department of Mathematical Sciences, Georgia Southern University, Statesboro, GA, United States*

## 4.1 Introduction

The development of new and generalized distributions to obtain greater flexibility when modeling data is of tremendous practical importance. Examples of recently generalized distributions include the type II generalized Topp–Leone family of distributions by Hassan et al. [10], the exponentiated generalized Topp–Leone-G family of distributions by Reyad et al. [17], the Topp–Leone generalized inverted Kumaraswamy distribution by Reyad et al. [18], the Marshall–Olkin generalized-G family of distributions by Yousof et al. [21], the Kumaraswamy generalized power Lomax distribution by Nagarjuna et al. [13] and the Marshall–Olkin generalized Erlang-truncated exponential distribution developed by Okorie et al. [13], to mention just a few.

An important family of distributions in the literature is the generalized-G distribution with recent extensions, which include new generalized-G class of distribution by Rasheed [15], the exponentiated generalized-G Poisson family of distributions by Aryal and Yousof [3], and the exponentiated generalized class of distributions by Cordeiro et al. [6], just to mention a few.

The following are the general motivations for developing this new family of distributions:

- To develop generalized distributions for modeling different real data sets, with different levels of skewness and kurtosis.
- To develop models that provide better fits than other generalized distributions with the same underlying model.
- To construct special models with different shapes (monotone and non-monotone) for the hazard rate function.

Mathematical Modeling in Bioscience. https://doi.org/10.1016/B978-0-44-315445-4.00010-X

Copyright © 2025 Elsevier Inc. All rights are reserved, including those for text and data mining, AI training, and similar technologies.

- To extend the generalized-G family of distributions, and the exponentiated half logistic-G family of distributions.

The rest of the paper is organized as follows: In Section 4.2, we present the new family of distributions. Section 4.3 contain the statistical properties of the new family of distributions, including quantile function, expansion of the probability density function, moments, generating function, probability weighted moments, stochastic orders, distribution of order statistics, and Rényi entropy. Maximum likelihood estimates of the parameters are given in Section 4.4. Some special cases of the new family of distributions are presented in Section 4.5. A Monte Carlo simulation study to examine the average bias and root mean square error of the maximum likelihood estimates are presented in Section 4.6. Section 4.7 contain applications of a special case of the new family of distributions to real data sets, and, lastly, concluding remarks in Section 4.8.

## 4.2 The new family

The new Topp–Leone exponentiated half logistic-generalized-G (TL-EHL-GG) family of distributions and its sub-families are presented in this section. The Topp–Leone generated family of distributions was introduced by Al-Shomrani et al. [2] with the cumulative distribution function (cdf) and probability density function (pdf) given as

$$F_{TL-G}(x;b,\Psi)=[1-\overline{G}^2(x;\Psi)]^b, \tag{4.1}$$

and

$$f_{TL-G}(x;b,\Psi)=2b[1-\overline{G}^2(x;\Psi)]^{b-1}\overline{G}(x;\Psi)g(x;\Psi), \tag{4.2}$$

respectively, for $b>0$, $\overline{G}(x;\Psi)=1-G(x;\Psi)$ and parameter vector $\Psi$. Cordeiro et al. [5] developed the exponentiated half logistic-G family of distributions with cdf given by

$$F_{EHL-G}(x;\alpha,\beta,\Psi)=\left[\frac{1-\overline{G}^{\alpha}(x;\Psi)}{1+\overline{G}^{\alpha}(x;\Psi)}\right]^{\beta}, \tag{4.3}$$

for $\alpha,\beta>0$ and parameter vector $\Psi$. Note that the cdf

$$F_{G-G}(x;\alpha,\Psi)=1-\overline{G}^{\alpha}(x;\Psi),$$

for $\alpha>0$ and parameter vector $\Psi$, is the generalized-G family of distributions. Now, by combining Eq. (4.1) and Eq. (4.3), we obtain the TL-EHL-GG family of distributions with the cdf

$$F_{TL-EHL-GG}(x;\alpha,\beta,b,\Psi)=\left[1-\left(1-\left(\frac{1-\overline{G}^{\alpha}(x;\Psi)}{1+\overline{G}^{\alpha}(x;\Psi)}\right)^{\beta}\right)^2\right]^b, \tag{4.4}$$

for $\alpha, \beta, b > 0$ and parameter vector $\Psi$. The corresponding pdf is given by

$$f_{TL-EHL-GG}(x;\alpha,\beta,b,\Psi)=4\alpha\beta b\left[1-\left(1-\left(\frac{1-\overline{G}^{\alpha}(x;\Psi)}{1+\overline{G}^{\alpha}(x;\Psi)}\right)^{\beta}\right)^{2}\right]^{b-1}$$
$$\times\left(1-\left(\frac{1-\overline{G}^{\alpha}(x;\Psi)}{1+\overline{G}^{\alpha}(x;\Psi)}\right)^{\beta}\right)\frac{(1-\overline{G}^{\alpha}(x;\Psi))^{\beta-1}}{(1+\overline{G}^{\alpha}(x;\Psi))^{\beta+1}}$$
$$\times[\overline{G}(x;\Psi)]^{\alpha-1}g(x;\Psi), \tag{4.5}$$

for $\alpha, \beta, b > 0$ and parameter vector $\Psi$.

The hazard rate function (hrf) for the new family of distribution is given by

$$h_{TL-EHL-GG}(x;\alpha,\beta,b,\Psi)=4\alpha\beta b\left[1-\left(1-\left(\frac{1-\overline{G}^{\alpha}(x;\Psi)}{1+\overline{G}^{\alpha}(x;\Psi)}\right)^{\beta}\right)^{2}\right]^{b-1}g(x;\Psi)$$
$$\times\left(1-\left(\frac{1-\overline{G}^{\alpha}(x;\Psi)}{1+\overline{G}^{\alpha}(x;\Psi)}\right)^{\beta}\right)\frac{(1-\overline{G}^{\alpha}(x;\Psi))^{\beta-1}}{(1+\overline{G}^{\alpha}(x;\Psi))^{\beta+1}}$$
$$\times[\overline{G}(x;\Psi)]^{\alpha-1}\left(1-\left[1-\left(1-\left(\frac{1-\overline{G}^{\alpha}(x;\Psi)}{1+\overline{G}^{\alpha}(x;\Psi)}\right)^{\beta}\right)^{2}\right]^{b}\right)^{-1}, \tag{4.6}$$

for $\alpha, \beta, b > 0$ and parameter vector $\Psi$. This new family generalized the Topp–Leone half logistic generalized-G distribution by Moakofi et al. [12].

## 4.2.1 Sub-families

We present some of the sub-families of the TL-EHL-GG family of distributions.

- If we let $b = 1$, we obtain a new family of distributions with the cdf given as

$$F(x;\alpha,\beta,\Psi)=1-\left(1-\left(\frac{1-\overline{G}^{\alpha}(x;\Psi)}{1+\overline{G}^{\alpha}(x;\Psi)}\right)^{\beta}\right)^{2},$$

  for $\alpha, \beta > 0$ and parameter vector $\Psi$.
- When $\beta = 1$, we obtain a family of distributions with the cdf

$$F(x;\alpha,b,\Psi)=\left[1-\left(1-\left(\frac{1-\overline{G}^{\alpha}(x;\Psi)}{1+\overline{G}^{\alpha}(x;\Psi)}\right)\right)^{2}\right]^{b},$$

  for $\alpha, b > 0$ and parameter vector $\Psi$. (See Moakofi et al. [12] for details.)

- By letting $\alpha = 1$, we have the Topp–Leone exponentiated half logistic-G (TL-EHL-G) family of distributions with the cdf given by

$$F(x;\beta,b,\Psi)=\left[1-\left(1-\left(\frac{G(x;\Psi)}{1+\overline{G}(x;\Psi)}\right)^{\beta}\right)^{2}\right]^{b},$$

for $\beta, b > 0$ and parameter vector $\Psi$.

- When $b = \beta = 1$, we obtain a new family of distribution with the cdf

$$F(x;\alpha,\Psi)=1-\left(1-\left(\frac{1-\overline{G}^{\alpha}(x;\Psi)}{1+\overline{G}^{\alpha}(x;\Psi)}\right)\right)^{2},$$

for $\alpha > 0$ and parameter vector $\Psi$.

- If we let $b = \beta = \alpha = 1$, we have a new family of distributions given by the cdf

$$F(x;\Psi)=1-\left(1-\left(\frac{G(x;\Psi)}{1+\overline{G}(x;\Psi)}\right)\right)^{2},$$

with parameter vector $\Psi$.

## 4.3 Properties

This section entails statistical properties of the TL-EHL-GG family of distributions. These properties include the quantile function, linear representation of the density function, moments, generating function, probability weighted moments, distribution of the $i$th-order statistics and Rényi entropy.

### 4.3.1 Quantile function

We compute the quantile function (qf) of the proposed family of distributions by inverting the non-linear equation $F_{TL-EHL-GG}(x;\alpha,\beta,b,\Psi)=u$, for $0 \leqslant u \leqslant 1$. That is,

$$\left[1-\left(1-\left(\frac{1-\overline{G}^{\alpha}(x;\Psi)}{1+\overline{G}^{\alpha}(x;\Psi)}\right)^{\beta}\right)^{2}\right]^{b}=u.$$

The above equation can be written as

$$(1-u^{\frac{1}{b}})^{\frac{1}{2}}=1-\left(\frac{1-\overline{G}^{\alpha}(x;\Psi)}{1+\overline{G}^{\alpha}(x;\Psi)}\right)^{\beta},$$

which is equivalent to

$$\frac{1-\overline{G}^{\alpha}(x;\Psi)}{1+\overline{G}^{\alpha}(x;\Psi)}=\left(1-(1-u^{\frac{1}{b}})^{\frac{1}{2}}\right)^{\frac{1}{\beta}}.$$

Therefore the qf of the TL-EHL-GG distribution is obtained as

$$Q_G(u;\alpha,\beta,b,\Psi)=G^{-1}\left[1-\left(\frac{1-\left(1-(1-u^{\frac{1}{b}})^{\frac{1}{2}}\right)^{\frac{1}{\beta}}}{1+\left(1-(1-u^{\frac{1}{b}})^{\frac{1}{2}}\right)^{\frac{1}{\beta}}}\right)^{\frac{1}{\alpha}}\right]. \tag{4.7}$$

Using R software, we can obtain random numbers from (4.7) for a specified baseline cdf G.

## 4.3.2 Series expansion of the density function

The series expansion of the density function of the TL-EHL-GG family of distributions is presented in this section. We derive the linear representation of the density function by considering the following generalized binomial expansion:

$$(1-x)^c=\sum_{s=0}^{\infty}(-1)^s\binom{c}{s}x^s, \quad \text{for } |x|<1.$$

Taking into consideration the following generalized binomial expansions:

$$\left[1-\left(1-\left(\frac{1-\overline{G}^{\alpha}(x;\Psi)}{1+\overline{G}^{\alpha}(x;\Psi)}\right)^{\beta}\right)^{2}\right]^{b-1}=\sum_{i=0}^{\infty}(-1)^i\binom{b-1}{i}\left(1-\left(\frac{1-\overline{G}^{\alpha}(x;\Psi)}{1+\overline{G}^{\alpha}(x;\Psi)}\right)^{\beta}\right)^{2i},$$

and

$$\left(1-\left(\frac{1-\overline{G}^{\alpha}(x;\Psi)}{1+\overline{G}^{\alpha}(x;\Psi)}\right)^{\beta}\right)^{2i+1}=\sum_{j=0}^{\infty}(-1)^j\binom{2i+1}{j}\left(\frac{1-\overline{G}^{\alpha}(x;\Psi)}{1+\overline{G}^{\alpha}(x;\Psi)}\right)^{\beta j},$$

we can write the pdf of the TL-EHL-GG family of distributions as

$$f_{TL-EHL-GG}(x;\alpha,\beta,b,\Psi)=4\alpha\beta b\sum_{i,j=0}^{\infty}(-1)^{i+j}\binom{b-1}{i}\binom{2i+1}{j}$$
$$\times\frac{(1-\overline{G}^{\alpha}(x;\Psi))^{\beta(j+1)-1}}{(1+\overline{G}^{\alpha}(x;\Psi))^{\beta(j+1)+1}}[\overline{G}(x;\Psi)]^{\alpha-1}g(x;\Psi).$$

Applying the generalized binomial expansion, we finally obtain the pdf of TL-EHL-GG family of distributions as

$$f_{TL-EHL-GG}(x;\alpha,\beta,b,\Psi)=4\alpha\beta b\sum_{i,j,k,l,m=0}^{\infty}(-1)^{i+j+k+l+m}\binom{b-1}{i}\binom{2i+1}{j}$$
$$\times\binom{\beta(j+1)-1}{k}\binom{-(\beta(j+1)+1)}{l}\binom{\alpha(k+l+1)-1}{m}$$

$$\times \frac{(m+1)}{(m+1)} G^m(x;\Psi) g(x;\Psi)$$
$$= \sum_{m=0}^{\infty} W_{m+1} g^*_{m+1}(x;\Psi),$$

where

$$W_{m+1} = 4\alpha\beta b \sum_{i,j,k,l=0}^{\infty} (-1)^{i+j+k+l+m} \binom{b-1}{i} \binom{2i+1}{j} (m+1)$$
$$\times \binom{\beta(j+1)-1}{k} \binom{-(\beta(j+1)+1)}{l} \binom{\alpha(k+l+1)-1}{m}, \tag{4.8}$$

and $g^*_{m+1}(x;\Psi) = (m+1)g(x;\Psi)G^m(x;\Psi)$ is the exponentiated-G (Exp-G) density with power parameter (m + 1). Therefore the pdf of TL-EHL-GG family of distributions can be written as an infinite linear combination of Exp-G densities. Consequently, the properties of the TL-EHL-GG family of distributions can be readily obtained from those of the Exp-G distribution.

### 4.3.3 Moments

We obtain the $n$th moment for the TL-EHL-GG family of distributions as follows:

$$E(X^n) = \int_{-\infty}^{\infty} x^n f_{TL-EHL-GG}(x;\alpha,\beta,b,\Psi)dx = \sum_{m=0}^{\infty} W_{m+1} E(Y^n_{m+1}),$$

where $E(Y^n_{m+1})$ is the $n$th moment of $Y_{m+1}$, which follows an exp-G distribution with power parameter $(m+1)$ and $W_{m+1}$ is given in Eq. (4.8). The moment generating function for the TL-EHL-GG family of distributions is given by

$$M_X(t) = E\left(e^{tX}\right) = \sum_{m=0}^{\infty} W_{m+1} E\left(e^{tY_{m+1}}\right),$$

where $E\left(e^{tY_{m+1}}\right)$ is the moment generating function of the exp-G family of distributions with power parameter $(m+1)$ and $W_{m+1}$ is given in Eq. (4.8).

### 4.3.4 Probability weighted moments

In this subsection, we present the probability weighted moments (PWMs) of the TL-EHL-GG family of distributions. The PWMs are used to estimate distribution parameters when the inverse form cannot be given explicitly (Greenwood et al. [9]). The PWMs for a random variable X is given by

$$\eta_{a,r} = E(X^a[F(X)]^r) = \int_{-\infty}^{\infty} x^a f(x)[F(x)]^r dx.$$

Note that, for a random variable X following the TL-EHL-GG family of distributions, we can write

$$f(x)[F(x)]^r = 4\alpha\beta b\left[1-\left(1-\left(\frac{1-\overline{G}^{\alpha}(x;\Psi)}{1+\overline{G}^{\alpha}(x;\Psi)}\right)^{\beta}\right)^2\right]^{b(r+1)-1}$$
$$\times\left(1-\left(\frac{1-\overline{G}^{\alpha}(x;\Psi)}{1+\overline{G}^{\alpha}(x;\Psi)}\right)^{\beta}\right)\frac{(1-\overline{G}^{\alpha}(x;\Psi))^{\beta-1}}{(1+\overline{G}^{\alpha}(x;\Psi))^{\beta+1}}$$
$$\times[\overline{G}(x;\Psi)]^{\alpha-1}g(x;\Psi).$$

By applying the generalized binomial expansion used in the expansion of the density function, we obtain

$$f(x)[F(x)]^r = 4\alpha\beta b\sum_{i,j,k,l,m=0}^{\infty}(-1)^{i+j+k+l+m}\binom{b(r+1)-1}{i}\binom{2i+1}{j}$$
$$\times\binom{\beta(j+1)-1}{k}\binom{-(\beta(j+1)+1)}{l}\binom{\alpha(k+l+1)-1}{m}$$
$$\times\frac{(m+1)}{(m+1)}G^m(x;\Psi)g(x;\Psi)$$
$$=\sum_{m=0}^{\infty}V_{m+1}g^*_{m+1}(x;\Psi),$$

where

$$V_{m+1} = 4\alpha\beta b\sum_{i,j,k,l=0}^{\infty}(-1)^{i+j+k+l+m}\binom{b(r+1)-1}{i}\binom{2i+1}{j}(m+1)$$
$$\times\binom{\beta(j+1)-1}{k}\binom{-(\beta(j+1)+1)}{l}\binom{\alpha(k+l+1)-1}{m},$$

and $g^*_{m+1}(x;\Psi)=(m+1)g(x;\Psi)G^m(x;\Psi)$ is the exp-G density with power parameter $(m+1)$. Therefore the PWMs of the TL-EHL-GG family of distributions can be obtained from the moments of the exp-G distribution and is given by

$$\eta_{a,r}=\sum_{m=0}^{\infty}V_{m+1}\int_{-\infty}^{\infty}x^a g^*_{m+1}(x;\Psi)dx. \tag{4.9}$$

## 4.3.5 Stochastic ordering

Stochastic orders for the TL-EHL-GG family of distributions are presented in this section. Let Y and Q be random variables with the distribution functions $F_Y(r)$ and $F_Q(r)$, respectively, and $\overline{F}_Y(r)=1-F_Y(r)$ the survival function. Then Y is stochastically smaller than

Q if $\overline{F}_Y(r) \le \overline{F}_Q(r)$ for all r or $F_Y(r) \ge F_Q(r)$ for all r, denoted by $Y <_s Q$. Hazard rate order and likelihood ratio order are given by $Y <_{hr} Q$ if $h_Y(r) \ge h_Q(r)$ for all r, and $Y <_{lr} Q$ if $\frac{f_Y(r)}{f_Q(r)}$ is decreasing in r, (Shaked and Shanthikumar [19]), respectively. It is well known that $Y <_{lr} Q \Rightarrow Y <_{hr} Q \Rightarrow Y <_s Q$.

**Theorem.** *Suppose* $X_1 \sim TL-EHL-GG(\alpha,\beta,b_1,\Psi)$ *and* $X_2 \sim TL-EHL-GG(\alpha,\beta,b_2,\Psi)$, *respectively. If* $b_1 \le b_2$, *then* $X_1$ *and* $X_2$ *are stochastically ordered.*

*Proof.* Note that we can write the pdfs of $X_1$ and $X_2$ as follows:

$$f_1(x;\alpha,\beta,b_1,\Psi) = 4\alpha\beta b_1\left[1-\left(1-\left(\frac{1-\overline{G}^{\alpha}(x;\Psi)}{1+\overline{G}^{\alpha}(x;\Psi)}\right)^{\beta}\right)^{2}\right]^{b_1-1}$$
$$\times\left(1-\left(\frac{1-\overline{G}^{\alpha}(x;\Psi)}{1+\overline{G}^{\alpha}(x;\Psi)}\right)^{\beta}\right)\frac{(1-\overline{G}^{\alpha}(x;\Psi))^{\beta-1}}{(1+\overline{G}^{\alpha}(x;\Psi))^{\beta+1}}$$
$$\times[\overline{G}(x;\Psi)]^{\alpha-1}g(x;\Psi)$$

and

$$f_2(x;\alpha,\beta,b_2,\Psi) = 4\alpha\beta b_2\left[1-\left(1-\left(\frac{1-\overline{G}^{\alpha}(x;\Psi)}{1+\overline{G}^{\alpha}(x;\Psi)}\right)^{\beta}\right)^{2}\right]^{b_2-1}$$
$$\times\left(1-\left(\frac{1-\overline{G}^{\alpha}(x;\Psi)}{1+\overline{G}^{\alpha}(x;\Psi)}\right)^{\beta}\right)\frac{(1-\overline{G}^{\alpha}(x;\Psi))^{\beta-1}}{(1+\overline{G}^{\alpha}(x;\Psi))^{\beta+1}}.$$
$$\times[\overline{G}(x;\Psi)]^{\alpha-1}g(x;\Psi).$$

Now,

$$\frac{f_1(x;\alpha,\beta,b_1,\Psi)}{f_2(x;\alpha,\beta,b_2,\Psi)} = \frac{b_1}{b_2}\left[1-\left(1-\left(\frac{1-\overline{G}^{\alpha}(x;\Psi)}{1+\overline{G}^{\alpha}(x;\Psi)}\right)^{\beta}\right)^{2}\right]^{b_1-b_2}, \tag{4.10}$$

and upon differentiating Eq. (4.10) with respect to x, we obtain

$$\frac{d}{dx}\left(\frac{f_1(x;\alpha,\beta,b_1,\Psi)}{f_2(x;\alpha,\beta,b_2,\Psi)}\right) = 4\alpha\beta\frac{b_1}{b_2}(b_1-b_2)\left[1-\left(1-\left(\frac{1-\overline{G}^{\alpha}(x;\Psi)}{1+\overline{G}^{\alpha}(x;\Psi)}\right)^{\beta}\right)^{2}\right]^{b_1-b_2-1}$$
$$\times\left(1-\left(\frac{1-\overline{G}^{\alpha}(x;\Psi)}{1+\overline{G}^{\alpha}(x;\Psi)}\right)^{\beta}\right)\left(\frac{1-\overline{G}^{\alpha}(x;\Psi)}{1+\overline{G}^{\alpha}(x;\Psi)}\right)^{\beta-1}$$

$$\times\,[\overline{G}(x;\Psi)]^{\alpha-1}g(x;\Psi), \tag{4.11}$$

which is $\leq 0$ if $b_1 \leq b_2$. Therefore $X_1 <_{lr} X_2$, $X_1 <_{hr} X_2$ and $X_1 <_s X_2$. Consequently, the random variables $X_1$ and $X_2$ are stochastically ordered. □

## 4.3.6 Distribution of order statistics

Let $X_1, X_2, \ldots, X_n$ be a random sample from the TL-EHL-GG family of distributions, and $X_{1:n} \leq X_{2:n} \leq \ldots \leq X_{n:n}$ the corresponding order statistics. Let $f(x;\alpha,\beta,b,\Psi) = f(x)$ and $F(x;\alpha,\beta,b,\Psi) = F(x)$. The pdf of the $i$th order statistic, $X_{i:n}$ from the TL-EHL-GG family of distributions is given by

$$f_{i:n}(x) = \frac{n!}{(i-1)!(n-1)!}\sum_{s=0}^{n-i}(-1)^s\binom{n-i}{s}f(x;\alpha,\beta,b,\Psi))[F(x;\alpha,\beta,b,\Psi)]^{s+i-1}.$$

Note that

$$\begin{aligned}f(x)[F(x)]^{s+i-1} &= 4\alpha\beta b\left[1-\left(1-\left(\frac{1-\overline{G}^{\alpha}(x;\Psi)}{1+\overline{G}^{\alpha}(x;\Psi)}\right)^{\beta}\right)^{2}\right]^{b(s+i)-1}\\ &\times\left(1-\left(\frac{1-\overline{G}^{\alpha}(x;\Psi)}{1+\overline{G}^{\alpha}(x;\Psi)}\right)^{\beta}\right)\frac{(1-\overline{G}^{\alpha}(x;\Psi))^{\beta-1}}{(1+\overline{G}^{\alpha}(x;\Psi))^{\beta+1}}\\ &\times[\overline{G}(x;\Psi)]^{\alpha-1}g(x;\Psi).\end{aligned}$$

Applying the following generalized binomial expansion, which was used in the density expansion, we can write

$$\begin{aligned}f(x)[F(x)]^{s+i-1} &= 4\alpha\beta b\sum_{r,j,k,l,m=0}^{\infty}(-1)^{r+j+k+l+m}\binom{b(s+i)-1}{r}\binom{2r+1}{j}\\ &\times\binom{\beta(j+1)-1}{k}\binom{-(\beta(j+1)+1)}{l}\binom{\alpha(k+l+1)-1}{m}\\ &\times\frac{(m+1)}{(m+1)}G^{m}(x;\Psi)g(x;\Psi)\\ &= \sum_{m=0}^{\infty}b_{m+1}g^{*}_{m+1}(x;\Psi),\end{aligned}$$

where

$$b_{m+1} = 4\alpha\beta b\sum_{r,j,k,l=0}^{\infty}(-1)^{r+j+k+l+m}\binom{b(s+i)-1}{r}\binom{2r+1}{j}(m+1)$$

$$\times \binom{\beta(j+1)-1}{k}\binom{-(\beta(j+1)+1)}{l}\binom{\alpha(k+l+1)-1}{m},$$

and $g^*_{m+1}(x;\Psi) = (m+1)g(x;\Psi)G^m(x;\Psi)$ is the Exp-G density with power parameter $(m+1)$. Consequently, the pdf of the $i$th order statistics from the TL-EHL-GG family of distributions is given by

$$f_{i:n}(x) = \frac{n!}{(i-1)!(n-1)!}\sum_{s=0}^{n-i}\sum_{m=0}^{\infty}(-1)^s\binom{n-i}{s}b_{m+1}g^*_{m+1}(x;\Psi). \tag{4.12}$$

## 4.3.7 Entropy

We present Rényi entropy of the TL-EHL-GG family of distributions. Rényi entropy [16] of a random variable X following TL-EHL-GG family of distributions is obtained as follows:

$$I_R(\nu) = (1-\nu)^{-1}\log\left[\int_0^{\infty} f^{\nu}(x;\alpha,\beta,b,\Psi)dx\right] \quad \text{for} \quad \nu > 0 \quad \text{and} \quad \nu \neq 1.$$

Note that

$$\begin{aligned} f^{\nu}(x;\alpha,\beta,b,\Psi) = (4\alpha\beta b)^{\nu}&\left[1-\left(1-\left(\frac{1-\overline{G}^{\alpha}(x;\Psi)}{1+\overline{G}^{\alpha}(x;\Psi)}\right)^{\beta}\right)^{2}\right]^{\nu(b-1)} \\ &\times\left(1-\left(\frac{1-\overline{G}^{\alpha}(x;\Psi)}{1+\overline{G}^{\alpha}(x;\Psi)}\right)^{\beta}\right)^{\nu}\frac{(1-\overline{G}^{\alpha}(x;\Psi))^{\nu(\beta-1)}}{(1+\overline{G}^{\alpha}(x;\Psi))^{\nu(\beta+1)}} \\ &\times[\overline{G}(x;\Psi)]^{\nu(\alpha-1)}g^{\nu}(x;\Psi). \end{aligned}$$

Applying the generalized binomial expansion, which was used in the density expansion, we obtain

$$\begin{aligned} f^{\nu}(x;\alpha,\beta,b,\Psi) = (4\alpha\beta b)^{\nu}&\sum_{i,j,k,l,m=0}^{\infty}(-1)^{i+j+k+l+m}\binom{\nu(b-1)}{i}\binom{2i+1}{j} \\ &\times\binom{\beta(j+\nu)-\nu}{k}\binom{-(\beta(j+\nu)+\nu)}{l}\binom{\alpha(k+l+\nu)-\nu}{m} \\ &\times G^m(x;\Psi)g^{\nu}(x;\Psi). \end{aligned}$$

Finally, we represent Rényi entropy of the TL-EHL-GG family of distributions as

$$\begin{aligned} I_R(\nu) = (1-\nu)^{-1}\log\Bigg[(4\alpha\beta b)^{\nu}&\sum_{i,j,k,l,m=0}^{\infty}(-1)^{i+j+k+l+m}\binom{\nu(b-1)}{i}\binom{2i+1}{j} \\ \times\binom{\beta(j+\nu)-\nu}{k}&\binom{-(\beta(j+\nu)+\nu)}{l}\binom{\alpha(k+l+\nu)-\nu}{m} \end{aligned}$$

$$\times \frac{1}{(\frac{m}{\nu}+1)^{\nu}} \int_0^{\infty} \left( \left(\frac{m}{\nu}+1\right) g(x;\Psi) G^{\frac{m}{\nu}}(x;\Psi) \right)^{\nu} dx \Bigg]$$

$$= (1-\nu)^{-1} \log \left[ \sum_{m=0}^{\infty} R_m^* \exp((1-\nu) I_{REG}) \right],$$

where $I_{REG} = (1-\nu)^{-1} \log \left[ \int_0^{\infty} \left( \left(\frac{m}{\nu}+1\right) g(x;\Psi) G^{\frac{m}{\nu}}(x;\Psi) \right)^{\nu} dx \right]$ is Rényi entropy of Exp-G densities with power parameter $(\frac{m}{\nu}+1)$ and

$$R_m^* = (4\alpha\beta b)^{\nu} \sum_{i,j,k,l=0}^{\infty} (-1)^{i+j+k+l+m} \binom{\nu(b-1)}{i} \binom{2i+1}{j}$$
$$\times \binom{\beta(j+\nu)-\nu}{k} \binom{-(\beta(j+\nu)+\nu)}{l} \binom{\alpha(k+l+\nu)-\nu}{m} \frac{1}{\left(\frac{m}{\nu}+1\right)^{\nu}}.$$

## 4.4 Parameter estimation

Suppose that $X \sim TL-EHL-GG(\alpha, \beta, b, \Psi)$ and that $\Delta = (\alpha, \beta, b, \Psi)^T$ is the parameter vector, then the log-likelihood function $(\ell_n)$ of a random sample of size $n$ from the $TL-EHL-GG(\alpha, \beta, b, \Psi)$ family of distributions is given by

$$\ell_n(\Delta) = n\log(4\alpha\beta b) + (b-1)\sum_{i=1}^{n} \log\left[1 - \left(1 - \left(\frac{1-\overline{G}^{\alpha}(x_i;\Psi)}{1+\overline{G}^{\alpha}(x_i;\Psi)}\right)^{\beta}\right)^2\right]$$
$$+ \sum_{i=1}^{n} \log\left[1 - \left(\frac{1-\overline{G}^{\alpha}(x_i;\Psi)}{1+\overline{G}^{\alpha}(x_i;\Psi)}\right)^{\beta}\right] + (\beta-1)\sum_{i=1}^{n}\log[1-\overline{G}^{\alpha}(x_i;\Psi)]$$
$$- (\beta+1)\sum_{i=1}^{n}\log[1+\overline{G}^{\alpha}(x_i;\Psi)] + (\alpha-1)\sum_{i=1}^{n}\log[\overline{G}(x_i;\Psi)]$$
$$+ \sum_{i=1}^{n}\log[g(x_i;\Psi)].$$

The maximum likelihood estimates (MLEs) of the parameters can be obtained by solving a system of non-linear equations $\left(\frac{\partial \ell_n}{\partial \alpha}, \frac{\partial \ell_n}{\partial \beta}, \frac{\partial \ell_n}{\partial b}, \frac{\partial \ell_n}{\partial \Psi_k}\right)^T = \mathbf{0}$ by numerical methods. Elements of the score vector $\Delta = (\alpha, \beta, b, \Psi)$ are given in the appendix.

The Fisher information matrix (FIM) of the TL-EHL-GG family of distributions is given by $\mathbf{I}(\Delta) = [\mathbf{I}_{\Delta_i,\Delta_j}]_{(3+q)\times(3+q)} = E\left(-\frac{\partial^2 \ell_n}{\partial \Delta_i \partial \Delta_j}\right)$, $i, j = 1, 2, ..., (3+q)$, and can be numerically obtained by mle2 package in R software. The total Fisher information matrix $n\mathbf{I}(\Theta)$ can be approximated by $\mathbf{J}_n(\hat{\Delta}) \approx \left[-\frac{\partial^2 \ell_n}{\partial \Delta_i \partial \Delta_j}\Big|_{\Delta=\hat{\Delta}}\right]_{(3+q)\times(3+q)}$, $i, j = 1, 2, ..., (3+q)$. The multivariate normal distribution $N_{(3+q)}(\underline{0}, \boldsymbol{J}(\hat{\boldsymbol{\Delta}})^{-1})$ can be used to construct confidence intervals and confidence regions for the individual model parameters. The approximate $100(1-\eta)\%$ two-sided confidence intervals for $\alpha$, $\beta$, $b$ and $\Psi_k$ are given by

$$\hat{\alpha}\pm Z_{\frac{\eta}{2}}\sqrt{I_{\alpha\alpha}^{-1}(\hat{\Delta})},\quad \hat{\beta}\pm Z_{\frac{\eta}{2}}\sqrt{I_{\beta\beta}^{-1}(\hat{\Delta})},\quad \hat{b}\pm Z_{\frac{\eta}{2}}\sqrt{I_{bb}^{-1}(\hat{\Delta})} \quad \text{and} \quad \hat{\Psi}_k\pm Z_{\frac{\eta}{2}}\sqrt{I_{\Psi_{kk}}^{-1}(\hat{\Delta})},$$

respectively, where $I_{\alpha\alpha}^{-1}(\hat{\Delta})$, $I_{\beta\beta}^{-1}(\hat{\Delta})$, $I_{bb}^{-1}(\hat{\Delta})$ and $I_{\Psi_{kk}}^{-1}(\hat{\Delta})$ are the diagonal elements of $I_n^{-1}(\hat{\Delta})=(nI(\hat{\Delta}))^{-1}$, and $Z_{\frac{\eta}{2}}$ is the upper $(\frac{\eta}{2})$th percentile of the standard normal distribution.

## 4.5 Specific cases

In this section, we specify the baseline cdf $G(x;\Psi)$ and pdf $g(x;\Psi)$ in Eqs. (4.4) and (4.5) to come up with some special cases of TL-EHL-GG family of distributions.

### 4.5.1 Topp–Leone-exponentiated half logistic generalized-log-logistic (TL-EHL-GLLo) distribution

Suppose we take the baseline distribution to be the log-logistic distribution with cdf and pdf given by $G(x;c)=1-(1+x^c)^{-1}$ and $g(x;c)=cx^{c-1}(1+x^c)^{-2}$, respectively, for $c,x>0$, then the cdf and pdf of the TL-EHL-GLLo distribution are given by

$$F_{TL-EHL-GLLo}(x;\alpha,\beta,b,c)=\left[1-\left(1-\left(\frac{1-(1+x^c)^{-\alpha}}{1+(1+x^c)^{-\alpha}}\right)^{\beta}\right)^2\right]^b \tag{4.13}$$

and

$$\begin{aligned} f_{TL-EHL-GLLo}(x;\alpha,\beta,b,c) &= 4\alpha\beta b\left[1-\left(1-\left(\frac{1-(1+x^c)^{-\alpha}}{1+(1+x^c)^{-\alpha}}\right)^{\beta}\right)^2\right]^{b-1} \\ &\times\left(1-\left(\frac{1-(1+x^c)^{-\alpha}}{1+(1+x^c)^{-\alpha}}\right)^{\beta}\right)\frac{(1-(1+x^c)^{-\alpha})^{\beta-1}}{(1+(1+x^c)^{-\alpha})^{\beta+1}} \\ &\times cx^{c-1}(1+x^c)^{-\alpha-1}, \end{aligned} \tag{4.14}$$

respectively. The corresponding hrf is given by

$$\begin{aligned} h_{TL-EHL-GLLo}(x;\alpha,\beta,b,c) &= \alpha\beta b\left[1-\left(1-\left(\frac{1-(1+x^c)^{-\alpha}}{1+(1+x^c)^{-\alpha}}\right)^{\beta}\right)^2\right]^{b-1}(1+x^c)^{-\alpha-1} \\ &\times\left(1-\left(\frac{1-(1+x^c)^{-\alpha}}{1+(1+x^c)^{-\alpha}}\right)^{\beta}\right)\frac{(1-(1+x^c)^{-\alpha})^{\beta-1}}{(1+(1+x^c)^{-\alpha})^{\beta+1}} \end{aligned}$$

$$\times\, cx^{c-1}\left(1-\left[1-\left(1-\left(\frac{1-(1+x^c)^{-\alpha}}{1+(1+x^c)^{-\alpha}}\right)^{\beta}\right)^{2}\right]^{b}\right)^{-1}, \tag{4.15}$$

for $\alpha, \beta, b, c > 0$, and $x > 0$.

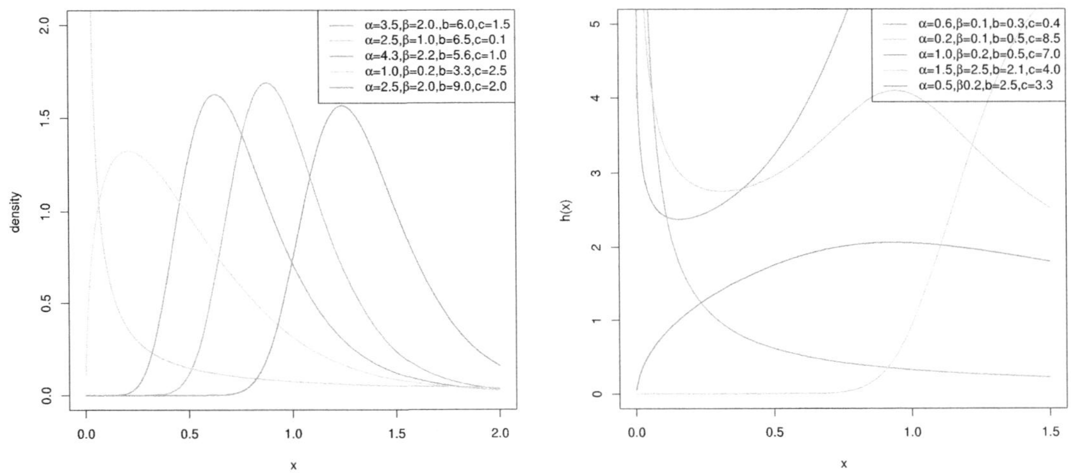


**FIGURE 4.1** Plots of the pdf and hrf for the TL-EHL-GLLo distribution.

Fig. 4.1 gives the plots of pdf and hrf of the TL-EHL-GLLo distribution. The shapes of the pdf of the TL-EHL-GLLo distribution include right-skewed, left-skewed, reverse-J, and almost symmetric shapes. The hrf on the other hand displays decreasing, increasing, bathtub, upside-down bathtub, and bathtub followed by upside-down bathtub shapes. Fig. 4.2 shows that TL-EHL-GLLo distribution can model data sets with different levels of skewness and kurtosis.

### 4.5.2 Topp–Leone-exponentiated half logistic generalized-Weibull (TL-EHL-GW) distribution

If we let the baseline distribution to be Weibull distribution with cdf and pdf given as $G(x;\lambda) = 1 - e^{-x^{\lambda}}$ and $g(x;\lambda) = \lambda x^{\lambda-1}e^{-x^{\lambda}}$, respectively, for $\lambda, x > 0$, we obtain the cdf, pdf and hrf of the TL-EHL-GW distribution as

$$F_{TL-EHL-GW}(x;\alpha,\beta,b,\lambda) = \left[1-\left(1-\left(\frac{1-e^{-\alpha x^{\lambda}}}{1+e^{-\alpha x^{\lambda}}}\right)^{\beta}\right)^{2}\right]^{b},$$

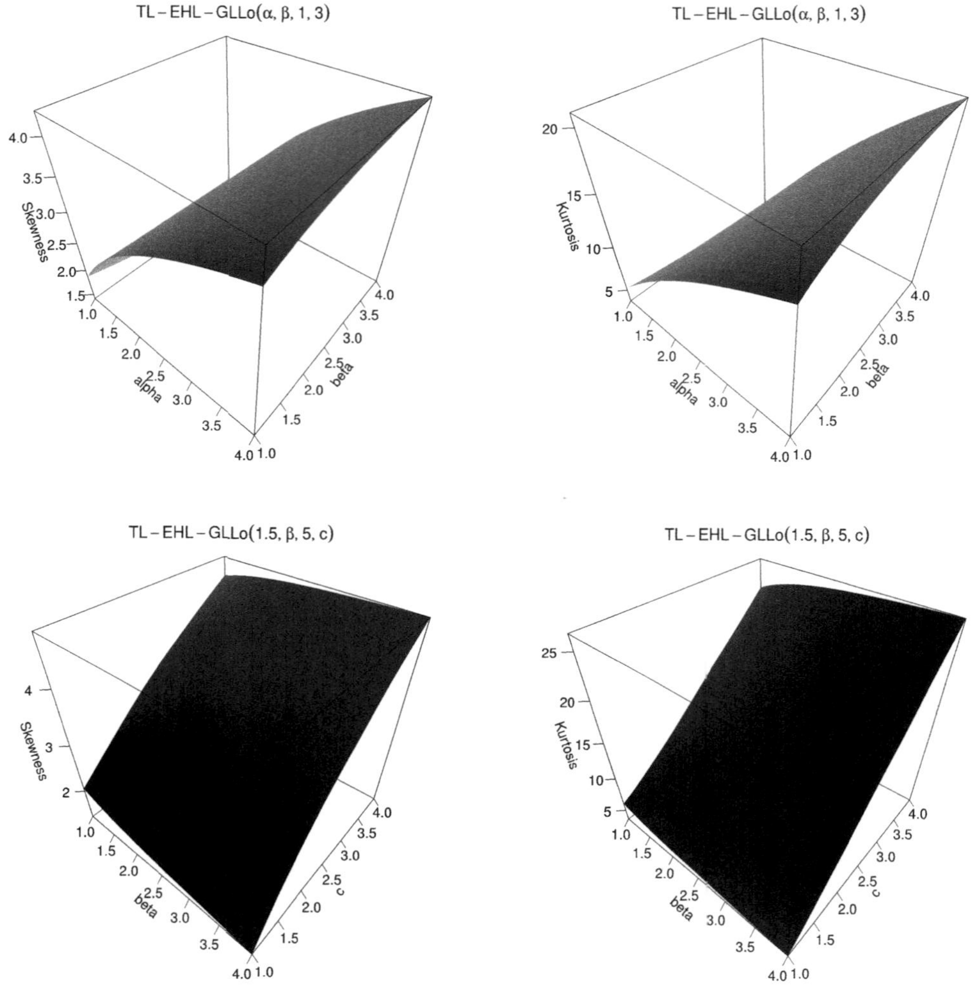


**FIGURE 4.2** Plots of the skewness and kurtosis of the TL-EHL-GLLo distribution for some selected parameter values.

$$f_{TL-EHL-GW}(x;\alpha,\beta,b,\lambda)=4\alpha\beta b\left[1-\left(1-\left(\frac{1-e^{-\alpha x^{\lambda}}}{1+e^{-\alpha x^{\lambda}}}\right)^{\beta}\right)^{2}\right]^{b-1}$$
$$\times\left(1-\left(\frac{1-e^{-\alpha x^{\lambda}}}{1+e^{-\alpha x^{\lambda}}}\right)^{\beta}\right)\frac{(1-e^{-\alpha x^{\lambda}})^{\beta-1}}{(1+e^{-\alpha x^{\lambda}})^{\beta+1}}$$
$$\times e^{-\alpha x^{\lambda}}\lambda x^{\lambda-1},$$

and

$$h_{TL-EHL-GW}(x;\alpha,\beta,b,\lambda) = 4\alpha\beta b\left[1-\left(1-\left(\frac{1-e^{-\alpha x^{\lambda}}}{1+e^{-\alpha x^{\lambda}}}\right)^{\beta}\right)^{2}\right]^{b-1}$$
$$\times\left(1-\left(\frac{1-e^{-\alpha x^{\lambda}}}{1+e^{-\alpha x^{\lambda}}}\right)^{\beta}\right)\frac{(1-e^{-\alpha x^{\lambda}})^{\beta-1}}{(1+e^{-\alpha x^{\lambda}})^{\beta+1}}$$
$$\times\left(1-\left[1-\left(1-\left(\frac{1-e^{-\alpha x^{\lambda}}}{1+e^{-\alpha x^{\lambda}}}\right)^{\beta}\right)^{2}\right]^{b}\right)^{-1}$$
$$\times e^{-\alpha x^{\lambda}}\lambda x^{\lambda-1},$$

for $\alpha, \beta, b, \lambda > 0$, and $x > 0$.

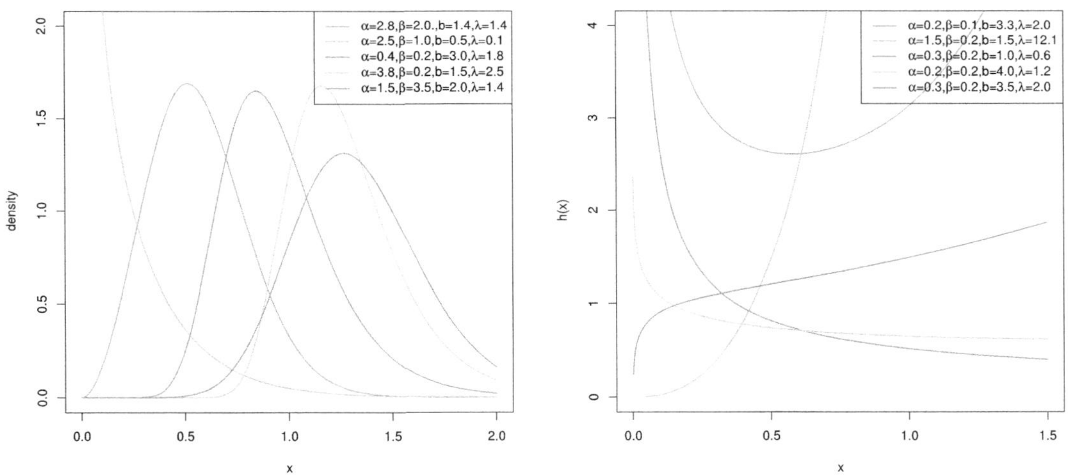


**FIGURE 4.3** Plots of the pdf and hrf for the TL-EHL-GW distribution.

Fig. 4.3 represents the plots of pdf and hrf of the TL-EHL-GW distribution. The pdf of the TL-EHL-GW distribution can be right-skewed, left-skewed, reverse-J, and almost symmetric shapes, whereas the hrf displays bathtub, decreasing, and increasing shapes. Fig. 4.4 indicates that TL-EHL-GW distribution can model data sets with different levels of skewness and kurtosis.

### 4.5.3 Topp–Leone-exponentiated half logistic generalized-power (TL-EHL-GP) distribution

If we take the baseline distribution to be power distribution with cdf and pdf given as $G(x;\lambda,\delta) = (\delta x)^{\lambda}$ and $g(x;\lambda,\delta) = \lambda\delta^{\lambda}x^{\lambda-1}$, respectively, for $\lambda, \delta > 0$ and $0 < x < \frac{1}{\delta}$. Then the cdf, pdf and hrf of the TL-EHL-GP distribution are given by

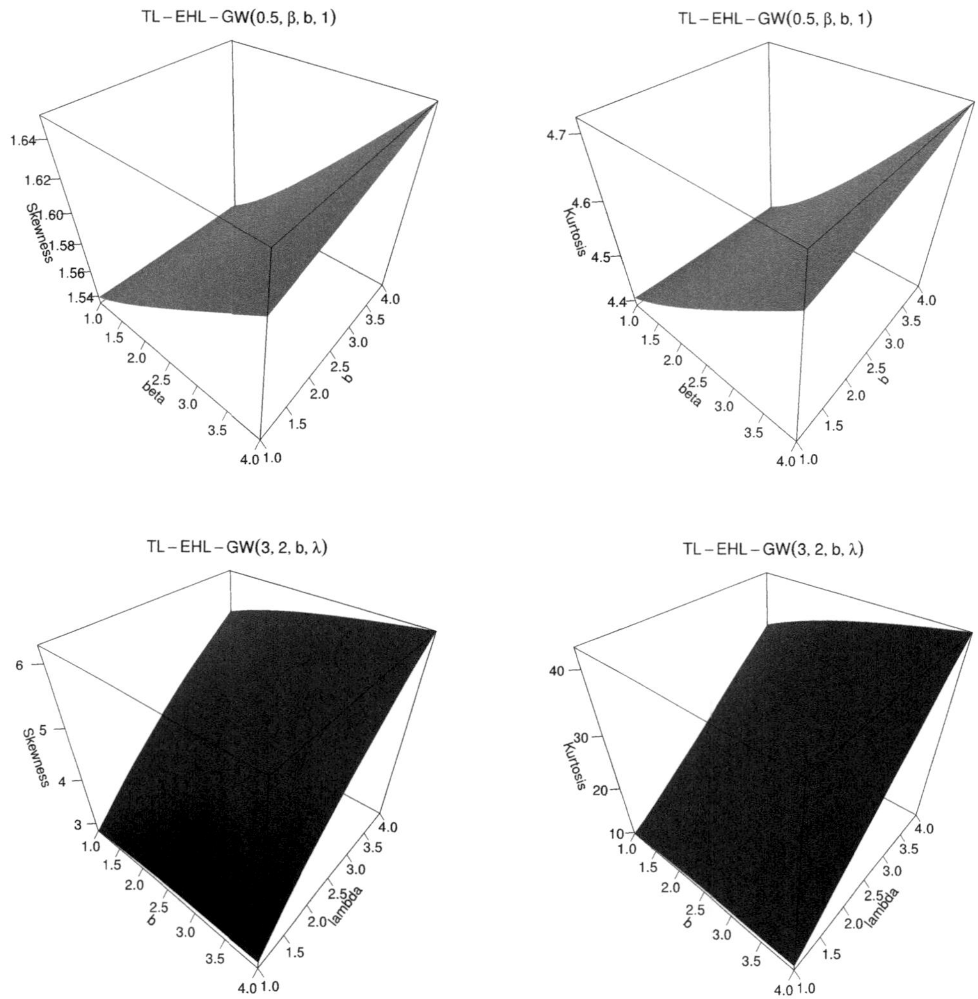


**FIGURE 4.4** Plots of the skewness and kurtosis of the TL-EHL-GW distribution for some selected parameter values.

$$F_{TL-EHL-GP}(x;\alpha,\beta,b,\lambda,\delta)=\left[1-\left(1-\left(\frac{1-(1-(\delta x)^{\lambda})^{\alpha}}{1+(1-(\delta x)^{\lambda})^{\alpha}}\right)^{\beta}\right)^{2}\right]^{b},$$

$$\begin{aligned}f_{TL-EHL-GP}(x;\alpha,\beta,b,\lambda,\delta)&=4\alpha\beta b\left[1-\left(1-\left(\frac{1-(1-(\delta x)^{\lambda})^{\alpha}}{1+(1-(\delta x)^{\lambda})^{\alpha}}\right)^{\beta}\right)^{2}\right]^{b-1}\\&\times\left(1-\left(\frac{1-(1-(\delta x)^{\lambda})^{\alpha}}{1+(1-(\delta x)^{\lambda})^{\alpha}}\right)^{\beta}\right)\frac{(1-(1-(\delta x)^{\lambda})^{\alpha})^{\beta-1}}{(1+(1-(\delta x)^{\lambda})^{\alpha})^{\beta+1}}\\&\times(\delta x)^{\lambda(\alpha-1)}\lambda\delta^{\lambda}x^{\lambda-1},\end{aligned}$$

and

$$h_{TL-EHL-GP}(x;\alpha,\beta,b,\lambda,\delta)=4\alpha\beta b\left[1-\left(1-\left(\frac{1-(1-(\delta x)^{\lambda})^{\alpha}}{1+(1-(\delta x)^{\lambda})^{\alpha}}\right)^{\beta}\right)^{2}\right]^{b-1}$$
$$\times\left(1-\left(\frac{1-(1-(\delta x)^{\lambda})^{\alpha}}{1+(1-(\delta x)^{\lambda})^{\alpha}}\right)^{\beta}\right)\frac{(1-(1-(\delta x)^{\lambda})^{\alpha})^{\beta-1}}{(1+(1-(\delta x)^{\lambda})^{\alpha})^{\beta+1}}$$
$$\times\left(1-\left[1-\left(1-\left(\frac{1-(1-(\delta x)^{\lambda})^{\alpha}}{1+(1-(\delta x)^{\lambda})^{\alpha}}\right)^{\beta}\right)^{2}\right]^{b}\right)^{-1}$$
$$\times(\delta x)^{\lambda(\alpha-1)}\lambda\delta^{\lambda}x^{\lambda-1}$$

for $\alpha,\beta,b,\lambda,\delta>0$.

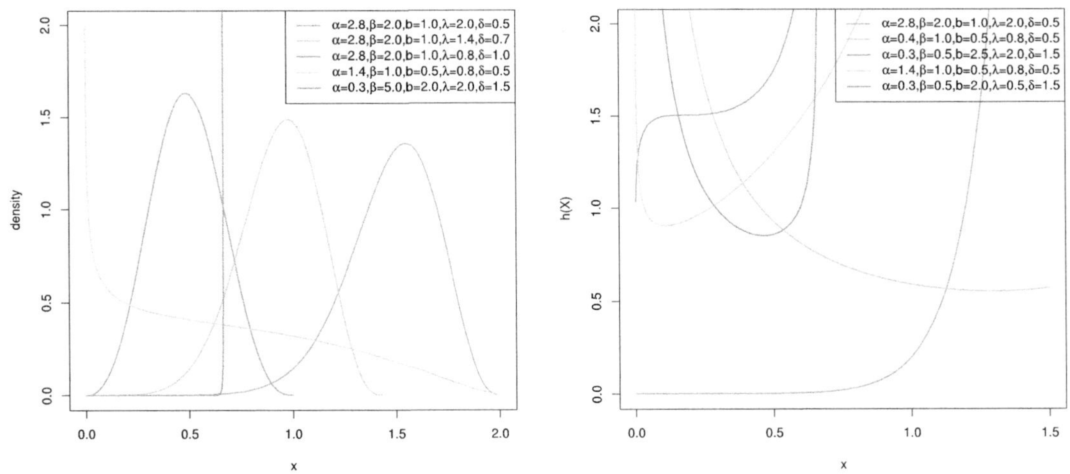


**FIGURE 4.5** Plots of the pdf and hrf for the TL-EHL-GP distribution.

Fig. 4.5 shows the plots of the pdf and hrf for the TL-EHL-GP distribution. The pdf displays shapes that are left-skewed, almost symmetric, reverse-J, and J-shapes, whereas the hrf shows bathtub, upside-down bathtub, followed by bathtub, decreasing, and increasing shapes. Fig. 4.6 gives plots of skewness and kurtosis for the TL-EHL-GP distribution.

## 4.6 Monte Carlo simulations

This section is devoted to assess the performance of the estimators of the parameters of the TL-EHL-GW distribution. We conduct simulations for different sample sizes ($n=50$, 100, 200, 400, 800, 1600). We simulate $N=3000$ samples for the selected parameters values of $(\alpha,\beta,b,\lambda)$, as given in Tables 4.1 and 4.2, respectively. The results were obtained using

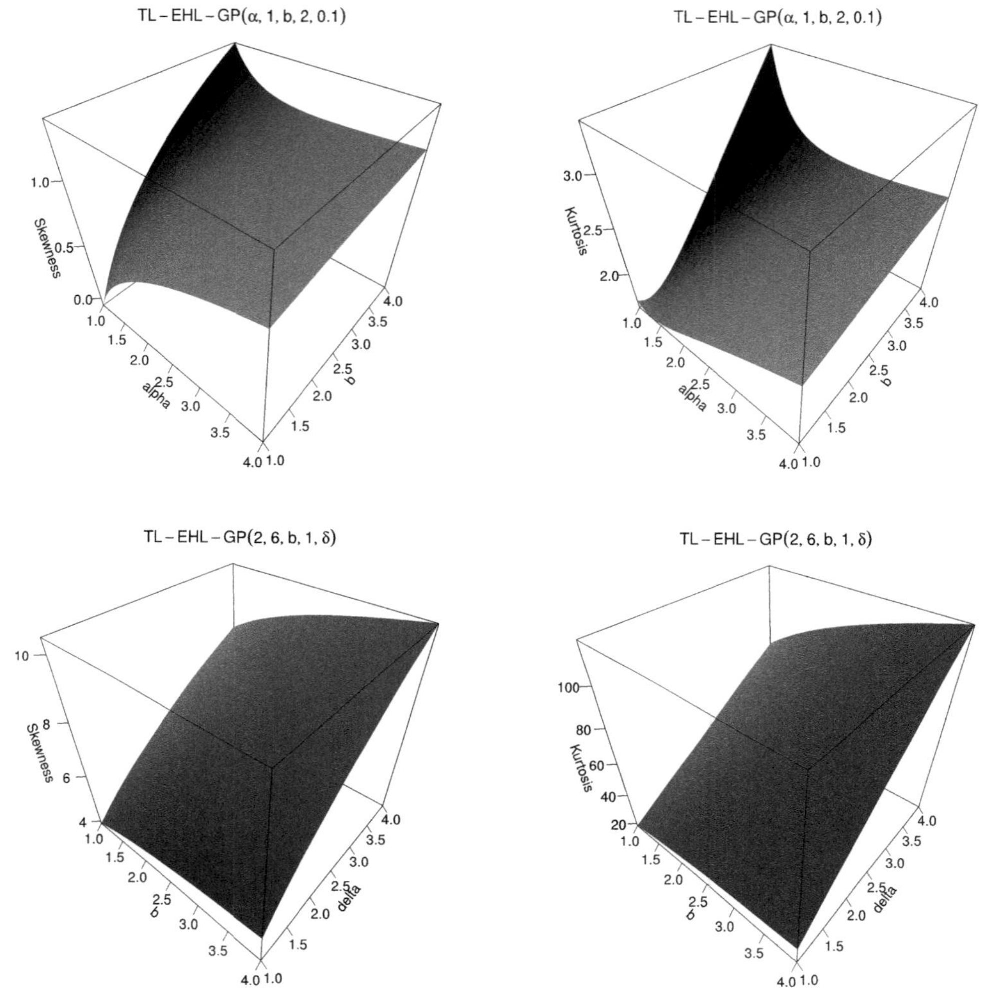


**FIGURE 4.6** Plots of the skewness and kurtosis of the TL-EHL-GP distribution for some selected parameter values.

R programming language. The average bias (ABIAS) and root mean square errors (RMSEs) for the estimated parameter, say, $\hat{\alpha}$, are given by

$$ABIAS(\hat{\alpha}) = \frac{\sum_{i=1}^{N} \hat{\alpha}_i}{N} - \alpha, \quad \text{and} \quad RMSE(\hat{\alpha}) = \sqrt{\frac{\sum_{i=1}^{N} (\hat{\alpha}_i - \alpha)^2}{N}},$$

respectively. From the results, we see that RMSEs decrease and ABIAS decreases towards zero as the sample size $n$ increases, and the mean estimates are getting closer to the true parameter values as the sample size increases.

**Table 4.1** Simulation results for TL-EHL-GW distribution: Mean estimates, RMSE, and ABIAS.

| | | (0.5, 1.0, 0.5, 2.5) | | | (0.1, 2.0, 1.0, 4.0) | | | (0.1, 0.7, 0.7, 3.0) | | |
|---|---|---|---|---|---|---|---|---|---|---|
| **Parameter** | $n$ | **Mean** | **RMSE** | **ABIAS** | **Mean** | **RMSE** | **ABIAS** | **Mean** | **RMSE** | **ABIAS** |
| $\alpha$ | 50 | 0.5850 | 0.4902 | 0.0850 | 0.1598 | 0.3240 | 0.0598 | 0.2089 | 0.2777 | 0.1089 |
| | 100 | 0.5342 | 0.4001 | 0.0342 | 0.1267 | 0.1259 | 0.0267 | 0.1674 | 0.2006 | 0.0674 |
| | 200 | 0.4987 | 0.3385 | −0.0013 | 0.1139 | 0.0685 | 0.0139 | 0.1378 | 0.1434 | 0.0379 |
| | 400 | 0.4771 | 0.2936 | −0.0229 | 0.1080 | 0.0474 | 0.0080 | 0.1228 | 0.1172 | 0.0228 |
| | 800 | 0.4771 | 0.2457 | −0.0229 | 0.1033 | 0.0328 | 0.0033 | 0.1154 | 0.0967 | 0.0154 |
| | 1600 | 0.4807 | 0.2018 | −0.0193 | 0.1020 | 0.0233 | 0.0020 | 0.1115 | 0.0753 | 0.0115 |
| $\beta$ | 50 | 1.5869 | 1.2846 | 0.5869 | 2.2705 | 2.7793 | 0.2705 | 1.3784 | 1.3184 | 0.6784 |
| | 100 | 1.4333 | 1.1795 | 0.4333 | 2.1514 | 1.0000 | 0.1514 | 1.2325 | 1.1972 | 0.5325 |
| | 200 | 1.3381 | 1.0934 | 0.3381 | 2.0938 | 0.5582 | 0.0938 | 1.1115 | 1.0382 | 0.4115 |
| | 400 | 1.2199 | 0.9503 | 0.2199 | 2.0709 | 0.5089 | 0.0709 | 0.9850 | 0.8730 | 0.2850 |
| | 800 | 1.1572 | 0.8274 | 0.1572 | 2.0481 | 0.3976 | 0.0481 | 0.9131 | 0.7308 | 0.2131 |
| | 1600 | 1.1059 | 0.6832 | 0.1059 | 2.0266 | 0.3057 | 0.0266 | 0.8294 | 0.5376 | 0.1294 |
| $b$ | 50 | 2.7004 | 8.9821 | 2.2004 | 2.9124 | 9.0313 | 1.9124 | 3.9604 | 3.4871 | 3.2604 |
| | 100 | 1.4469 | 3.8795 | 0.9469 | 1.6445 | 3.5929 | 0.6445 | 1.6764 | 3.4703 | 0.9764 |
| | 200 | 0.8400 | 1.1394 | 0.3400 | 1.2774 | 2.0256 | 0.2774 | 1.0252 | 1.2787 | 0.3252 |
| | 400 | 0.6889 | 0.5624 | 0.1889 | 1.0895 | 0.6098 | 0.0895 | 0.8672 | 0.6843 | 0.1671 |
| | 800 | 0.6345 | 0.4015 | 0.1345 | 1.0471 | 0.4086 | 0.0471 | 0.7897 | 0.4470 | 0.0897 |
| | 1600 | 0.5815 | 0.2768 | 0.0815 | 1.0245 | 0.2086 | 0.0245 | 0.7508 | 0.3087 | 0.0508 |
| $\lambda$ | 50 | 3.3701 | 2.2041 | 0.8701 | 4.3493 | 1.3711 | 0.3493 | 3.3312 | 1.4174 | 0.3312 |
| | 100 | 3.1792 | 1.7922 | 0.6792 | 4.1642 | 1.0253 | 0.1642 | 3.3279 | 1.2457 | 0.3279 |
| | 200 | 3.0361 | 1.4232 | 0.5361 | 4.0200 | 0.6594 | 0.0200 | 3.2881 | 1.0466 | 0.2881 |
| | 400 | 2.9237 | 1.1094 | 0.4237 | 4.0045 | 0.4968 | 0.0045 | 3.2571 | 0.9051 | 0.2571 |
| | 800 | 2.7828 | 0.8051 | 0.2828 | 4.0087 | 0.3631 | 0.0087 | 3.1839 | 0.7461 | 0.1839 |
| | 1600 | 2.6845 | 0.5833 | 0.1845 | 3.9977 | 0.2519 | −0.0023 | 3.0918 | 0.5705 | 0.0918 |

## 4.7 Some applications

In this section, we give examples of applications to real data sets to assess the usefulness of the TL-EHL-GW distribution. The TL-EHL-GW distribution is compared to several competing models with the same number of parameters, including Topp–Leone Weibull–Lomax (TLWLx) distribution by Jamal et al. [11], Topp–Leone generalized inverted Kumaraswamy (TLGIKw) distribution by Reyad et al. [18], Weibull–Lomax (WL) distribution by Tahir et al. [20], the beta odd Lindley-exponential (BOL-E) distribution by Chipepa et al. [8] and odd log-logistic exponentiated Weibull (OLLEW) distribution by Afify et al. [1]. The pdf's of the models listed above are given by

$$f_{TLWLx}(x; a, b, \alpha, \theta) = 2\theta\alpha ab(1+bx)^{a\alpha-1}\{1-(1+bx)^{-a}\}^{a-1}$$

**Table 4.2** Simulation results for TL-EHL-GW distribution: Mean estimates, RMSE, and ABIAS.

| | | (2.0, 2.0, 0.5, 1.5) | | | (0.5, 1.0, 0.3, 3.5) | | | (0.1, 1.0, 2.0, 4.0) | | |
|---|---|---|---|---|---|---|---|---|---|---|
| **Parameter** | $n$ | **Mean** | **RMSE** | **ABIAS** | **Mean** | **RMSE** | **ABIAS** | **Mean** | **RMSE** | **ABIAS** |
| $\alpha$ | 50 | 1.6522 | 0.9609 | −0.3478 | 0.5230 | 0.4537 | 0.0230 | 0.2072 | 0.4051 | 0.1072 |
| | 100 | 1.6256 | 0.8252 | −0.3744 | 0.4855 | 0.3797 | −0.0145 | 0.1397 | 0.1594 | 0.0397 |
| | 200 | 1.6340 | 0.7388 | −0.3659 | 0.4719 | 0.3316 | −0.0281 | 0.1189 | 0.0924 | 0.0189 |
| | 400 | 1.7082 | 0.6256 | −0.2918 | 0.4839 | 0.2930 | −0.0161 | 0.1199 | 0.0716 | 0.0199 |
| | 800 | 1.7879 | 0.5069 | −0.2121 | 0.4946 | 0.2528 | −0.0054 | 0.1108 | 0.0503 | 0.0108 |
| | 1600 | 1.8445 | 0.4243 | −0.1555 | 0.5067 | 0.2111 | 0.0047 | 0.1071 | 0.0376 | 0.0071 |
| $\beta$ | 50 | 1.7451 | 2.8126 | −0.2549 | 1.5562 | 1.4652 | 0.5562 | 1.7435 | 4.2186 | 0.7435 |
| | 100 | 1.7161 | 1.8882 | −0.2839 | 1.4893 | 1.4088 | 0.4893 | 1.2514 | 1.1209 | 0.2514 |
| | 200 | 1.7125 | 1.6476 | −0.2875 | 1.4346 | 1.3457 | 0.4346 | 1.1925 | 0.8149 | 0.1925 |
| | 400 | 1.8049 | 1.5528 | −0.1951 | 1.4307 | 1.2971 | 0.4307 | 1.2279 | 0.7963 | 0.2279 |
| | 800 | 1.8759 | 1.4283 | −0.1241 | 1.3875 | 1.1913 | 0.3875 | 1.1553 | 0.6491 | 0.1553 |
| | 1600 | 1.9319 | 1.3308 | −0.0681 | 1.3228 | 1.0474 | 0.3228 | 1.1010 | 0.5256 | 0.1010 |
| $b$ | 50 | 1.7699 | 5.2073 | 1.4699 | 1.5273 | 3.9444 | 1.2273 | 6.4417 | 9.2306 | 4.4417 |
| | 100 | 1.0355 | 2.2889 | 0.7355 | 0.9374 | 2.3473 | 0.6374 | 3.8541 | 6.9104 | 1.8541 |
| | 200 | 0.7475 | 1.0467 | 0.4475 | 0.6036 | 1.1441 | 0.3036 | 3.4697 | 6.1498 | 1.4697 |
| | 400 | 0.6298 | 0.6894 | 0.3298 | 0.4536 | 0.6455 | 0.1536 | 2.4469 | 2.8508 | 0.4469 |
| | 800 | 0.5767 | 0.4976 | 0.2767 | 0.3654 | 0.2677 | 0.0654 | 2.1883 | 1.5549 | 0.1883 |
| | 1600 | 0.5244 | 0.3985 | 0.2244 | 0.3451 | 0.1763 | 0.0451 | 2.0934 | 0.9089 | 0.0934 |
| $\lambda$ | 50 | 1.8914 | 0.8928 | 0.3914 | 4.9874 | 3.3345 | 1.4874 | 4.2569 | 1.5032 | 0.2569 |
| | 100 | 1.8861 | 0.8176 | 0.3861 | 4.7101 | 2.7727 | 1.2101 | 4.2426 | 1.2029 | 0.2426 |
| | 200 | 1.8638 | 0.7341 | 0.3638 | 4.4446 | 2.2434 | 0.9446 | 4.1193 | 0.8713 | 0.1193 |
| | 400 | 1.8036 | 0.6249 | 0.3036 | 4.0902 | 1.5939 | 0.5902 | 3.9822 | 0.6428 | −0.0178 |
| | 800 | 1.7494 | 0.4981 | 0.2494 | 3.8435 | 1.0876 | 0.3435 | 3.9883 | 0.4804 | −0.0117 |
| | 1600 | 1.6662 | 0.3621 | 0.1662 | 3.6637 | 0.7082 | 0.1637 | 3.9829 | 0.3584 | −0.0170 |

$$\times \exp\left\{-2\left(\frac{1-(1+bx)^{-a}}{(1+bx)^{-a}}\right)^{\alpha}\right\}$$
$$\times \left[1-\exp\left\{-2\left(\frac{1-(1+bx)^{-a}}{(1+bx)^{-a}}\right)^{\alpha}\right\}\right]^{\theta-1},$$

for $x > 0, a, b, \alpha, \theta > 0$,

$$f_{TLGIKw}(x; a, b, \alpha, \theta) = 2\alpha\beta\lambda\theta x^{\lambda-1}\left[1-\{1-[1-(1+x^{\lambda})^{-\alpha}]^{\beta}\}^{2}\right]^{\theta-1}$$
$$\times (1+x^{\lambda})^{-\alpha-1}[1-(1+x^{\lambda})^{-\alpha}]^{\beta-1}\{1-[1-(1+x^{\lambda})^{-\alpha}]^{\beta}\},$$

for $x > 0, \alpha, \lambda, \beta, \theta > 0$,

$$f_{WL}(x) = \frac{ab\alpha}{\beta}(1+(x/\beta))^{b\alpha-1}(1-(1+(x/\beta)))^{b-1}\exp\{-a[(1+(x/\beta))^{\alpha}-1]^{b}\},$$

for $x > 0, a, b, \alpha, \beta > 0$,

$$f_{BOL-E}(x; a, b, \lambda, \theta) = \frac{1}{B(a,b)} \left[1 - \frac{\lambda + e^{-\theta x}}{(1+\lambda)e^{-\theta x}} \exp\left\{-\lambda \frac{(1-e^{-\theta x})}{e^{-\theta x}}\right\}\right]^{a-1}$$
$$\times \left[\frac{\lambda + e^{-\theta x}}{(1+\lambda)e^{-\theta x}} \exp\left\{-\lambda \frac{(1-e^{-\theta x})}{e^{-\theta x}}\right\}\right]^{b-1}$$
$$\times \frac{\lambda^2}{(1+\lambda)} \frac{(\theta e^{-\theta x})}{e^{-3\theta x}} \exp\left\{-\lambda \frac{(1-e^{-\theta x})}{e^{-\theta x}}\right\},$$

for $x > 0, a, b, \lambda, \theta > 0$, and

$$f_{OLLEW}(x; \alpha, \beta, \gamma, \delta) = \frac{\theta\beta\gamma x^{\beta-1} e^{-(x/\alpha)^\beta} [1 - e^{-(x/\alpha)^\beta}]^{\gamma\theta-1} \{1 - [1 - e^{-(x/\alpha)^\beta}]^\gamma\}^{\theta-1}}{\alpha^\beta \left\{[1 - e^{-(x/\alpha)^\beta}]^{\gamma\theta} + \{1 - [1 - e^{-(x/\alpha)^\beta}]^\gamma\}^\theta\right\}^2},$$

for $x > 0, \alpha, \beta, \gamma, \delta > 0$.

The estimated values of parameters of the distributions with standard errors (in parentheses) are given in Tables 4.3 and 4.4, respectively. The goodness-of-fit statistics, including −2log-likelihood statistic (−2log(L)), Akaike information criterion (AIC), consistent Akaike information criterion (AICC), Bayesian information criterion (BIC), Cramér-von Mises ($W^*$), Anderson–Darling ($A^*$) statistic, Kolmogorov–Smirnov (KS) statistic and its P-value are also presented in the tables.

Plots of the fitted densities, the histogram of the data, probability plots, estimated cdf, Kaplan–Meier (K-M) survival plots, total time on test (TTT), and estimated hrf plots are given in Figs. 4.7, 4.8, 4.9 and 4.10. For the probability plots, we plotted $F(x_{(p)}; \hat{\alpha}, \hat{\beta}, \hat{b}, \hat{\lambda})$ against $\frac{p-0.375}{n+0.25}$, $p = 1, 2, \cdots, n$, where $x_{(p)}$ are the ordered values of the observed data. The measure of closeness to the diagonal line is given by the sum of squares (SS):

$$SS = \sum_{j=1}^{n} \left[F(x_{(p)}; \hat{\alpha}, \hat{\beta}, \hat{b}, \hat{\lambda}) - \left(\frac{p-0.375}{n+0.25}\right)\right]^2,$$

(see Chambers et al. [7] for additional details). For the scaled total time on test (TTT)-transform plot, we plotted $\Phi_F(u)$ vs $u$, where

$$\Phi_F(u) = \frac{H_F^{-1}(u)}{H^{-1}(1)} \quad \text{and} \quad H^{-1}(u) = \int_0^{F^{-1}(u)} (1 - F(x))dx, \quad \text{for} \quad 0 \leqslant u \leqslant 1.$$

## 4.7.1 Waiting times of bank customers data

The following data set corresponds to the waiting times (in minutes) of 100 bank customers. The data was also analyzed by Oluyede et al. [14]. The observations are as follows: 0.8, 0.8, 1.3, 1.5, 1.8, 1.9, 1.9, 2.1, 2.6, 2.7, 2.9, 3.1, 3.2, 3.3, 3.5, 3.6, 4.0, 4.1, 4.2, 4.2, 4.3, 4.3, 4.4, 4.4, 4.6, 4.7, 4.7, 4.8, 4.9, 4.9, 5.0, 5.3, 5.5, 5.7, 5.7, 6.1, 6.2, 6.2, 6.2, 6.3, 6.7, 6.9, 7.1, 7.1,

**Table 4.3** Parameter estimates and goodness-of-fit statistics of models for bank customers data.

| Model | Estimates | | | | Statistics | | | | | | | |
|---|---|---|---|---|---|---|---|---|---|---|---|---|
| | $\alpha$ | $\beta$ | $b$ | $\lambda$ | $-2\log L$ | $AIC$ | $AICC$ | $BIC$ | $W^*$ | $A^*$ | $K-S$ | p-value |
| TL-EHL-GW | 0.0921 | 0.1412 | 32.9208 | 0.9614 | 633.9 | 641.9 | 642.4 | 652.4 | 0.0179 | 0.1322 | 0.0369 | 0.9992 |
| | (0.1186) | (0.0649) | (0.0013) | (0.3423) | | | | | | | | |
| | $a$ | $b$ | $\alpha$ | $\theta$ | | | | | | | | |
| TLWLx | 18.1630 | 0.0041 | 0.5389 | 4.1174 | 636.6 | 644.6 | 645.1 | 655.1 | 0.0543 | 0.3508 | 0.0642 | 0.8050 |
| | $(2.8594\times10^{-05})$ | $(2.9104\times10^{-04})$ | $(4.0835\times10^{-02})$ | $(2.9813\times10^{-03})$ | | | | | | | | |
| | $\alpha$ | $\lambda$ | $\beta$ | $\theta$ | | | | | | | | |
| TLGIKw | 1.6288 | 0.7865 | 52.8214 | 0.2792 | 645.8 | 653.8 | 654.2 | 664.2 | 0.1254 | 0.8942 | 0.07389 | 0.6458 |
| | (0.4939) | (0.1814) | (40.0881) | (0.1697) | | | | | | | | |
| | $a$ | $b$ | $\alpha$ | $\beta$ | | | | | | | | |
| WL | $2.3751\times10^{-05}$ | 1.3714 | 1.0629 | $7.3688\times10^{-03}$ | 637.5 | 645.5 | 645.9 | 655.9 | 0.0629 | 0.3961 | 0.0577 | 0.8920 |
| | $(2.3750\times10^{-06})$ | $(4.5634\times10^{-10})$ | $(5.8906\times10^{-10})$ | $(1.1148\times10^{-08})$ | | | | | | | | |
| | $a$ | $b$ | $\lambda$ | $\theta$ | | | | | | | | |
| BOL-E | 2.0015 | 21.3750 | 24.7340 | 0.0004 | 634.6 | 642.6 | 643.0 | 653.04 | 0.0279 | 0.1839 | 0.0434 | 0.9917 |
| | $(5.0593\times10^{-09})$ | $(4.8510\times10^{-10})$ | $(4.4519\times10^{-10})$ | $(2.7384\times10^{-05})$ | | | | | | | | |
| | $\alpha$ | $\beta$ | $\gamma$ | $\theta$ | | | | | | | | |
| OLLEW | $2.6990\times10^{-05}$ | 0.1261 | 90.1420 | 2.5614 | 640.3 | 648.3 | 648.7 | 658.7 | 0.0693 | 0.4930 | 0.0551 | 0.9214 |
| | $(2.0210\times10^{-05})$ | $(7.5511\times10^{-03})$ | $(5.9157\times10^{-07})$ | $(1.9911\times10^{-04})$ | | | | | | | | |

7.1, 7.1, 7.4, 7.6, 7.7, 8.0, 8.2, 8.6, 8.6, 8.6, 8.8, 8.8, 8.9, 8.9, 9.5, 9.6, 9.7, 9.8, 10.7, 10.9, 11.0, 11.0, 11.1, 11.2, 11.2, 11.5, 11.9, 12.4, 12.5, 12.9, 13.0, 13.1, 13.3, 13.6, 13.7, 13.9, 14.1, 15.4, 15.4, 17.3, 17.3, 18.1, 18.2, 18.4, 18.9, 19.0, 19.9, 20.6, 21.3, 21.4, 21.9, 23.0, 27.0, 31.6, 33.1, 38.5.

The estimated variance-covariance matrix for TL-EHL-GW distribution on bank customers data set is given by

$$\begin{pmatrix} 0.0141 & 7.6168\times10^{-03} & 1.5592\times10^{-04} & -0.0404 \\ 0.0076 & 4.2077\times10^{-03} & 8.4272\times10^{-05} & -0.0217 \\ 0.0002 & 8.4272\times10^{-05} & 1.7348\times10^{-06} & -0.0005 \\ -0.0404 & -2.1749\times10^{-05} & -4.5053\times10^{-04} & 0.1172 \end{pmatrix}.$$

From Table 4.3, the values of the goodness-of-fit statistics related to AIC, AICC, and BIC are smallest for the TL-EHL-GW distribution compared to other non-nested models. Again, the values of the goodness-of-fit statistics related to K-S, $W^*$, and $A^*$ are smallest for the proposed distribution. Therefore we conclude that the TL-EHL-GW distribution performs better for the waiting times of customers data than non-nested distributions listed in the table. Fig. 4.7 indicates that our model outperforms the competing non-nested models on waiting times of customers data.

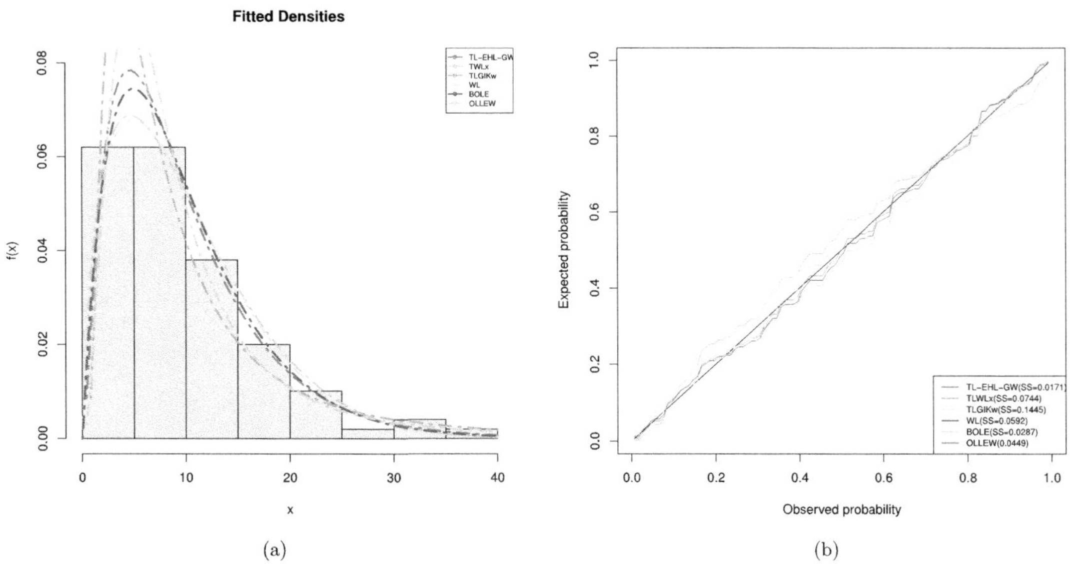


**FIGURE 4.7** Fitted densities and probability plots for waiting times of customers data.

In Fig. 4.8, the survival function is close to the K-M survival curve, demonstrating that our model is effective at describing the waiting times of customers data. The cdf line for the TL-EHL-GW distribution aligns closely with the empirical cdf, and the TTT-transform plot suggests an increasing hrf(c), which is in agreement with the estimated hrf of the TL-EHL-GW distribution, as depicted in (d).

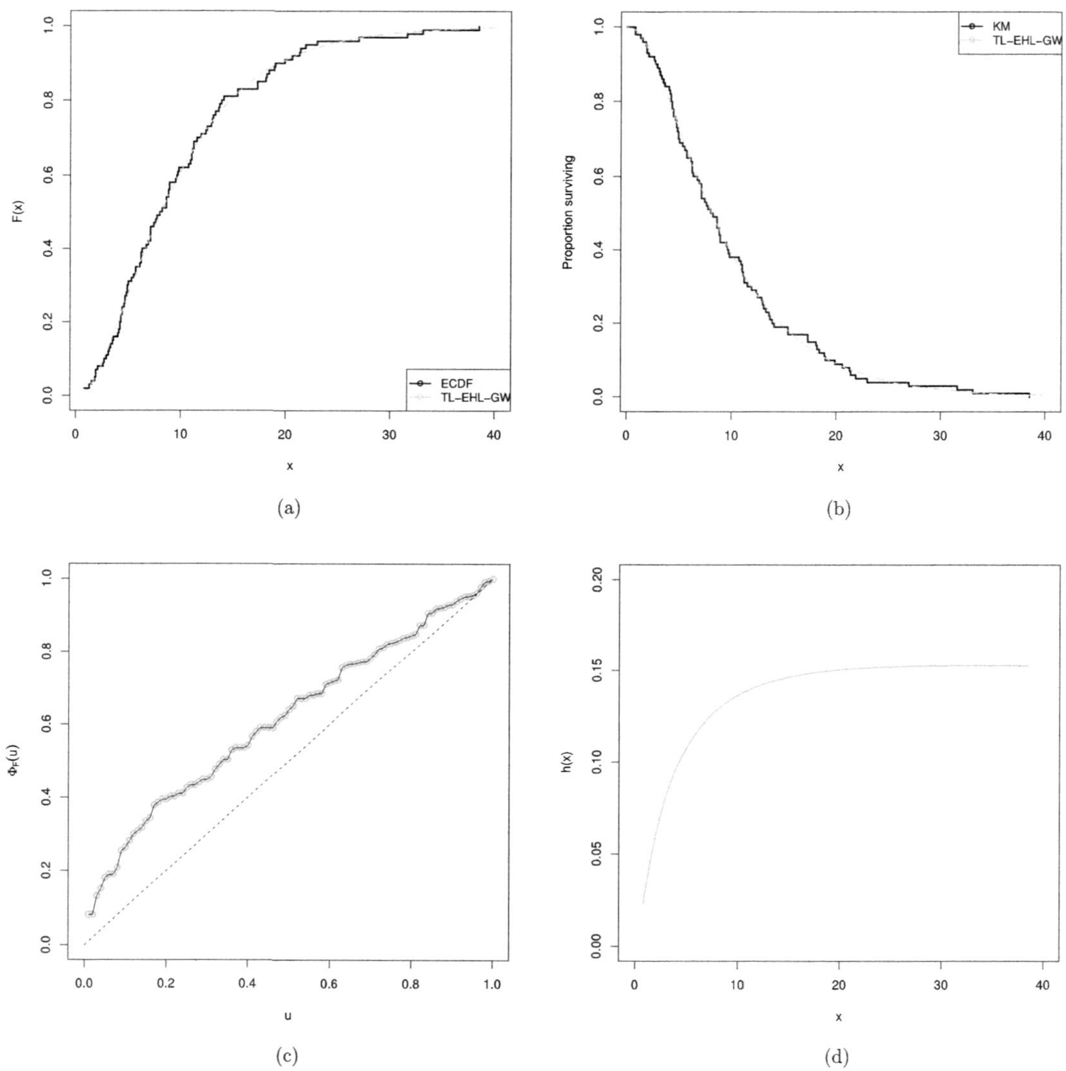


**FIGURE 4.8** Ecdf, K-M survival, TTT-transform and estimated hrf plots for customers data.

## 4.7.2 Remission times of cancer patients data

The second data set corresponds to remission times (months) of 128 bladder cancer patients and was analyzed by [4]. The observations are the following: 0.08, 4.98, 25.74, 3.7, 10.06, 2.69, 7.62, 1.26, 7.87, 4.4, 2.02, 21.73, 2.09, 6.97, 0.5, 5.17, 14.77, 4.18, 10.75, 2.83, 11.64, 5.85, 3.31, 2.07, 3.48, 9.02, 2.46, 7.28, 32.15, 5.34, 16.62, 4.33, 17.36, 8.26, 4.51, 3.36, 4.87, 13.29, 3.64, 9.74, 2.64, 7.59, 43.01, 5.49, 1.4, 11.98, 6.54, 6.93, 6.94, 0.4, 5.09, 14.76, 3.88, 10.66, 1.19, 7.66, 3.02, 19.13, 8.53, 8.65, 8.66, 2.26, 7.26, 26.31, 5.32, 15.96, 2.75, 11.25, 4.34, 1.76, 12.03, 12.63, 13.11, 3.57, 9.47, 0.81, 7.39, 36.66, 4.26, 17.14, 5.71, 3.25, 20.28, 22.69,

**Table 4.4** Parameter estimates and goodness-of-fit statistics of models for cancer patients data.

| | Estimates | | | | Statistics | | | | | | | |
|---|---|---|---|---|---|---|---|---|---|---|---|---|
| **Model** | $\alpha$ | $\beta$ | $b$ | $\lambda$ | $-2\log L$ | $AIC$ | $AICC$ | $BIC$ | $W^*$ | $A^*$ | $K-S$ | p-value |
| TL-EHL-GW | 0.3526 | 0.4202 | 8.8054 | 0.5723 | 821.4 | 829.4 | 829.7 | 840.8 | 0.0435 | 0.2873 | 0.0492 | 0.9159 |
| | (0.2075) | (0.1984) | (4.9118) | (0.1409) | | | | | | | | |
| | $a$ | $b$ | $\alpha$ | $\theta$ | | | | | | | | |
| TLWLx | 0.3221 | 0.3487 | 1.9375 | 0.6780 | 823.5 | 831.5 | 831.8 | 842.9 | 0.0691 | 0.4268 | 0.0633 | 0.6833 |
| | (0.0998) | (0.2494) | (0.5162) | (0.1718) | | | | | | | | |
| | $\alpha$ | $\lambda$ | $\beta$ | $\theta$ | | | | | | | | |
| TLGIKw | 1.8070 | 0.5660 | 21.2119 | 0.4521 | 836.0 | 844.0 | 844.3 | 855.4 | 0.1805 | 1.2019 | 0.0790 | 0.4010 |
| | (0.7101) | (0.1707) | (17.9377) | (0.3089) | | | | | | | | |
| | $a$ | $b$ | $\alpha$ | $\beta$ | | | | | | | | |
| WL | 66.9267 | 6.0094 | 0.0573 | 0.0079 | 823.4 | 831.4 | 831.8 | 842.8 | 0.0691 | 0.4443 | 0.0568 | 0.8030 |
| | (0.0089) | (2.4934) | (0.0122) | (0.0246) | | | | | | | | |
| | $a$ | $b$ | $\lambda$ | $\theta$ | | | | | | | | |
| BOL-E | 1.1720 | 65.4980 | 28.4930 | 0.0001 | 826.8 | 834.8 | 835.1 | 846.2 | 0.1202 | 0.7211 | 0.0748 | 0.4716 |
| | $(3.2846\times10^{-10})$ | $(5.9050\times10^{-12})$ | $(1.4052\times10^{-11})$ | $(5.6323\times10^{-06})$ | | | | | | | | |
| | $\alpha$ | $\beta$ | $\gamma$ | $\theta$ | | | | | | | | |
| OLLEW | 62.3937 | 4.7172 | 0.0657 | 2.5396 | 833.5 | 841.5 | 841.8 | 852.9 | 0.1522 | 0.9281 | 0.0579 | 0.7825 |
| | (8.6099) | (1.2416) | (0.0211) | (0.2870) | | | | | | | | |

23.63, 5.06, 14.24, 2.62, 10.34, 1.05, 5.41, 79.05, 7.93, 4.5, 2.02, 0.2, 7.09, 25.82, 3.82, 14.83, 2.69, 7.63, 1.35, 11.79, 6.25, 3.36, 2.23, 9.22, 0.51, 5.32, 34.26, 4.23, 17.12, 2.87, 18.1, 8.37, 6.76, 3.52, 13.8, 2.54, 7.32, 0.9, 5.41, 46.12, 5.62, 1.46, 12.02, 12.07.

The estimated variance-covariance matrix for TL-EHL-GW distribution on cancer patients data set is given by

$$\begin{pmatrix} 0.0430 & 0.0282 & 0.0703 & -0.0288 \\ 0.0282 & 0.0394 & -0.6524 & -0.0174 \\ 0.0703 & -0.6524 & 24.1259 & -0.0903 \\ -0.0288 & -0.0174 & -0.0903 & 0.0199 \end{pmatrix}.$$

It is quite clear from the results in Table 4.4 that the TL-EHL-GW distribution has the smallest goodness-of-fit statistics compared to other non-nested distributions. Therefore we can conclude that the TL-EHL-GW model performs better than the non-nested distributions listed in the table for the remission times of cancer patients data. Moreover, the TL-EHL-GW model has the highest K-S p-value and smallest sum of squares value from the probability plots in Fig. 4.9.

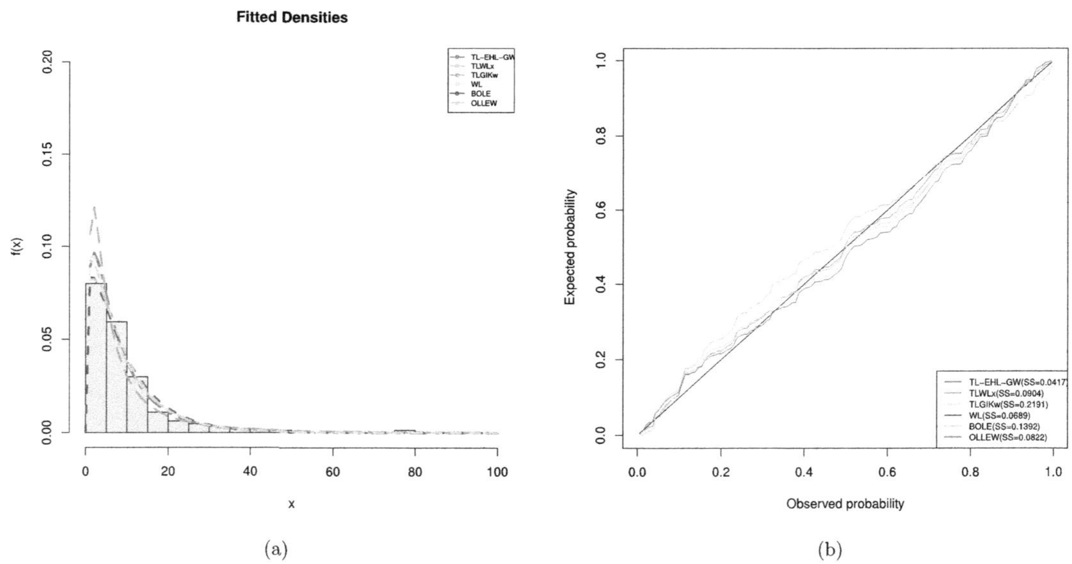


**FIGURE 4.9** Fitted densities and probability plots for cancer patients data.

Fig. 4.10 shows the estimated cdf, K-M survival curve and the scaled TTT plots. From the plots we see that the estimated cdf line for the TL-EHL-GW distribution is closer to the empirical cdf while the survival function is also close to the K-M survival curve, hence our model is best in explaining the remission times of cancer patients data. The TTT-transform plot and estimated hrf of the TL-EHL-GW distribution on the cancer data are compatible, showing that our model can fit data with upside-down bathtub hrf.

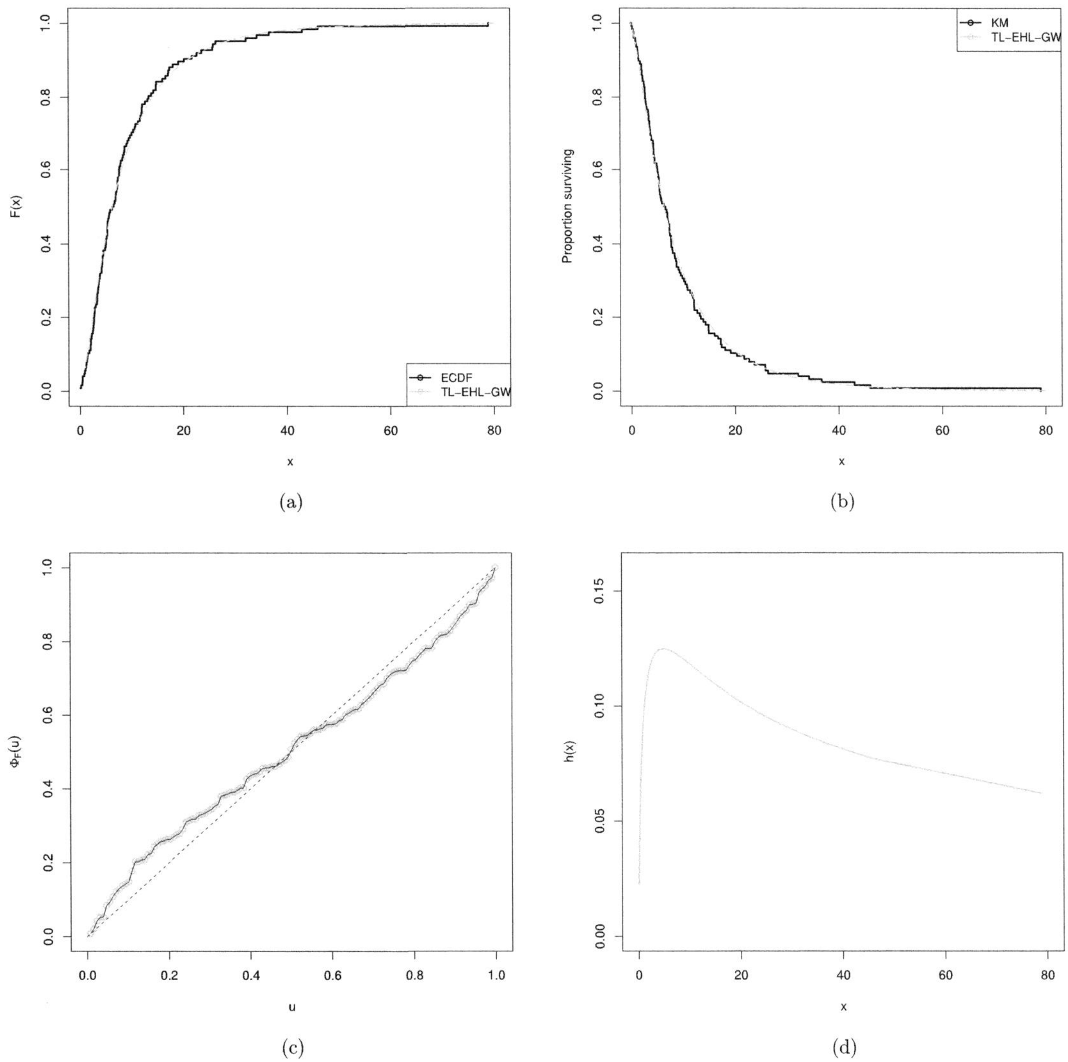


**FIGURE 4.10** Ecdf, K-M survival, TTT-transform and estimated hrf plots for cancer patients data.

# 4.8 Concluding remarks

We proposed and studied a new family of distributions named, the Topp–Leone-exponentiated half logistic-generalized-G distribution. Some important statistical properties of the new family of distributions were derived. Some special cases of the new family of distributions were also studied in detail. We conducted a simulation study to examine the performance of the maximum likelihood estimates. Lastly, two real data examples were given to show the importance and flexibility of the proposed family of distributions, where we established that indeed the model is useful and performs better than some selected competing models. We plan to take into account additional statistical features, including other parameter estimation techniques, such as weighted least squares, maximum product of spacing, and Bayesian estimators, which are not included in this study, in our future research.

# Appendix 4.A

The elements of the score vector $U(\Delta)$ are given by

$$\frac{\partial \ell_n(\Delta)}{\partial \alpha} = \frac{n}{\alpha} - 2\beta(b-1)\sum_{i=1}^{n} \frac{\left(\frac{1-\overline{G}^{\alpha}(x_i;\Psi)}{1+\overline{G}^{\alpha}(x_i;\Psi)}\right)^{\beta-1}\left[\frac{1-\overline{G}^{\alpha}(x_i;\Psi)}{1+\overline{G}^{\alpha}(x_i;\Psi)}\right]\ln\left[\frac{1-\overline{G}(x_i;\Psi)}{1+\overline{G}(x_i;\Psi)}\right]}{\left[1-\left(1-\left(\frac{1-\overline{G}^{\alpha}(x_i;\Psi)}{1+\overline{G}^{\alpha}(x_i;\Psi)}\right)^{\beta}\right)^{2}\right]}$$

$$-\beta\sum_{i=1}^{n} \frac{\left(\frac{1-\overline{G}^{\alpha}(x_i;\Psi)}{1+\overline{G}^{\alpha}(x_i;\Psi)}\right)^{\beta-1}\left[\frac{1-\overline{G}^{\alpha}(x_i;\Psi)}{1+\overline{G}^{\alpha}(x_i;\Psi)}\right]\ln\left[\frac{1-\overline{G}(x_i;\Psi)}{1+\overline{G}(x_i;\Psi)}\right]}{\left(1-\left(\frac{1-\overline{G}^{\alpha}(x_i;\Psi)}{1+\overline{G}^{\alpha}(x_i;\Psi)}\right)^{\beta}\right)} + \sum_{i=1}^{n}\ln[\overline{G}^{\alpha}(x_i;\Psi)]$$

$$+(\beta-1)\sum_{i=1}^{n}\frac{\overline{G}^{\alpha}(x_i;\Psi)\ln[\overline{G}(x_i;\Psi)]}{[1-\overline{G}^{\alpha}(x_i;\Psi)]} - (\beta+1)\sum_{i=1}^{n}\frac{\overline{G}^{\alpha}(x_i;\Psi)\ln[\overline{G}(x_i;\Psi)]}{[1+\overline{G}^{\alpha}(x_i;\Psi)]},$$

$$\frac{\partial \ell_n(\Delta)}{\partial \beta} = \frac{n}{\beta} - 2(b-1)\sum_{i=1}^{n} \frac{\left(\frac{1-\overline{G}^{\alpha}(x_i;\Psi)}{1+\overline{G}^{\alpha}(x_i;\Psi)}\right)^{\beta}\ln\left[\frac{1-\overline{G}^{\alpha}(x_i;\Psi)}{1+\overline{G}^{\alpha}(x_i;\Psi)}\right]}{\left[1-\left(1-\left(\frac{1-\overline{G}^{\alpha}(x_i;\Psi)}{1+\overline{G}^{\alpha}(x_i;\Psi)}\right)^{\beta}\right)^{2}\right]}$$

$$-\sum_{i=1}^{n} \frac{\left(\frac{1-\overline{G}^{\alpha}(x_i;\Psi)}{1+\overline{G}^{\alpha}(x_i;\Psi)}\right)^{\beta}\ln\left[\frac{1-\overline{G}^{\alpha}(x_i;\Psi)}{1+\overline{G}^{\alpha}(x_i;\Psi)}\right]}{\left[1-\left(\frac{1-\overline{G}^{\alpha}(x_i;\Psi)}{1+\overline{G}^{\alpha}(x_i;\Psi)}\right)^{\beta}\right]} + \sum_{i=1}^{n}\ln[1-\overline{G}^{\alpha}(x_i;\Psi)]$$

$$-\sum_{i=1}^{n}\ln[1+\overline{G}^{\alpha}(x_i;\Psi)],$$

$$\frac{\partial \ell_n(\Delta)}{\partial b} = \frac{n}{b} + \sum_{i=1}^{n}\ln\left[1-\left(1-\left(\frac{1-\overline{G}^{\alpha}(x_i;\Psi)}{1+\overline{G}^{\alpha}(x_i;\Psi)}\right)^{\beta}\right)^{2}\right],$$

and

$$\frac{\partial \ell_n(\Delta)}{\partial \Psi_k} = 2\beta\alpha\sum_{i=1}^{n} \frac{\left(\frac{1-\overline{G}^{\alpha}(x_i;\Psi)}{1+\overline{G}^{\alpha}(x_i;\Psi)}\right)^{\beta}\left(\frac{1-\overline{G}^{\alpha}(x_i;\Psi)}{1+\overline{G}^{\alpha}(x_i;\Psi)}\right)^{\beta-1}\frac{1-[\overline{G}(x_i;\Psi)]^{\alpha-1}}{1+[\overline{G}(x_i;\Psi)]^{\alpha-1}}\frac{\partial \overline{G}(x_i;\Psi)}{\partial \Psi_k}}{\left[1-\left(1-\left(\frac{1-\overline{G}^{\alpha}(x_i;\Psi)}{1+\overline{G}^{\alpha}(x_i;\Psi)}\right)^{\beta}\right)^{2}\right]}$$

$$
\begin{aligned}
&+\beta\alpha\sum_{i=1}^{n}\frac{\left(\frac{1-\overline{G}^{\alpha}(x_i;\Psi)}{1+\overline{G}^{\alpha}(x_i;\Psi)}\right)^{\beta-1}\frac{1-[\overline{G}(x_i;\Psi)]^{\alpha-1}}{1+[\overline{G}(x_i;\Psi)]^{\alpha-1}}\frac{\partial\overline{G}(x_i;\Psi)}{\partial\Psi_k}}{\left[1-\left(\frac{1-\overline{G}^{\alpha}(x;\Psi)}{1+\overline{G}^{\alpha}(x_i;\Psi)}\right)^{\beta}\right]}-\alpha(\beta-1)\sum_{i=1}^{n}\frac{[\overline{G}(x_i;\Psi)]^{\alpha-1}\frac{\partial\overline{G}(x_i;\Psi)}{\partial\Psi_k}}{[1-\overline{G}^{\alpha}(x_i;\Psi)]}\\
&-\alpha(\beta+1)\sum_{i=1}^{n}\frac{[\overline{G}(x_i;\Psi)]^{\alpha-1}\frac{\partial\overline{G}(x_i;\Psi)}{\partial\Psi_k}}{[1+\overline{G}^{\alpha}(x_i;\Psi)]}+(\alpha-1)\sum_{i=1}^{n}\frac{\frac{\partial\overline{G}(x_i;\Psi)}{\partial\Psi_k}}{[\overline{G}(x_i;\Psi)]}+\frac{\frac{\partial g(x_i;\Psi)}{\partial\Psi_k}}{g(x_i;\Psi)}.
\end{aligned}
$$

# References


[1] A.Z. Afify, M. Alizadeh, M. Zayed, T.G. Ramires, F. Louzada, The odd log-logistic exponentiated Weibull distribution: regression modeling, properties, and applications, Iranian Journal of Science and Technology. Transaction A, Science 42 (4) (2017) 2273–2288.

[2] A. Al-Shomrani, O. Arif, A. Shawky, S. Hanif, M.Q. Shahbaz, Topp-Leone family of distributions: some properties and application, Pakistan Journal of Statistics and Operation Research 12 (3) (2016) 443–451.

[3] G.R. Aryal, H.M. Yousof, The exponentiated generalized-G Poisson family of distributions, Stochastics and Quality Control 32 (1) (2017) 7–23.

[4] S. Chamunorwa, B. Makubate, B. Oluyede, F. Chipepa, The exponentiated half logistic-log-logistic Weibull distribution: model, properties and applications, Journal of Statistical Modelling: Theory and Applications 2 (1) (2021) 101–120.

[5] G.M. Cordeiro, M. Alizadeh, E.M. Ortega, The exponentiated half logistic family of distributions: properties and applications, Journal of Probability and Statistics 2014 (2014).

[6] G.M. Cordeiro, E.M. Ortega, D.C. da Cunha, The exponentiated generalized class of distributions, Journal of Data Science 11 (1) (2013) 1–27.

[7] J. Chambers, W. Cleveland, B. Kleiner, J. Tukey, Graphical Methods for Data Analysis, Chapman and Hall, London, 1983.

[8] F. Chipepa, B. Oluyede, B. Makubate, A. Fagbamigbe, The beta odd Lindley-G family of distributions with applications, Journal of Probability and Statistical Science 17 (1) (2019) 51–83.

[9] J.A. Greenwood, J.M. Landwehr, N.C. Matalas, J.R. Wallis, Probability weighted moments: definition and relation to parameters of several distributions expressible in inverse form, Water Resources Research 15 (5) (1979) 1049–1054.

[10] A.S. Hassan, M. Elgarhy, Z. Ahmad, Type II generalized Topp-Leone family of distributions: properties and applications, Journal of Data Science 17 (4) (2019) 638–658.

[11] F. Jamal, H.M. Reyad, M.A. Nasir, C. Chesneau, M.A.A. Shah, S.O. Ahmed, Topp-Leone Weibull-Lomax Distribution: Properties, Regression Model and Applications, hal-02270561f, 2019.

[12] T. Moakofi, B. Oluyede, F. Chipepa, The Topp-Leone half logistic generalized-G family of distributions: model, properties and applications, Eurasian Bulletin of Mathematics 4 (2) (2021) 107–133.

[13] I.E. Okorie, A.C. Akpanta, J. Ohakwe, Marshall-Olkin generalized Erlang-truncated exponential distribution: properties and applications, Cogent Mathematics 4 (1) (2017) 1285093.

[14] B. Oluyede, T. Moakofi, F. Chipepa, B. Makubate, A new power generalized Weibull-G family of distributions: properties and applications, Journal of Statistical Modelling: Theory and Applications 1 (2) (2020) 167–191.

[15] N. Rasheed, A new generalized-G class of distributions and its applications with Dagum distribution, Research Journal of Mathematical and Statistical Science 8 (3) (2020) 1–13.

[16] A. Rényi, On measures of entropy and information, in: Proceedings of the Fourth Berkeley Symposium on Mathematical Statistics and Probability, 1, 1960, pp. 547–561.

[17] H.M. Reyad, M. Alizadeh, F. Jamal, S. Othman, G.G. Hamedani, The exponentiated generalized Topp Leone-G family of distributions: properties and applications, Pakistan Journal of Statistics and Operation Research XV (1) (2019) 1–24.

[18] H. Reyad, F. Jamal, S. Othman, N. Yahia, The Topp Leone generalized inverted Kumaraswamy distribution: properties and applications, Asian Research Journal of Mathematics 13 (3) (2019) 1–15.

[19] M. Shaked, J.G. Shanthikumar, Stochastic Orders, Springer, New York, 2007.
[20] M. Tahir, G. Cordeiro, M. Mansoor, M. Zubair, The Weibull-Lomax distribution: properties and applications, Hacettepe University Bulletin of Natural Sciences and Engineering Series B: Mathematics and Statistics (2014), https://doi.org/10.15672/HJMS.2014147465.
[21] H.M. Yousof, A.Z. Afify, S. Nadarajah, G. Hamedani, G.R. Aryal, The Marshall-Olkin generalized-G family of distributions with applications, Statistica 78 (3) (2018) 273–295.

5

# A mathematical model for the dynamics of visceral leishmaniasis disease with time delay

Fahad Al Basir[a] and Amar Nath Chatterjee[b]
[a]Department of Mathematics, Asansol Girls' College, Asansol, West Bengal, India [b]Department of Mathematics, K.L.S. College, Nawada, Magadh University, Bodh Gaya, Bihar, India

## 5.1 Introduction

Leishmania parasites are the main cause of leishmaniasis, which is primarily spread by sand flies. The disease is spread through the bite of female *phlebotomine* sandflies that are infected. Leishmaniasis primarily affects the poor, the malnourished, those with weakened immune systems, and those who reside in contaminated environments [1]. Deforestation, urbanization, and climate change all contribute to the disease's spread. Each year, there are one million new cases of leishmaniasis according to the World Health Organization [2].

There are three types of leishmaniasis. These are VL (visceral leishmaniasis), CL (cutaneous leishmaniasis), and ML (mucocutaneous leishmaniasis). If not treated in time, VL, also referred to as kala-azar, is fatal. Irregular fever, weight loss, a fatty liver, and anemia are the most common symptoms of VL [3]. India, Brazil, and South Africa are the most affected regions by VL. Worldwide, approximately 50 000–90 000 new cases are found annually. It is one of the top parasitic diseases with a high mortality effect [4]. CL is the most common form of leishmaniasis. The most common symptoms of CL are skin lesions, primarily ulcers, on exposed body parts. These lesions leave permanent scars and can severely impair one's ability to function or cause social stigma. America, the Mediterranean Basin, the Middle East, and Central Asia are the most affected regions by CL. In this region, approximately 95% of these cases occur annually. One of the main causes of ML is the partial or complete destruction of the nose and mouth mucous membranes. Approximately 90% of ML cases occur in Bolivia, Brazil, Ethiopia, and Peru [5].

Infections can be categorized as symptomatic, CL, or VL types based on the parasite, host, and other factors. After consuming infected blood cells, sandflies become infected. The promastigotes that are formed from the vector amastigotes develop in the gastrointestinal tract before floating to the proboscis. It has been discovered that about 70 animal species, including humans, serve as leishmania parasites' natural reservoir hosts [6,7].

The movement of non-immune people into areas with outbreaks and migration both play a significant part in the spread of the CL and VL epidemics. Breaking the transmission

Mathematical Modeling in Bioscience. https://doi.org/10.1016/B978-0-44-315445-4.00011-1

Copyright © 2025 Elsevier Inc. All rights are reserved, including those for text and data mining, AI training, and similar technologies.

cycle plays a pivotal role in controlling epidemics [8]. Proper diagnosis and parasitological and serological tests can help detect the VL. Clinical manifestations confirmed by parasitological tests confirm the detection of CL and ML. The patients suffering from VL require quick and comprehensive treatment [9]. Leishmaniasis control requires comprehensive intervention strategies, which include a human or animal reservoir host, the parasite, and sandflies. Early diagnosis with prompt treatment, vector control, effective disease surveillance, control of animal reservoir hosts, and social mobilization are the key strategies for prevention of the disease [10].

To control the leishmaniasis, WHO organizes various initiatives and national level programs. Regular surveillance and monitoring the trend of the disease help to get rid of the epidemic. The awareness about leishmaniasis world wide and developing health services help to control the disease. The standard control procedure is established by the advancement of leishmaniasis research and the application of its findings. Various governmental organizations, including WHO, support research by funding it.

In 2002, Courtenay et al. published their findings from their investigation into the ineffectiveness of dog scrapping as a ZVL prevention measure [1]. To determine the threshold value, a mathematical model was used by Maroli et al. to investigate a range of preventative measures to control sandflies in 2004 [3]. 2004 saw a rise in sandflies [3]. Alexender et al. [4] developed the sandfly management strategy. Sharma et al. looked into how humans and vectors interact and developed a number of control measures [10].

Neghina et al. [11] contributed a report on the various leishmaniasis cases that occurred in 2010. A VL vaccination model was the subject of research by Elmojtaba et al. (2012) [12]. In 2013, Ribas et al. looked into the ZVL mathematical model for ideal control. Chatterjee et al. studied IRS and DDT this year to control sandflies in VL disease [13]. The best control model and the backward bifurcation of VL transmission were analyzed by Zhao et al. in 2016 [14]. To comprehend the dynamics of the illness, El Mojtaba conducted research on the VL-malaria coinfection theory [15]. In 2016, Rock KS et al. worked to integrate biology for VL control to better understand VL dynamics [16,17]. Hussaini et al. [18] developed a mathematical model of VL infection dynamics to describe the dynamics of VL infection spread in human and other animal reservoir populations. Zou et al. took into account the populations of dogs, people, and sandflies to create a mathematical model and investigate the dynamics of VL transients. In 2018 [19], Boukhalfa et al. presented a mathematical model of VL [20]. To analyze the 2018 CL outbreak in Madrid and Tolima, Mubayi et al. [21] used the reproduction number ($R_0$). Agyingi et al. [22] presented a model of AA leishmaniasis transmission, which takes into account the dynamics of the disease. 2019's diagnosis and treatment of VL were covered by Griensren et al. [23]. In 2021, Ghosh et al. looked at the prevalence of post-kala-azar dermal leishmaniasis (PKDL) in various Indian regions. They came to the conclusion that PKDL was the root of many social problems [24].

By taking the delay dynamics of the disease transmission into account, the study of disease dynamics becomes more realistic and close to nature. Recent research on the COVID-19 disease has shown that effective delaying strategies, such as lockdowns, travel

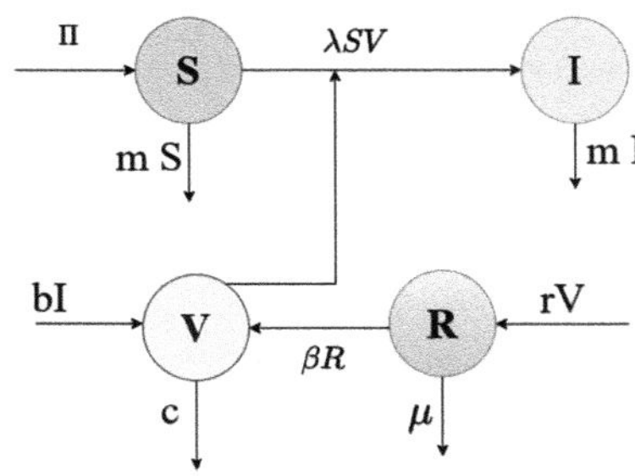


**FIGURE 5.1** Schematic diagram of the model (5.1).

restrictions, facemasks, hand sanitizer, and social seclusion, are crucial in stopping the disease's chain of transmission.

Even though several mathematical models have been developed, the delay model is one of the most efficient methods for containing the epidemic. The disease's progression and its restraint can both be achieved with the aid of various delay tactics. The study of the delay differential equation model of leishmaniasis can help to advance and limit the disease.

The chapter is organized as follows: In Section 5.2, we propose the mathematical model and discuss its basic properties. In Section 5.3, we study the delayed model with mathematical stability theories and stability switches. In Section 5.4, we provide the numerical simulations of analytical outcomes. Finally, in Section 5.5, we conclude the main results and significance of this study.

## 5.2 The mathematical model

Here, we consider a SI-type model with respect to the history of infection to describe human population. The model contains four population, namely the susceptible human $S(t)$, infected human $I(t)$, infective reservoir $R(t)$, and infective sandfly $V(t)$ populations. The total human population is thus $N(t) = S(t) + I(t)$. Transmission dynamics of this disease is shown in Fig. 5.1.

We assume that all the humans initially grow at a constant birth rate $\Pi$ and death rate $m$. They remain susceptible to the infection. After being bitten by an infectious sandfly or susceptible humans are considered to have become infected, a force of infection $\frac{e^{-m\tau}\lambda S(t-\tau)V(t-\tau)}{S(t-\tau)+I(t-\tau)}$, where $\lambda$ is the mean rate of bites per sandfly, is realized

From the above assumption the following system is obtained:

$$\frac{dS}{dt} = \Pi - \frac{\lambda V S}{S+I} - mS, \tag{5.1}$$

$$\frac{dI}{dt} = \frac{e^{-m\tau}(\lambda S(t-\tau)V(t-\tau)}{S(t-\tau)+I(t-\tau)} - mI,$$

$$\frac{dR}{dt} = rV - \mu R,$$
$$\frac{dV}{dt} = bI + \beta R - cV;$$

the initial conditions are

$$S(\theta) > 0,\ I(\theta) > 0,\ R(\theta) > 0,\ V(\theta) > 0, \text{where } \theta \in [-\tau, 0]. \tag{5.2}$$

Here $e^{-m\tau}$ is the survival probability of human through the latent time $\tau$.

Let $C$ denotes the Banach space of continuous functions $\psi : [-\tau, 0] \to \mathbb{R}^4_+$ with the norm

$$\|\psi\| = \sup_{-\tau<\theta<0} \{|\psi_1(\theta)|, |\psi_2(\theta)|, |\psi_3(\theta)|, |\psi_4(\theta)|\},$$

where $\psi = (\psi_1, \psi_2, \psi_3, \psi_4)$. The initial conditions of (5.1) are given as

$$S(\theta) = \psi_1(\theta),\ I(\theta) = \psi_2(\theta), R(\theta) = \psi_3(\theta),\ V(\theta) = \psi_4(\theta), \theta \in [-\tau, 0], \tag{5.3}$$

where the initial function $\psi = (\psi_1, \psi_2, \psi_3, \psi_4)$ belongs to the Banach space $C = C([-\tau, 0], \mathbb{R}^4_+)$ of continuous functions mapping the initial $[-\tau, 0]$ into $\mathbb{R}^4$. For biological reasons, the initial functions are assumed as

$$\psi_i(\theta) \geq 0, \theta \in [-\tau, 0], \psi_i(0) > 0, i = 1, 2, 3, 4.$$

System (5.1) can be written as follows:

$$\frac{dX}{dt} = f(X), \tag{5.4}$$

with

$$X(\theta) = (S(\theta),\ I(\theta),\ R(\theta),\ V(\theta))^T \in C, \tag{5.5}$$
$$\psi_i(0) > 0, i = 1, 2, 3, 4, \text{ and } f = (f_1, f_2, f_3, f_4)^T, i = 1, 2, 3, 4. \tag{5.6}$$

Here, $f_i, i = 1, 2, 3, 4$ are the right sides of system (5.1). For example, $f_1 = \Pi - \frac{(\lambda_1 V + \lambda_2 R)S}{N} - mS$, etc.

By the fundamental theory of functional differential equations [27], we know that there is a unique solution $(S(t), I(t), R(t),\ V(t))$ to system (5.1) with initial conditions given in (5.2).

### 5.2.1 Some basic properties

In this section, some basic properties, such as positive invariance and boundedness of the solutions of the delayed system (5.1) are illustrated.

#### *5.2.1.1 Positive invariance*

Biologically, positivity implies the survival of the populations. To proof this, the method of Bodnar [28] and Yang et al. [29] has been followed, and provide the following theorem:

**Theorem 5.1.** *All the solution of (5.1) with initial conditions (5.2) are positive.*

*Proof.* Using the results from [28,29] on existence and negativity of solutions, we show that any solution of system (5.1) exists in the region $\mathbb{R}^4_+$, and all solutions remain non-negative for all $t > 0$.

It is easy to check in system (5.1) that whenever choosing $X(\theta) \in R_+$ such that $S = 0$, $I = 0$, $R = 0$, $V = 0$, then

$$f_i(X)|_{x_i=0, X\in R^4_+} \geq 0,$$

with $x_1(t) = S(t)$, $x_2(t) = I(t)$, $x_3(t) = R(t)$, $x_4(t) = V(t)$.

Using Lemma 1 in [29], Theorem 1 in [28], we can conclude that any solution of (5.1) with $X(\theta) \in C$, say $X(t) = X(t, X(\theta))$, is such that $X(\theta) \in R^4_+$ for all $t \geq 0$. Hence the solution of the system of system (5.1) exist in the region $R^4_+$. Thus all solutions remain non-negative for all $t > 0$, and hence the positive octant $\mathbb{R}^4_+$ is an invariant region for the system (5.1). □

#### *5.2.1.2 Boundedness*

Boundedness of a system indicates that the system is well behaved. We have the following result based on the boundedness of the model populations:

**Theorem 5.2.** *All the solutions of (5.1) with initial conditions (5.2) are ultimately bounded.*

*Proof.* Let $N$ be the total human population at any time $t$, then

$$\begin{aligned} \frac{dN}{dt} &\leq \Pi - mN, \\ &\Rightarrow \limsup_{t\to\infty} N(t) \leq \frac{\Pi}{m}. \end{aligned} \tag{5.7}$$

From the second equation of (5.1), we write

$$\begin{aligned} \frac{dR}{dt} &\leq \frac{r\Pi}{m} - \mu R, \\ &\Rightarrow \limsup_{t\to\infty} R \leq \frac{r\Pi}{\mu m}. \end{aligned} \tag{5.8}$$

Thus the region of attraction is given by the set

$$\mathcal{D} = \{(S, I, R, V) \in \mathbb{R}^4_+ : 0 \leq S + I \leq \frac{\Pi}{m}, 0 \leq R \leq \frac{r\Pi}{\mu m},\ 0 \leq V \leq \frac{r\Pi}{cm}\},$$

which is positively invariant, and it attracts all solutions initiating inside the interior of positive octant. □

### 5.2.2 Basic reproduction number

The basic reproduction number is a key factor of an infection. It can be described as the average number of new cases generated by an infective individual in susceptible population over the course of its infectious period. Basically, it let us know the transmission dynamics of an epidemic.

The basic reproduction number $\mathcal{R}_0$ for the system (5.1) is given by

$$\mathcal{R}_0 = \frac{\lambda e^{-m\tau} b\mu}{m(\mu c - \beta r)}. \tag{5.9}$$

This can be obtained using next generation matrix method [25,26].

## 5.3 Equilibrium points

There are only three possible equilibria for system (5.1):

**i)** the disease-free equilibrium $E_0 = (\frac{\Pi}{m}, 0, 0, 0)$;
**ii)** the endemic equilibrium $E^* = (S^*,\ I^*,\ R^*,\ V^*)$, where,

$$I^* = \frac{(c\mu - \beta r)V^*}{b}, \quad R^* = \frac{rV^*}{\mu},$$
$$S^* = \frac{b\Pi - m(c\mu - \beta r)V^*}{m}, \tag{5.10}$$

and $V^*$ satisfies

$$(S+I)\Pi - \lambda SV - mS(S+I) = 0. \tag{5.11}$$

### 5.3.1 Characteristic equation

The characteristic equation of system (5.1), at any equilibrium point $E(S, I, R, V)$, is given by

$$\triangle(\xi) = \mid \xi I - J - e^{-\xi\tau} G \mid = 0,$$

where $J, G$ are the $4 \times 4$ matrices as given below:

$$J = [J_{ij}] = \begin{bmatrix} -\frac{\lambda I}{(S+I)^2} - m & \frac{\lambda SV}{(S+I)^2} & 0 & -\frac{\lambda S}{(S+I)} \\ 0 & -m & 0 & 0 \\ 0 & r & -\mu & 0 \\ 0 & b & 0 & -c \end{bmatrix},$$

and

$$G = [G_{ij}] = \begin{bmatrix} 0 & 0 & 0 & 0 \\ \frac{e^{-m\tau}\lambda I}{(S+I)^2} & \frac{(e^{-m\tau}\lambda SV}{(S+I)^2} & 0 & \frac{\lambda S}{(S+I)} \\ 0 & 0 & 0 & 0 \\ 0 & 0 & 0 & 0 \end{bmatrix}.$$

The above equation can be written as

$$\xi^4 + a_1\xi^3 + a_2\xi^2 + a_3\xi + a_4 + e^{\xi\tau}[b_1\xi^3 + b_2\xi^2 + b_3\xi + b_4] = 0, \tag{5.12}$$

where

$$\begin{aligned} a_1 &= -(J_{11} + J_{22} + J_{33} + J_{44}), \\ a_2 &= J_{11}J_{33} + J_{11}J_{44} + J_{33}J_{44}, \\ a_3 &= -J_{11}J_{33}J_{44} - J_{22}J_{33}J_{44} - J_{11}J_{22}J_{33} - J_{11}J_{22}J_{44}, \\ a_4 &= J_{11}J_{22}J_{33}J_{44}, \end{aligned}$$

and

$$\begin{aligned} b_1 &= -G_{22}, \quad b_2 = -J_{12}G_{21} + J_{11}G_{22} - G_{23}J_{32} + G_{22}J_{33} - G_{24}J_{42} + G_{22}J_{44}, \\ b_3 &= J_{11}G_{23}J_{32} + J_{12}G_{21}J_{33} - J_{11}G_{22}J_{33} - J_{14}G_{21}J_{42} + J_{11}G_{24}J_{42} + G_{24}J_{33}J_{42} \\ &\quad + J_{12}G_{21}J_{44} - J_{11}G_{22}J_{44} + G_{23}J_{32}J_{44} - G_{22}J_{33}J_{44}, \\ b_4 &= J_{14}G_{21}J_{33}J_{42} - J_{11}G_{24}J_{33}J_{42} - J_{11}G_{23}J_{32}J_{44} - J_{12}G_{21}J_{33}J_{44} + J_{11}G_{22}J_{33}J_{44}. \end{aligned} \tag{5.13}$$

## 5.4 Stability of the system for $\tau = 0$

We have the following result on the stability of disease-free equilibrium:

For $\tau = 0$, at the disease-free equilibrium $E_0$, the Jacobian matrix has one eigenvalue $-m$; rest eigenvalues satisfy

$$\xi^3 + l_1\xi^2 + l_2\xi + l_3 = 0, \tag{5.14}$$

where $l_1 = m + \mu + c$, $l_2 = \mu c - \beta r = b\mu$, $l_3 = m(\mu c - \beta r)$.

Using Routh–Hurwitz criteria, we can say that the equilibrium $E_0$ is stable for $\mathcal{R}_0 < 1$ and unstable otherwise. Based on the foregoing, we have the following theorem:

**Theorem 5.3.** *For $\mathcal{R}_0 < 1$, the disease-free system is stable, and unstable when $\mathcal{R}_0 > 1$. A forward transcritical bifurcation occurs at $\mathcal{R}_0 = 1$.*

For the stability of endemic equilibrium of system (5.1), we get the following theorem:

**Theorem 5.4.** *The endemic equilibria is stable if the following conditions are satisfied:*

$$\sigma_2 > 0, \quad \sigma_3 > 0, \quad \sigma_4 > 0,$$

$$\sigma_1\sigma_2 - \sigma_3 > 0,$$
$$\sigma_1\sigma_2\sigma_3 - \sigma_3^2 - \sigma_4\sigma_1^2 > 0. \tag{5.15}$$

*Proof.* For $\tau = 0$, the characteristic equation for $E^*$ is given by

$$\xi^4 + \sigma_1\xi^3 + \sigma_2\xi^2 + \sigma_3\xi + \sigma_4 = 0, \tag{5.16}$$

where,

$$\sigma_1 = a_1 + b_1, \quad \sigma_2 = a_2 + b_2, \quad \sigma_3 = a_3 + b_3, \quad \sigma_4 = a_4 + b_4. \tag{5.17}$$

According to the Routh–Hurwitz criterion, the characteristic equation (5.16) has negative roots or roots with negative real parts if the conditions of Theorem 5.4 are satisfied. Thus the endemic equilibria $E^*$ is stable when the conditions in Theorem 5.4 are fulfilled. □

## 5.5 Stability of the system for $\tau > 0$

In this section, the local stability of the delayed system (5.1) has been studied.

The characteristic equation (5.16) is a transcendental equation in $\xi$. It is known that $E^*$ is locally asymptotically stable if all the roots of the corresponding characteristic equation have negative real parts. Hopf bifurcation occurs when a pair of purely imaginary roots exist.

Let $i\varsigma$ be a purely imaginary root of Eq. (5.16), from which we get

$$\varsigma^4 - a_2\varsigma^2 + a_4 = [\varsigma^2 b_2 - b_4]\cos\varsigma\tau - [\varsigma b_3]\sin\varsigma\tau,$$
$$a_1\varsigma^3 - a_3\varsigma = [\varsigma^2 b_2 - b_4]\sin\varsigma\tau + [\varsigma b_3]\cos\varsigma\tau. \tag{5.18}$$

Squaring and adding above two equations, we get the following result:

$$\varsigma^8 + \alpha_1\varsigma^6 + \alpha_2\varsigma^4 + \alpha_3\varsigma^2 + \alpha_4 = 0. \tag{5.19}$$

Substituting $\varsigma^2 = l$ in Eq. (5.19), and then simplifying, we get the following relation:

$$l^4 + \alpha_1 l^3 + \alpha_2 l^2 + \alpha_3 l + \alpha_4 = 0, \tag{5.20}$$

where,

$$\begin{aligned}
\alpha_1 &= {a_1}^2 - 2a_2,\\
\alpha_2 &= 2a_4 + a_2^2 - 2a_1a_3 - b_2^2,\\
\alpha_3 &= -2a_4a_2 + a_3^2 + 2b_2b_4 - b_3^2,\\
\alpha_4 &= a_4^2 - b_4^2.
\end{aligned} \tag{5.21}$$

Let us define $H(l)$ as

$$H(l) = l^4 + \alpha_1 l^3 + \alpha_2 l^2 + \alpha_3 l + \alpha_4.$$

We see that $H(0) = \alpha_4 < 0$, and $\lim_{l\to\infty} H(l) = \infty$. Hence, there exists a $l_0 \in (0, \infty)$ so that $H(l_0) = 0$, and this completes the proof.

Thus we have the following lemma:

**Lemma 5.1.** *If* $\alpha_4 < 0$, *then* (5.20) *has at least one positive root.*

From the definition of $H(l)$, we have

$$H'(l) = 4l^3 + 3\alpha_1 l^2 + 2\alpha_2 z + \alpha_3. \tag{5.22}$$

Assume $H'(l) = 0$, and let $z = l + \frac{3\alpha_1}{4}$, then (5.22) becomes

$$z^3 + \Lambda_1 z^2 + \Lambda_2 = 0, \tag{5.23}$$

where

$$\Lambda_1 = \frac{\alpha_2}{2} - \frac{3\alpha_1^2}{16}, \tag{5.24}$$
$$\Lambda_2 = \frac{\alpha_1^3}{32} - \frac{\alpha_1\alpha_2}{8} + \alpha_3.$$

We now define

$$\Delta = \frac{\Lambda_2^2}{4} + \frac{\Lambda_1^3}{27}, \quad \gamma = \frac{-1+\sqrt{3}i}{2},$$
$$z_1 = \sqrt[3]{-\frac{\Lambda_2}{2} + \sqrt{\Delta}} + \sqrt[3]{-\frac{\Lambda_2}{2} - \sqrt{\Delta}},$$
$$z_2 = \sqrt[3]{-\frac{\Lambda_2}{2} + \sqrt{\Delta}\gamma} + \sqrt[3]{-\frac{\Lambda_2}{2} - \sqrt{\Delta}\gamma^2},$$
$$z_3 = \sqrt[3]{-\frac{\Lambda_2}{2} + \sqrt{\Delta}\gamma^2} + \sqrt[3]{-\frac{\Lambda_2}{2} - \sqrt{\Delta}\gamma},$$
$$l_i = z_i - \frac{3\Lambda_1}{4}, i = 1, 2, 3. \tag{5.25}$$

If $\Delta \geq 0$, then (5.23) has unique real root $z_1$ (say), that is, (5.22) has unique real root $l_1$. Since $H(l)$ is the derivable function and $\lim_{l\to\infty} H(l) = \infty$, we have $l_1$ is the unique stationary point of $H(l)$ and the minimum point of $H(l)$. The sufficient condition follows directly. So we prove the necessary condition only.

Here, either $l_1 \leq 0$ or $l_1 > 0$, and $H(l_1) > 0$. If $H(0) = \alpha_4 > 0$ is the minimum of $H(l)$ for $l \geq 0$, then $H(l)$ has no positive real zeros. If $l_1 > 0$ and $H(l_1) > 0$, then $min_{l>0} H(l) = H(l_1) > 0$. Thus $H(l)$ will have no positive real zeros.

Again, if $\Delta < 0$, then (5.23) has only three roots $l_1, l_2, l_3$, that is (5.22) has only three roots $l_1, l_2, l_3$ at least one of them is real. We assume that $l_1, l_2, l_3$ are all real, i.e., $H(l)$ has at most three stationary points at $l_1, l_2, l_3$.

Thus we have the following lemma:

**Lemma 5.2.** *Suppose that* $\alpha_4 < 0$,

**(i)** *For* $\Delta \geq 0$, *Eq.* (5.20) *has positive roots if and only if* $l_1 > 0$ *and* $H(l_1) < 0$.
**(ii)** *For* $\Delta < 0$, *Eq.* (5.20) *has positive root if and only if there exists a positive* $l^* \in \{l_1, l_2, l_3\}$ *such that* $H(l^*) \leq 0$.

We assume that (5.20) has four positive roots, namely $l_i^*$, i = 1, 2, 3, 4. Then (5.19) has four purely imaginary roots, $\omega_k = \sqrt{z_i^*}$, k = 1, 2, 3, 4.

From the relation (5.18), we determine $\tau$ as follows:

$$\tau_k^n = \frac{1}{\omega_k} cos^{-1}\left[\frac{\omega_0^2 b_4[a_1\omega_0^2 - a_3] + (b_2\omega_0^2 - b_4)[\omega_0^4 - a_2\omega_0^2 + a_4]}{(b_2\omega_0^2 - b_4)^2 + b_3^2\omega_0^2}\right] + \frac{2\pi n}{\omega_0}, \tag{5.26}$$

$$k = 1, 2, 3, 4 \text{ and } n = 0, 1, 2, 3, \ldots \tag{5.27}$$

Therefore $\pm i\omega_k$ are the required pair of purely imaginary roots. We now determine the smallest value of $\tau$, for which stability switch occurs.

Let

$$\tau^* = \tau_{k_0}^{n_0} = \min_{n \geq 0, 1 \leq k \leq 4} \{\tau_k^n\}, \omega_0 = \omega_{k_0}, l_0 = l_{k_0}^*. \tag{5.28}$$

Thus we have the following lemma:

**Lemma 5.3.** *Suppose that* (5.4) *holds, i.e., the* $E^*$ *of system* (5.1) *is stable for* $\tau = 0$.

**(A)** *If one of following holds:*
  **(a)** $\alpha_4 < 0$,
  **(b)** $\alpha_4 \geq 0, \Delta \geq 0, l_1 > 0$ *and* $H(l_1) \leq 0$,
  **(c)** $\alpha_4 \geq 0$, $\Delta < 0$, *and there exists a* $l^* \in \{l_1, l_2, l_3\}$ *such that* $l^* > 0$ *and* $H(l^*) \leq 0$,
  *then all roots of* (5.19) *have negative real parts when* $\tau \in [0, \tau^*)$.
**(B)** *If the conditions (a)–(c) of (A) are not satisfied, then all roots of* (5.19) *have negative real parts for all* $\tau \geq 0$, *where* $l_0$ *is defined in (5.28).*

From the above lemmas, we can say that if the conditions (a)–(c) are not satisfied, then (5.19) has no roots with zero real part for all $\tau \geq 0$. If one of (a), (b), or (c) holds, when $\tau \neq \tau^*$, Eq. (5.19) has no roots with zero real part, and $\tau^*$ is the minimum value of $\tau$ so that (5.19) has purely imaginary roots.

**Lemma 5.4.** *Suppose* $H'(l_0) \neq 0$. *If* $\tau = \tau^*$, *then* $\pm i\omega_0$ *is a pair of purely imaginary roots of* (5.19). *Moreover, if the conditions of Lemma 5.3(A) holds, then the sign of* $\frac{d(Re\xi)}{d\tau} \neq 0$ *is same as* $H'(l_0)$.

*Proof.* Differentiating (5.19) with respect to $\tau$, we get

$$\frac{d\tau}{d\xi} = \frac{4\xi^3 + 3(a_1\xi^2 + 2a_2\xi + a_3)}{b_2\xi^3 + b_3\xi^2 + b_4\xi} e^{\xi\tau} + \frac{2b_2\xi + b_3}{b_2\xi^3 + b_3\xi^2 + b_4\xi} - \frac{\tau}{\xi}.$$

We use the relation (5.18) to get

$$Sgn\left[\frac{d(Re\xi}{d\tau}\right]_{\tau=\tau^*} = Sgn[Re(\frac{d\xi}{d\tau})^{-1}]_{\xi=i\omega_0},$$
$$= Sgn\left[\frac{4\omega_0^6 + A_1\omega_0^4 + A_2\omega_0^2 + A_3}{b_3^2\omega_0^2 + [-b_2\omega_0^2 + b_4]^2}\right], \tag{5.29}$$

where

$$\alpha_1 = 3a_1 - 6a_2,$$
$$\alpha_2 = 2a_2 + 4a_4 - 4a_1a_3 - 2b_2^2,$$
$$\alpha_3 = a_3^2 - 2a_2a_4 - b_3^2 + 2b_2b_4.$$

As $b_3^2\omega_0^2 + [-b_2\omega_0^2 + b_4]^2 \neq 0$, one can conclude that

$$Sgn[\frac{d(Re\xi}{d\tau}]_{\tau=\tau^*} > 0$$

when

$$4\omega_0^6 + A_1\omega_0^4 + A_2\omega_0^2 + A_3 > 0, \text{ i.e., when } H'(l_0) > 0.$$

□

Finally, from the above three lemmas, we have the following theorem:

**Theorem 5.5.**

**(i)** *If the conditions (a)–(c) of Lemma 5.3 are not satisfied, then* (5.19) *has no roots with zero real part for all* $\tau \geq 0$*; hence stability switch will not occur.*
**(ii)** *If one of three conditions in (a), (b), and (c) of Lemma 5.3 holds, then Eq.* (5.19) *has roots with negative real parts for* $\tau \in [0, \tau^*)$. *For* $\tau = \tau^*$ *and* $[\frac{d(Re\xi}{d\tau}]_{\tau=\tau^*} > 0$, *then* $\pm i\omega_0$ *are the roots of Eq.* (5.19), *and all other roots have negative real parts; hence the system undergoes a Hopf bifurcation at* $\tau = \tau^*$.

## 5.6 Sensitivity analysis

In this section, a local sensitivity analysis of the model parameters of (5.1) is studied using the method studied in [30,31]. This analysis evaluates effectiveness of a parameter belonging to the model under investigation. The sensitivity of a population $P$ (say) with respect to a generic parameter $\eta$ is obtained by plotting the values of the term $\frac{\partial P}{\partial \eta}$ as function of time. To do this, we use the Matlab® software that calculates sensitivities by combining the original model system with the auxiliary differential equations for the sensitivities. These additional equations are obtained by differentiating the original equations with respect to the parameter $\eta$.

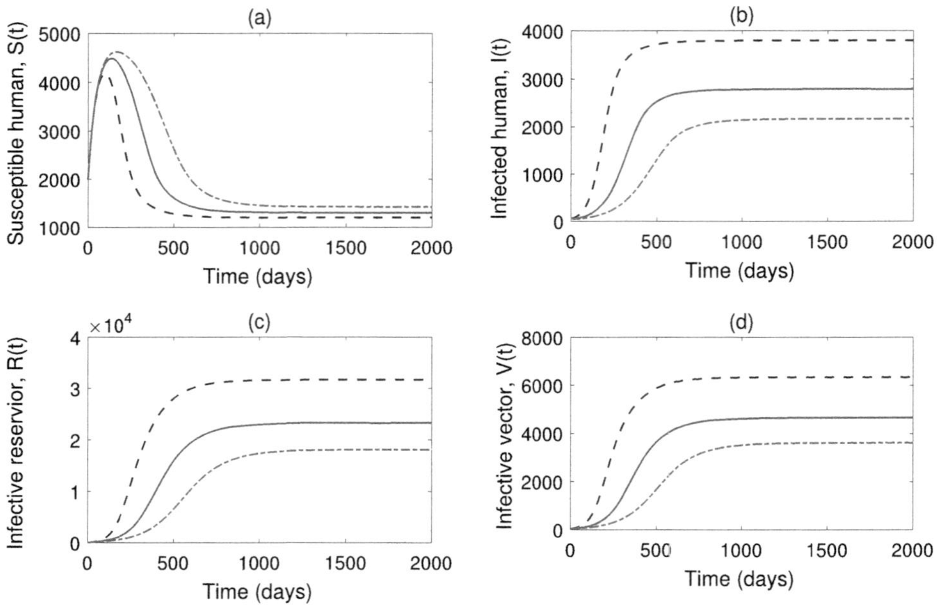


**FIGURE 5.2** Numerical solution of the model (5.1) for the set of parameters: $r = 0.1$; $m = 0.02$, $\Pi = 100$, $\lambda = 0.02$, $b = 0.1$, $\mu = 0.02$; $c = 0.12$ for three different values of the delay $\tau$.

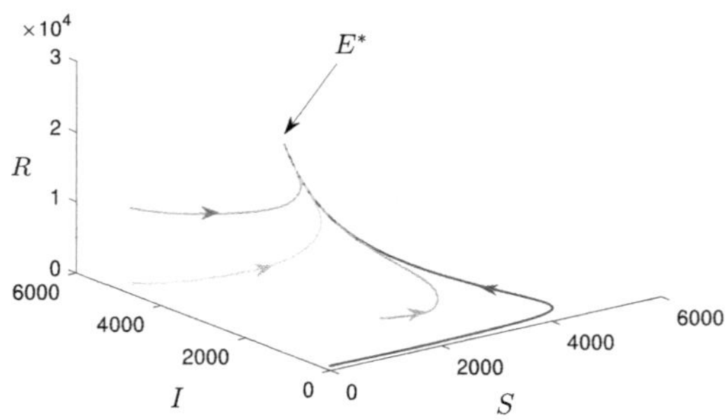


**FIGURE 5.3** Numerical solution of the model (5.1) for the set of parameters as in Fig. 5.2 using three different initial conditions.

## 5.7 Numerical simulations

Based on the results of the previous sections, we have provided the numerical results of system (5.1) in the following figures:

In Fig. 5.2, we have shown the impact of incubation periods in form of the time delay. Solution trajectory is plotted for different initial conditions in Fig. 5.3. It shows the global stability of endemic equilibrium point $E^*$. As the delay increases, the infection reduces significantly; no stability switches occur in the system, rather we observe the forward bifurcation, as shown in Fig. 5.4. Generally, higher resistance towards the disease can indicate a disease-free situation. Such situation can be achieved by using vaccination.

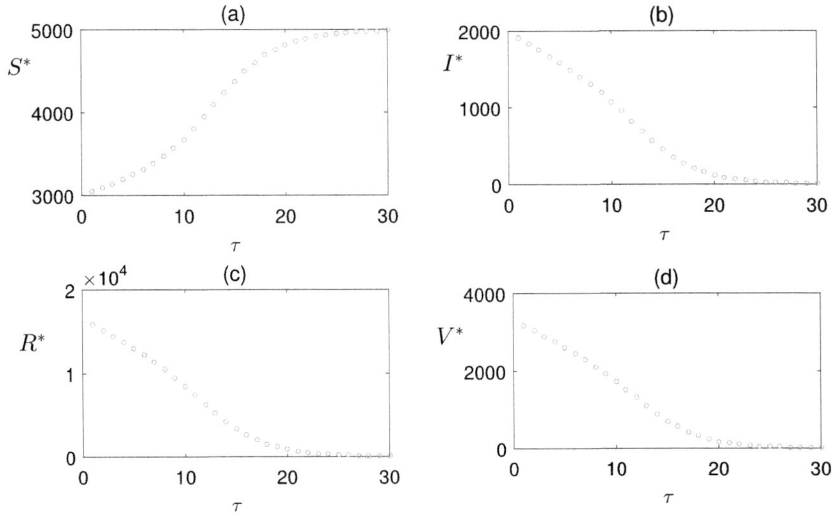


**FIGURE 5.4** Populations of model (5.1) is plotted as function of $\tau$ for the set of parameters as in Fig. 5.2.

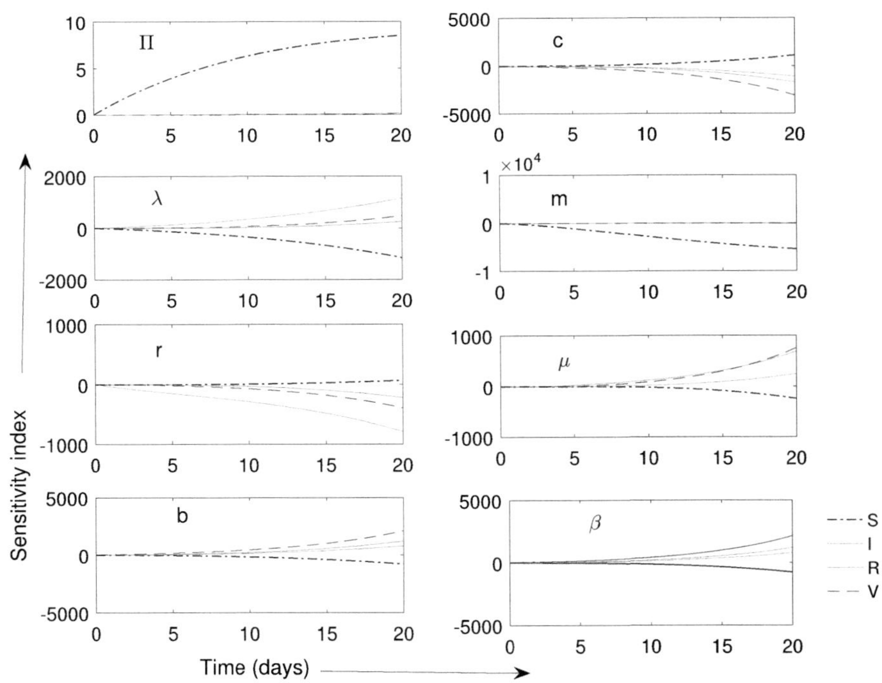


**FIGURE 5.5** Sensitivity of the parameters of the model (5.1) is plotted using the set of parameters as in Fig. 5.2.

The sensitivity index plotted in Fig. 5.5 shows that each parameter is sensitive for the dynamics of model (5.1).

## 5.8 Discussion and conclusion

In this research, we have developed a new mathematical model of visceral leishmaniasis based on population dynamics, considering four populations that are interrelated to the transmission of the disease, namely the susceptible human, infected human, infective reservoir, and the infected vector population. Furthermore, we develop the model using delay differential equations, because there are time lags for developing and transmitting the disease in the vector and human population due to resistance to a disease by humans. Our primary focus is to investigate the effectiveness of disease dynamics on infected populations in the presence of the said delay.

We see that as the delay is increased, the equilibrium value of uninfected populations and also disease-free state can be obtained if delay is higher than three weeks. This research established that delay can obtain a disease-free state when human population has strong immunity power, which is established here as time delay parameter. We noted that for a larger value of the delay, the endemic state can switch to the disease-free state. The results obtained here can be helpful to the policymakers for controlling visceral leishmaniasis.

## References


[1] O. Courtenay, R.J. Quinnell, L.M. Garcez, J.J. Shaw, C. Dye, Infectiousness in a cohort of Brazilian dogs: why culling fails to control visceral Leishmaniasis in area of high transmission, The Journal of Infectious Diseases 186 (2001) 1314–1320.

[2] J. Alvar, I.D. Vélez, C. Bern, M. Herrero, P. Desjeux, J. Cano, J. Jannin, M.D. Boer, WHO Leishmaniasis Control Team, Leishmaniasis worldwide and global estimates of its incidence, PLoS ONE 7 (5) (2012 May 31) e35671.

[3] H. Alatif, Burden trends of leishmaniasis over the last one decade across the globe: trend analysis of WHO regions, Integrative Journal of Medical Sciences 8 (2021).

[4] R. Rocha, A. Pereira, C. Maia, Non-endemic Leishmaniases reported globally in humans between 2000 and 2021–a comprehensive review, Pathogens 11 (8) (2022) 921.

[5] S. Sasidharan, P. Saudagar, Leishmaniasis: where are we and where are we heading?, Parasitology Research 120 (5) (2021) 1541–1554.

[6] T.D. Serafim, I.V. Coutinho-Abreu, R. Dey, R. Kissinger, J.G. Valenzuela, F. Oliveira, S. Kamhawi, Leishmaniasis: the act of transmission, Trends in Parasitology 37 (11) (2021) 976–987.

[7] S. Mann, K. Frasca, S. Scherrer, A.F. Henao-Martínez, S. Newman, P. Ramanan, J.A. Suarez, A review of leishmaniasis: current knowledge and future directions, Current Tropical Medicine Reports 8 (2021) 121–132.

[8] M. Maroli, C. Khoury, Prevention and control of leishmaniasis vectors: current approaches, Parasitology 46 (2004) 211–216.

[9] B. Alexander, M. Maroli, Control of phlebotomine sandflies, Medical and Veterinary Entomology 17 (2003) 1–18.

[10] U. Sharma, S. Singh, Insect vectors of Leishmania: distribution, physiology and their control, Journal of Vector Borne Diseases 45 (2008) 255–272.

[11] R. Neghina, A.M. Neghina, Leishmaniasis, a global concern for travel medicine, Scandinavian Journal of Infectious Diseases 42 (2010) 563–570.

[12] I.M. ELmojtaba, J.Y.T. Mugisha, M.H.A. Hashim, Vaccination model for visceral Leishmaniasis with infective immigrants, Mathematical Methods in the Applied Sciences 36 (2013) 216–226.

[13] A.N. Chatterjee, P.K. Roy, J. Mandal, Mathematical model for suppression of sand flies through IRS with DDT in visceral Leishmaniasis, American Journal of Mathematics and Sciences 2 (1) (Jan 2013) 105–112.

[14] S. Zhao, Y. Kuang, C. Wu, D. Ben-Arieh, M. Ramalho Ortigao, et al., Zoonotic visceral leishmaniasis transmission: modeling, backward bifurcation, and optimal control, Journal of Mathematical Biology 73 (2016) 1525–1560.
[15] I.M. ELmojtaba, Mathematical model for the dynamics of visceral Leishmaniasis–malaria co-infection, Mathematical Methods in the Applied Sciences 39 (2016) 4334–4353.
[16] K.S. Rock, R.J. Quinnell, G.F. Medley, O. Courtenay, Chapter two - progress in the mathematical modelling of visceral Leishmaniasis, Advances in Parasitology 94 (2016) 49–131.
[17] K.S. Rock, E.A.L. Rutte, S.J.D. Vlas, E.R. Adams, G.F. Medley, et al., Uniting mathematics and biology for control of visceral Leishmaniasis, Trends in Parasitology 31 (2015) 251–259.
[18] N. Hussaini, K. Okuneye, A.B. Gumel, Mathematical analysis of a model for zoonotic visceral Leishmaniasis, Infectious Disease Modelling 2 (2017) 455–474.
[19] L. Zou, J. Chen, S. Ruan, Modeling and analyzing the transmission dynamics of visceral Leishmaniasis, Mathematical Biosciences 14 (2017) 1585–1604.
[20] F. Boukhalfa, M. Helal, A. Lakmeche, Mathematical analysis of visceral leishmaniasis model, Results in Applied Mathematics 1 (2017) 01.
[21] A. Mubayi, M. Paredes, J. Ospina, A comparative assessment of epidemiologically different cutaneous leishmaniasis outbreaks in Madrid, Spain and Tolima, Colombia: an estimation of the reproduction number via a mathematical model, Tropical Medicine and Infectious Diseases 3 (2018) 01.
[22] E.O. Agyingi, D.S. Ross, K. Bathena, A model of the transmission dynamics of Leishmaniasis, Journal of Biological Systems 19 (2011) 237–250.
[23] J.V. Griensven, E. Diro, Visceral Leishmaniasis: recent advances in diagnostics and treatment regimens, Infectious Disease Clinics of North America 33 (2019) 79–99.
[24] P. Ghosh, P. Roy, S.J. Choudhuri, N. Kanti Das, Epidemiology of post-kala-azar dermal Leishmaniasis, Indian Journal of Dermatology (2021) 12–23.
[25] H.M. Yang, The basic reproduction number obtained from Jacobian and next generation matrices–a case study of Dengue transmission modelling, Biosystems 126 (2014) 52–75.
[26] O. Diekmann, J.A.P. Heesterbeek, M.G. Roberts, The construction of next-generation matrices for compartmental epidemic models, Journal of the Royal Society Interface 7 (47) (2010) 873–885.
[27] J. Hale, Theory of Functional Differential Equations, Springer-Verlag, NY, Heidelberg, Berlin, 1977.
[28] M. Bodnar, The nonnegativity of solutions of delay differential equations, Applied Mathematics Letters 13 (6) (2000) 91–95.
[29] X. Yang, L. Chen, J. Chen, Permanence and positive periodic solution for the single species nonautonomus delay diffusive model, Computers & Mathematics with Applications 32 (1996) 109–116.
[30] A. Saltelli, M. Scott, Guest editorial: the role of sensitivity analysis in the corroboration of models and its link to model structural and parametric uncertainty, Reliability Engineering & System Safety 57 (1) (1997) 1–4.
[31] A. Saltelli, S. Tarantola, K.S. Chan, A quantitative model-independent method for global sensitivity analysis of model output, Technometrics 41 (1) (1999) 39–56.

6

# Fractalization through stochasticization processes in epileptic dynamics

Maricel Agop[a,b], Alina Gavriluţ[c], Vlad Ghizdovăţ[d], and Lucian Eva[e]

[a]Physics Department, "Gheorghe Asachi" Technical University, Iaşi, Romania [b]Academy of Romanian Scientists, Bucureşti, Romania [c]Faculty of Mathematics, Alexandru Ioan Cuza University from Iaşi, Iaşi, Romania [d]Biophysics and Medical Physics Department, "Grigore T. Popa" University of Medicine and Pharmacy, Iaşi, Romania [e]Nicolae Oblu Hospital, Iaşi, Romania

## 6.1 Introduction

The brain is, by far, the most complex organ in the human body, being made of roughly 100 billion neurons, which communicate with each other through an even greater number of connections, called synapses. The brain is part of the central nervous system, together with the spinal cord, which together with the peripheral nervous system, made of nerves and spinal ganglions, is responsible for the majority of the functions carried out by a living organism: from breathing and hearing to learning and memory [1,2]. The information is acquired by nerves from the other organs of the body, is processed at the brain level, and then the response is sent back through the nerves as well.

The morpho-functional unit of the nervous system is thus represented by the neuron. According to their function, the neurons can be [3–6]:

- **i)** receptors: they receive stimuli from the external environment or from inside the organism;
- **ii)** motors: they are linked with the effector organs;
- **iii)** intercalary: they link the receptor neurons and the motor neurons.

The neuron is made of the cellular body and one or several extensions, which are of two types: dendrites and axons. The dendrites are short, numerous and have the role of receiving the nervous impulse from the receptors or from the extensions of other neurons. They then pass on the impulse to the neuronal body. The neuronal body will establish if the information should be passed on or not. The axon is unique, longer; it can reach up to 1 meter in length, and it has the role of leading the impulse from the neuronal body to the effectors or to other neurons. The extremity of the axon is ramified into several axonic terminations, each of them having a bulge at their end, called a terminal button.

Mathematical Modeling in Bioscience. https://doi.org/10.1016/B978-0-44-315445-4.00012-3

Copyright © 2025 Elsevier Inc. All rights are reserved, including those for text and data mining, AI training, and similar technologies.

The information received from the internal or external medium of the organism is later translated into a nervous impulse at receptor level. This impulse is further transmitted to the effectors by means of the neurons and the synapses.

The cytoplasmic membrane of the neurons is similar to the cytoplasmic membrane of other cells and, in resting conditions, it is found in a polarized state. Due to environmental stimuli, the cytoplasmic membrane becomes permeable for $K^+$ ions and non-permeable for $Na^+$ ions. The $Na^+$ ions will diffuse from the extracellular to the intracellular space. In this moment, the concentrations of the ions from the two faces of the membrane are modified; it is associated with a depolarization taking place together with the commencement of the so-called action potential [7]. The appearance of an action potential in an area of the neuronal membrane determines the apparition of a new action potential in a neighboring region, as such, along the entire membrane, successive depolarizations take place [8].

There exist two types of nervous fibers: myelinic and amyelinic ones. Correspondingly, there exist two types of nervous impulse transmission: through jumps or along the entire length of the axonic membrane. In the case of amyelinic axons, the action potential can appear in any region of the membrane, and transmission takes place in one way, with a speed of 10 m/s. The properties of the myelin play a very important part in the leading of the nervous impulse, thus the action potential at the level of the myelinic axons propagates in jumps, with a speed of 100 m/s [9].

The propagation of the nervous impulse is made under the following laws:

- **i)** the law of nerve integrity: the nervous fiber must not be damaged, compressed, or sectioned;
- **ii)** the law of isolated leading: the nervous fibers are completely independent and do not influence each other in the transmission of the impulse;
- **iii)** the law of bilateral leading: inside the organism, the sense of impulse propagation is unique: from the dendrites to the body of the cell and from the cellular body to the axon. This law is in effect due to the synapses.

A normal organism activity assumes that the nervous system is in equilibrium with the internal factors (the concentration of the molecular oxygen and blood glucose level, the alternation of intellectual activities with physical activities, soliciting neuron activities, etc.) and the external ones (temperature, diet, oxygen and light degree in a room etc.) [7,8]. Some studies appreciate that somatic diseases are approximately 70% of psychoneurological origin. Trauma, chemical substances, insufficient oxygenation, dietary deficiencies, stress—all these can lead to nervous system disfunctions.

Convulsions represent one of the most important illnesses that affect the nervous system. These are spontaneous, intermittent, and abnormal electrical discharges. They can be of two types [4,5]:

- **i)** isolated (non-recurrent): they manifest only in certain situations, such as brain trauma, febrile diseases, etc.
- **ii)** recurrent: they are characterized by attacks with sudden onset and uncontrolled motor activity, represented by epilepsy.

Epilepsy is a chronic neurological disorder, characterized by the emergence of convulsive seizures, and they are always accompanied by loss of conscience. The convulsions manifest through a contraction of the limbs, followed by their stretching, and by the arching of the back. These are often accompanied by a unison trembling of all the limbs. In most cases, the skin of the person acquires a blue tint, because breathing is cut off. During the seizure, a loss of control over the bladder or the bowel functions is possible. After the trembling ceases, the necessary recovery time for a person is usually 10–30 minutes, but sometimes it can be up to several hours. Recovery is accompanied by tiredness, headache, speaking difficulties, and abnormal behavior. Psychosis appears in 6–10% of cases. Most people have no recollection of what occurred during the seizure.

The exact mechanism of what triggers epilepsy is still unknown. The normal electric activity of the brain is not synchronized. During epileptic seizures, a group of neurons discharge abnormally, excessively but in a synchronized manner, causing the emergence of a depolarization wave. Normally, after the discharge, an excitatory neuron becomes resistant to discharges for a certain period of time [5]. In the case of epilepsy, the resistance of excitatory neurons to discharges decrease. This leads to the emergence of an area predisposed to the triggering of epileptic seizures, known as the seizure focus [6–8]. Another epilepsy mechanism could be the amplifying of the sensitivity of the receptors or the diminishing of the response of inhibitory circuits due to brain lesions [6–8]. The malfunctioning of the hematoencephalic barrier may represent a causal mechanism, because it can allow the substances found in blood to reach the brain [8].

A classification of epilepsies is very difficult to be made, because there are many types of epilepsy, having different causes, effects, and symptoms. In 1981, a classification system based on the observation of clinical phenomena was proposed [2,3], which contains 15 different sub-categories, grouped into 3 main categories: focal or partial epilepsies, non-focal or general epilepsies, continuous epilepsies.

The focal or partial epileptic seizure is triggered by the abnormal functioning or destruction of a group of neurons located in a specific brain region. The seizure does not necessarily start from this brain region, but it is the result of its presence and influence upon the entire neural network. These neurons cannot participate in the normal brain activity and are kept under control through a powerful inhibitory activity of the surrounding neural network. This is the reason why the brain does not exhibit a normal behavior, even in-between epileptic seizures [4]. Epileptic seizures are the result of a blockage of the inhibitory mechanism through which the affected neurons are kept under control [5]. These become hyperactive, triggering massive explosions of action potentials, which encompass the healthy neurons found in their vicinity or radius of action. If a focal seizure extends over a large portion of the brain, then it becomes secondarily generalized. The initiation and the expansion of the seizure is progressively made, sometimes very fast. Regarding the effects of focal epileptic seizures, the simple ones will produce light sensory and psycho-cognitive symptoms, without the loss of conscience, whereas complex focal seizures will lead to convulsions and the change of the conscience state [4–7]. The effects also depend on the brain location of the affected neurons; as such, certain senses might be lost (smell, taste, etc.) and hallucinations may be experienced.

Non-focal or generalized seizures are characterized by the fact that the patient exhibits seizures, which immediately manifest throughout the entire brain. This type of seizure is far less understood than the focal one, the emergence and evolution mechanism being still unknown [9]. As far as seizure onset goes, most probably there exists a combination of chemical (neurotransmitters, neuromodulators) and physiological (histology, neural network connections, resonances) imbalances. In the scientific community, the opinion that the origin of a generalized epileptic seizure can be found in the thalamus is sometimes expressed, as this is very well connected to all the parts of the brain and can ensure a rapid extension of the seizure [10]. Some types of generalized epileptic seizures appear mostly in children and disappear once adolescence sets in, whereas other types develop in maturity. It has been attempted to link these with hormonal changes or with the restructuring of neural connections due to aging [6,7].

Continuous epilepsies describe states in which there are no observable recoveries between seizures. They are by far the most dangerous types of epilepsy, because they can be life-threatening, even after five minutes from the onset of convulsions, due to the inability to ensure metabolic needs (oxygen, glucose, etc.) [9] A period of convulsions greater than 30 minutes leads to death in most cases.

The electroencephalogram is the base method in the diagnosis and monitoring of epilepsy. It is a non-invasive, electrophysiological method of exploring the nervous system by registering the electric activity of the brain. The method consists of placing electrodes on the surface of the head, which register the fluctuations of electric potential due to ionic bio-currents, which travel at neuron level. The registered signals are transmitted to a computer by means of a specialized acquisition plate, where they are analyzed and can be printed. The temporal resolution is high, allowing the characterization of rapid modifications of current fluxes, but the spatial resolution is rather low, being limited by the number of electrodes, their positioning, and the properties of the head.

In the case of scalp electroencephalogram, the signal must travel through several layers of non-neural tissue (cerebrospinal fluid, cranium, scalp), each of said layers affecting the signal in a different manner. The signal is damped and, due to this, to create the recording, it is necessary that greater regions of the brain must be simultaneously active. The procedure is inexpensive and easy to realize. Sometimes, it is capable of delivering sufficient information; other times it is only a preliminary step, to be followed by a detailed intracranial investigation. The registering of electric potentials is carried out by means of a helmet, on which the electrodes are fixed in standard positions, to make a comparison between the measurements made for the same patient, or for different patients.

In Fig. 6.1, the international system 10–20 of placing electrodes for scalp electroencephalogram is explained. On the left side of Figure (a), the positions and the labels of each electrode are presented, whereas on the right side (b), (c), two standard modes of signal measurement are depicted.

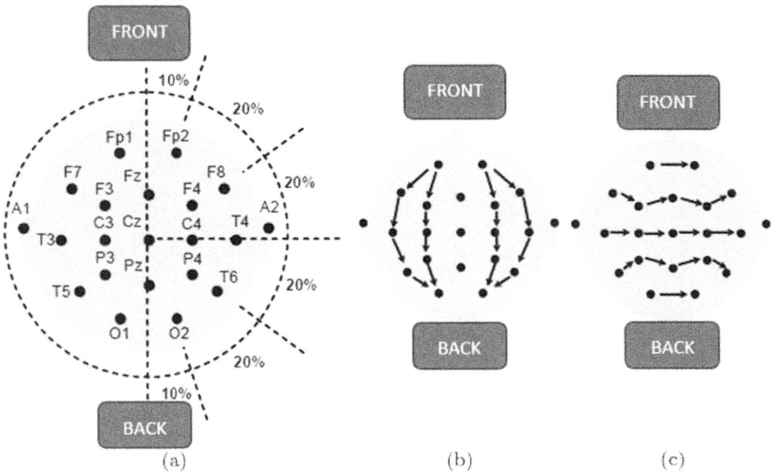


**FIGURE 6.1** The international 10–20 system for placing scalp EEG electrodes.

In the case of intracranial electroencephalogram, the recordings are made on electrodes placed on the cortex or on depth electrodes, which penetrate towards subcortical system (such as the thalamus) [10]. The technique is usually employed before surgery to accurately determine the brain region that is to be removed. It is a procedure that is rarely carried out, and the standard setup of the electrodes is difficult to achieve, being emplaced depending on the patient.

The analogic signals generated by the brain and recorded by the electrodes are transmitted to an interface responsible for the amplification, filtration, and digitalization of the signals before they are introduced in the computer for analysis and storage.

In Fig. 6.2, a typical electroencephalogram is represented. This contains, in a stacked manner, the time-dependent variations of electric tension, measured between two standard points, the points being specified in the left part of each signal.

The electroencephalogram measures the differences of potential induced by the currents of post-synaptic excitation and inhibition potentials. To generate signals with a good enough amplitude so that they may be registered, it is necessary for $10^4$–$10^7$ neurons to have a synchronous evolution. Subcortical neural networks have a negligible effect in the scalp electroencephalography [10–12]. An electroencephalogram will reflect the global dynamic of electric activity of a high population of neurons; this is the very reason why it is so useful in diagnosing epilepsy.

The interpretation of an electroencephalogram is a very difficult task; it requires a lot of experience, because there are no standards referring to "how it should look." There is no identical electroencephalogram, as modifications may appear as a function of conscience level (awake, or different sleep stages), operating mode of the sensing organs (opened or closed eyes, blocked ears etc.), age, etc. However, there are several models that tend to repeat; they appear in most persons, and the electroencephalograph can recognize them.

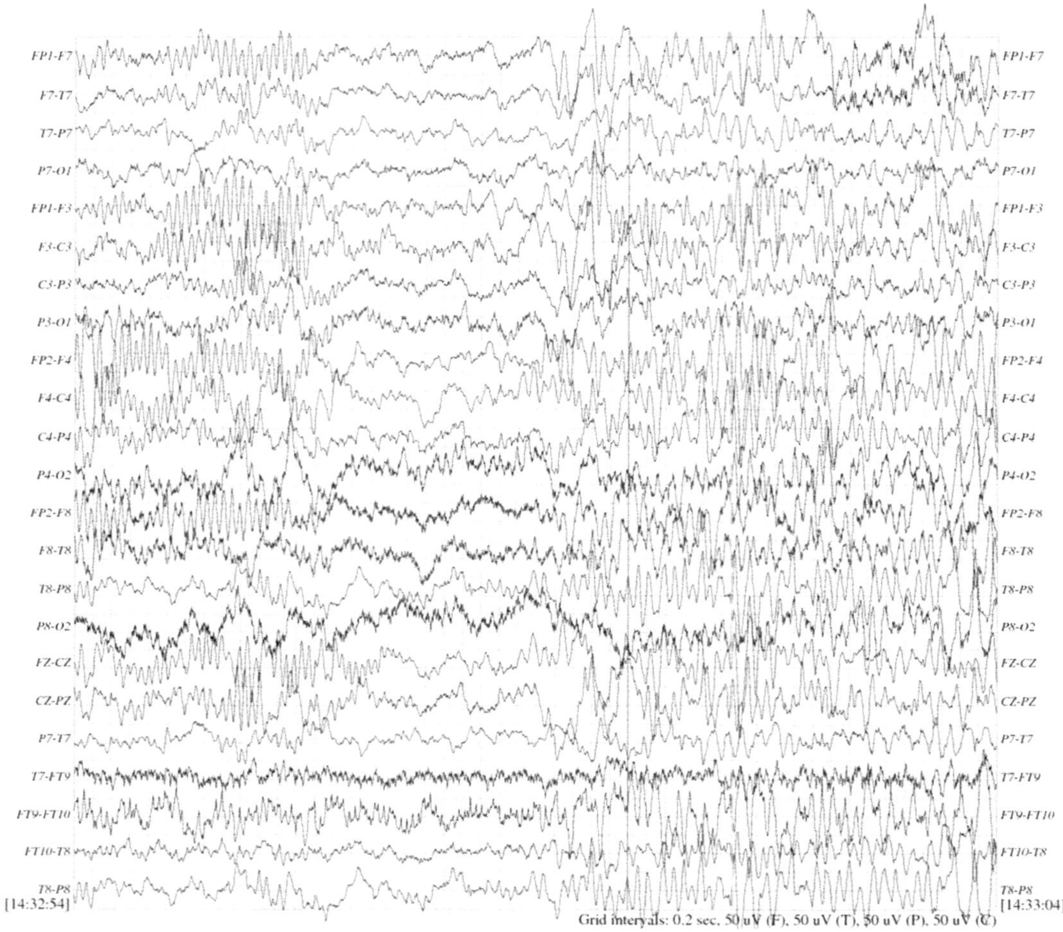


**FIGURE 6.2** A typical electroencephalogram.

The main modifications that appear in the case of an epileptic patient are

**i)** synchronization: the signals registered on different channels are similar, because the neural activity is similar on a large cortex region. Synchronization can be observed on several channels, or on a majority of them.

**ii)** high signal amplitude: after the triggering of an epileptic seizure, the amplitude of registered signals is much higher than before the seizure. This happens because, on one hand, a hyperexcitability of the cortex appears, and, on the other hand, a synchronization between channels emerges.

**iii)** the appearance of oscillations: it takes place once the epileptic activity has sufficiently extended.

It is not mandatory for all epileptic seizures to exhibit said characteristics.

The electroencephalogram is very useful in the study of epilepsy, because it quantitatively highlights time-dependent modifications of the electric activity of the brain. Thus it is possible to distinguish the transitions from a seizure to the interval between seizures and, more importantly, it is possible to identify the state before the seizure, which leads to the realization of predictions regarding the triggering moment of the epileptic seizure. However, the experience of tens of years of research has demonstrated that the identification of several characteristics based on which pre-seizure, seizure, and inter-seizure states can be distinguished remains an extremely difficult task. These characteristics have to be extracted from the electroencephalogram by means of signal processing, these being often affected by noise and are extremely complex, apparently chaotic. In the last years, important progress has been made in the interpretation of electroencephalograms, by means of non-linear dynamics developments, which allow the obtainment of supplementary information. Powerful methods and algorithms for signal processing are now available, which can deliver very valuable quantitative information. These can then be linked with qualitative information obtained through clinical observation of epileptic patients. The obtained data represents a basis for the development of models of brain dynamics or for the development of existing models.

## 6.2 Operational procedure 1: analysis of epileptic and eclamptic seizures by applying non-linear dynamics methods

Although the first electroencephalograms were registered more than 140 years ago, the progress relating to their interpretation has been extremely slow. Even nowadays, there is no classification of the structures that appear in the electroencephalogram so that a correspondence between them and the brain activity can be made. The clinical interpretation of electroencephalograms is mainly carried out by visual recognition of certain structures and through correspondences made by the specialist medic [6]. Fouriér analysis cannot be applied, because the signals associated with electroencephalograms are not stationary. The signals are extremely weak, in the domain of microvolts, "sunk in a sea of noise" [10–12]. This is why particular attention must be paid to the quality of employed electrodes, as well as to their positioning. Furthermore, the identification and analysis of artifacts should not be underestimated, as these can emerge due to slight electrode movements or due to the contraction of sub-electrode muscles.

The analyzed electroencephalograms were downloaded from PhysioNet database. This offers the possibility of accessing a free collection of physiological signals (PhysioBank), registered from a wide array of patients, as well as specialized software for their visualization and analysis. It is supported by the National Institute of General Medical Science (NIGMS) and the National Institute of Biomedical Imaging and Bioengineering (NIBIB), and the free access is realized according to ODC Public Domain Dedication and License v. 1.0. These resources are available for the encouragement of current research in the study of complex biomedical and physiological signals.

In the present work, an electroencephalogram, registered from an 11 year old patient with epilepsy, was analyzed. The characteristics of the electroencephalogram are listed below:

- The signals were recorded on 23 channels;
- The resolution of each signal was 16 bit;
- The sampling time is 4 ms;
- The signal duration is 60 minutes;
- The seizure duration is 40 seconds.

In Fig. 6.3, the signal registered on FP9-F10 channel is depicted. It can be observed that the neuronal activity does not have a regular dynamic.

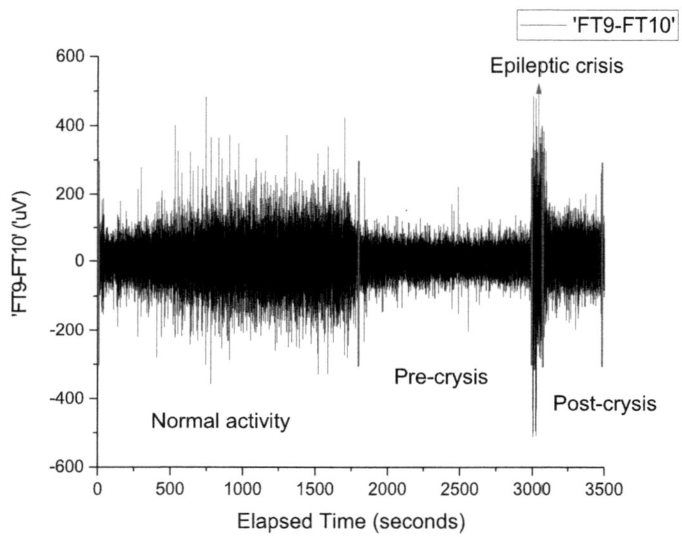


**FIGURE 6.3** Graphical representation of the signal recorded on channel FP9-F10.

The period of brain functioning can be split in four interest areas:

**a)** zone of normal brain activity (0–1800 s interval), characterized by a chaotic dynamic, with a relatively high signal amplitude;
**b)** zone of pre-seizure (1800–3000 s interval), characterized by a decrease of signal amplitude;
**c)** zone of epileptic seizure (3000–3040 s interval), in which the signal amplitude reaches its maximum value in a very short period of time, having a more regulated behavior following the synchronization of neuronal activities;
**d)** zone of post-seizure (3040–3600 s interval), in which the amplitude of the signal decreases to a relatively low value, but then increases to the value corresponding to the normal brain activity zone.

In Fig. 6.4 a–d, the electroencephalograms corresponding to the four zones previously described are illustrated. The corresponding signals have been analyzed by a series of statistic and non-linear dynamic methods, in what follows being described only those results that permitted the extraction of valuable information.

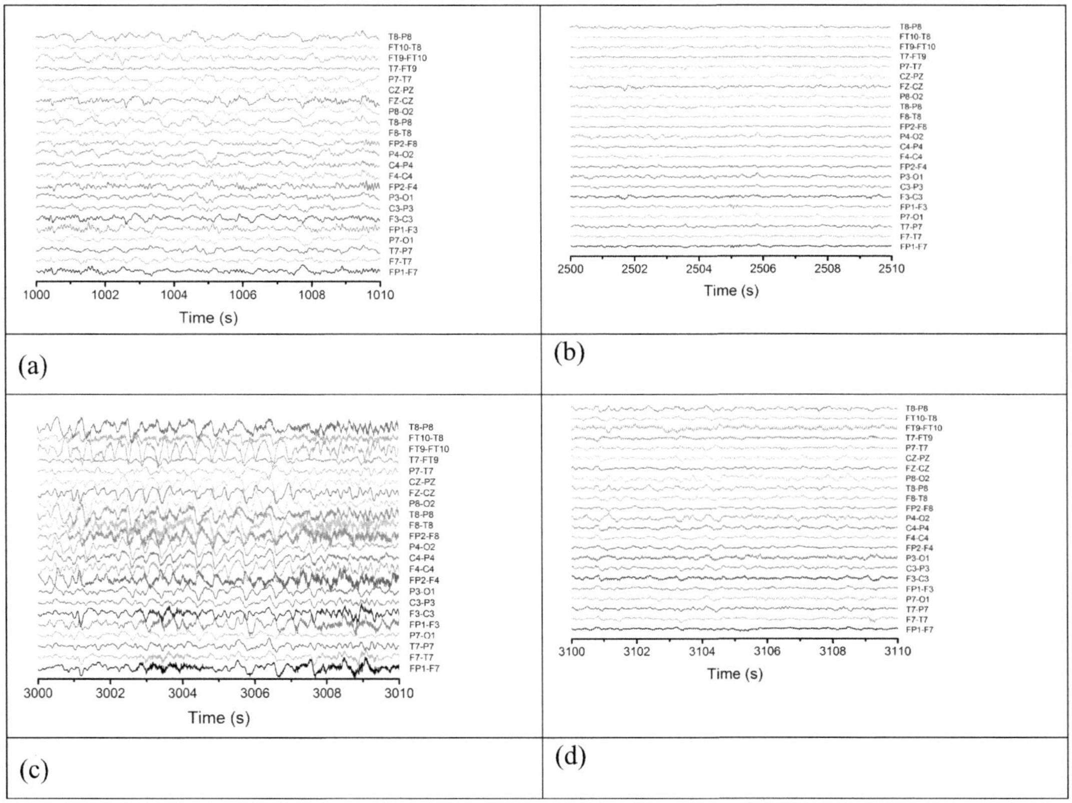


**FIGURE 6.4** Electroencephalogram corresponding to the normal functioning of the brain (a), the pre-seizure period (b), the epileptic seizure (c), and the post-seizure period (d).

The graphical representation of the standard deviation (Fig. 6.5a) shows that before the pre-seizure, its value suddenly drops (approximately to 1800 s), and then stays roughly constant until seizure onset (3000 s). During the epileptic seizure, the standard deviation presents a pronounced maximum. Because the standard deviation is an indicator of data dispersion, the fact that it remains at a small, approximately constant value during pre-seizure period, denotes that the registered potentials have small, relatively equal values. Thus the nervous impulses at the neural level are of small amplitude and "tranquil" dynamic. During a seizure, the values of the potentials strongly deviate from the average value.

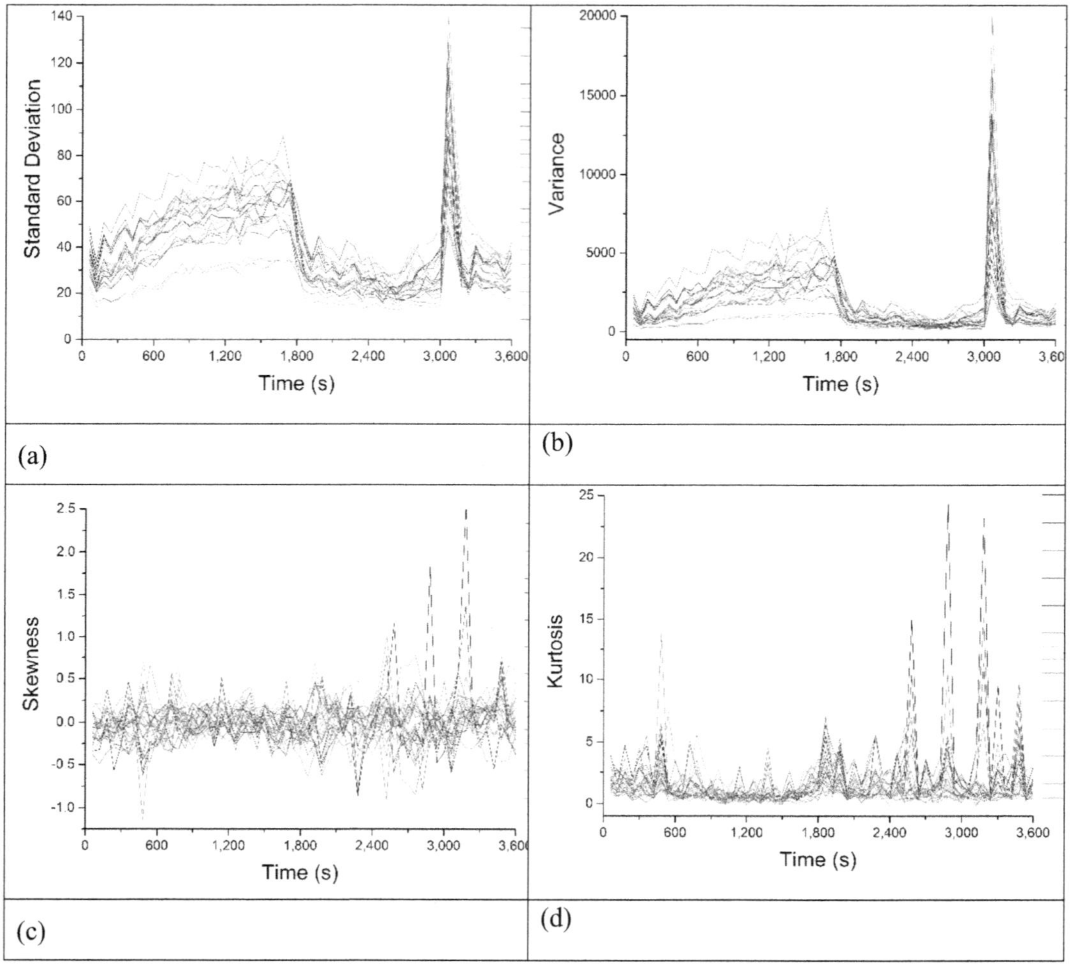


**FIGURE 6.5** Variation in time of the standard deviation (a), variance (b), skewness (c), and kurtosis (d), calculated for the component signals of the electroencephalogram.

The same result, but much better highlighted and with smaller errors, can be obtained through the graphical representation of the time-dependent variation of the variance (Fig. 6.5b).

In Fig. 6.5c,d, represented are time-dependent variations of the skewness and kurtosis, parameters that indicate the deviation from a Gauss-type normal distribution. From Fig. 6.5d, it is possible to observe that the skewness has an average value close to zero, with the exception of several pronounced positive maximums, which emerge in the pre-seizure and seizure regions, but only on a few channels (FP1-F7 and FP1-F3), which is an indication of the fact that, most probably, the epileptic seizure is a focal one, localized in the brain area near the electrode FP1.

Regarding kurtosis, this has positive average values, but lower than 3, with the exception of several pronounced positive maximums on channels FP1-F7, FP1-F3, and FP2-F4, correlated with the maximums observed for skewness. The behavior of this parameter confirms the fact that, most probably, the epileptic seizure is of a focal type.

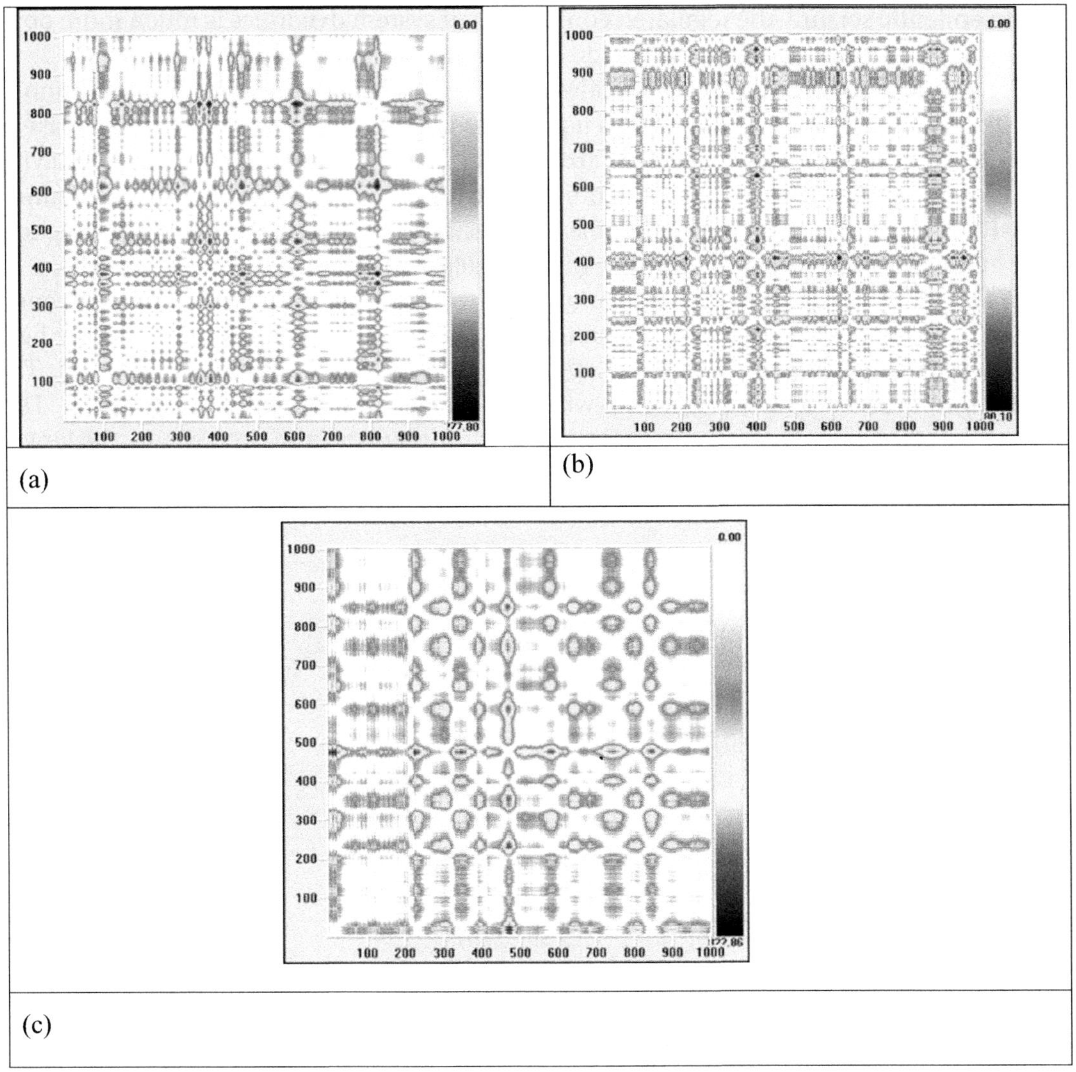


**FIGURE 6.6** The recurrence map corresponding to the normal functioning of the brain (a), the pre-seizure period (b), and the epileptic seizure (c).

The recurrence map gives global information about the dynamic of the brain and, because of this, no information can be gleaned about the focal or global character of the epileptic seizure. In Fig. 6.6a—c, the recurrence mappings for the signal registered on channel FP1-F7 are represented. They correspond to a normal brain function, pre-seizure, and seizure states, respectively. They were obtained with the help of visual recurrence analysis v.4.7.

The lack of homogeneity for the mappings indicates the existence of a non-stationary signal and the singular; isolated points indicate powerful fluctuations in the system. During an epileptic seizure, the regulated component of system dynamics is much more obvious, in accordance with previously-made observations.

For a more detailed quantitative analysis, in Fig. 6.7a, the time-dependent variation of space-time entropy for five channels is represented. It is possible to observe a decrease of the entropy until the onset of the pre-seizure, after which a sharp increase occurs, staying at a high value throughout the entire pre-seizure and seizure period. On some channels (FP1-F7, FP1-F3, and F7-T7), it is possible to observe some minimums, the value of space-time entropy decreasing to values close to the regularity limit of seizure.

In Fig. 6.7b, the time-dependent variation of the largest Lyapunov exponent is represented. It has been computed for 10 channels of the electroencephalogram, with the help of the subroutine "largest Lyapunov exponent" from Santis computer application. The largest Lyapunov exponent is positive, with an average value of approximately 0.09. This means that the brain dynamic is a chaotic one. During seizure and pre-seizure, the largest Lyapunov exponent showcases several sudden decreases to close to zero values, i.e., towards the regularity limit.

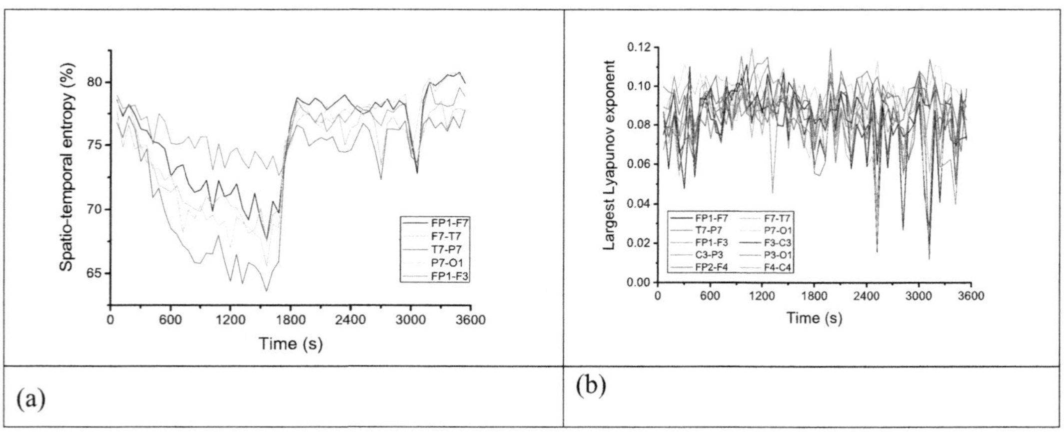


**FIGURE 6.7** Evolution in time of the spatial–temporal entropy for 5 component signals of the electroencephalogram (a) and the variation of the largest Lyapunov exponent corresponding to the 10-channel signals of the electroencephalogram (b).

The analyses carried out on signals corresponding to the electroencephalogram of an epileptic patient prove that some statistic parameters, such as standard deviation or variance, as well as space-time entropy, can be utilized to foresee, in a timely manner (approximately 20 minutes before the case of the currently investigated electroencephalogram), the onset of the epileptic seizure. For this, these parameters (or at least one of them) must be permanently monitored, and the warning system must me coupled to an automated intervention system upon the patient (for example, using drugs) such that the onset of the seizure can be prevented. Thus it is possible to lay the foundation for a functional electronic device, which can be permanently worn by the epileptic patient. As soon as a sensor detects the apparition of a pre-seizure type of dynamic behavior, it can administer, in a controlled manner, a treatment that can avoid the onset of an epileptic seizure.

In the future, statistical analysis must be extended to other types of electroencephalograms, in which multiple epileptic seizures appear at short time intervals. Furthermore, far more complex methods for signal analysis should be trialed, specific for non-stationary signals, such as the wavelet transform or the Hilbert–Huang transform.

## 6.3 Operational procedure 2: the reconstruction of EEG signals through scale relativity theory

In the multifractal theory of motion [13,14] two scenarios have been outlined in describing the neural dynamics of the brain:

**a.** The multifractal Schrödinger scenario explained by the differential equation

$$\lambda^2 (dt)^{\left[\frac{4}{f(\alpha)}-2\right]} \partial^l \partial_e \Psi + i\lambda (dt)^{\left[\frac{2}{f(\alpha)}\right]-1} \partial_t \Psi - U\Psi = 0; \tag{6.1}$$

**b.** The Madelung multifractal scenario explained by the system of differential equations

$$\partial_e V_D^i + \left(V_D^e \partial_e\right) V_D^i = -\rho^{-1} \partial_e \hat{\sigma}^{ie} + \partial^i U \tag{6.2}$$

$$\partial_t \rho + \partial_i \left(\rho V_D^i\right) = 0, \tag{6.3}$$

where

$$\partial_e \partial^e = \frac{\partial^2}{\partial x_e^2}, \quad \partial_t = \frac{\partial}{\partial t}, \quad i, e = 1, 2, 3.$$

In relations (6.1)–(6.3) the sizes have the following physical meanings:

- $\Psi$ is the multifractal state function;
- $\rho = \Psi\overline{\Psi}$, where $\overline{\Psi}$is the complex conjugate of $\Psi$, the density of multifractal states;
- $U$ is the scalar potential of the extended field;
- $x^e$ is the multifractal spatial coordinate;
- $t$ is the non-multifractal temporal coordinate,

$$V_D^i = 2\lambda (dt)^{\left[\frac{2}{f(\alpha)}-1\right]} \partial^i ln\Psi \tag{6.4}$$

is the non-multifractal speed,

$$\sigma^{ie} = -2\lambda^2 (dt)^{\left[\frac{4}{f(\alpha)}-2\right]} \rho \partial_i \partial_e ln\rho \tag{6.5}$$

is the multifractal tension tensor,

- $\lambda$ is the parameter associated with the multifractal-non-multifractal scale transition,
- $dt$ is the scale resolution,
- $f(\alpha)$ is the singularity spectrum of order $\Psi\alpha = \alpha(D_F)$, and
- $D_F$ is the fractal dimension of the motion curves.

Recall that in multifractal theory of motion, the dynamics of any neural structure are described by continuous and undifferentiable curves (fractal curves/multifractal curves); for further details, see references [13,14]. Note that the two scenarios for describing the dynamics of neural structures are not disjoint, but complementary (the brain can be considered to have two distinct behaviors, according to the functionality of the two hemispheres: digital behaviors, described by (6.1) and analogic behaviors, described by (6.2) and (6.3)). They allow, by an appropriate choice of scale resolution, either local or global descriptions, or specific local-global transition descriptions. In such a context, in references [13,14], we have shown that the presence of a multifractal tension tensor can be correlated by the relation

$$\partial^i \sigma_{ie} + \rho \partial_e Q = 0 \tag{6.6}$$

of the existence of multifractal scalar potential:

$$Q = -2\lambda^2 (dt)^{\left[\frac{4}{f(\alpha)}-2\right]} \frac{\partial^e \partial_e \sqrt{\rho}}{\sqrt{\rho}} = V_F^e V_F^e - \lambda (dt)^{\left[\frac{2}{f(\alpha)}-1\right]} \partial_i V_F^i, \tag{6.7}$$

where

$$V_F^i = \lambda (dt)^{\left[\frac{2}{f(\alpha)}-1\right]} \partial^i ln\rho \tag{6.8}$$

is the multifractal velocity.

Since in any of these behaviors, the SL(2R) group structure

$$t' \to \frac{\alpha t + \beta}{\gamma t + \delta}, \quad x' \to \frac{x}{\gamma t + \delta}, \tag{6.9}$$

where $\alpha$, $\beta$, $\gamma$, and $\delta$ are real elements, is invariant with respect to the set of equations (6.1)–(6.3), then, according to the general mathematical procedures from [15], a harmonic map from the Euclidean space to the hyperbolic plane, in the Beltrami–Poincaré representation, becomes functional. In such a context, through [14], the Euler–Lagrange equation

$$(h - \hbar)\, \partial_l \partial^l h = 2 \partial_l h \partial^l h \tag{6.10}$$

admits the solution

$$h = i \frac{\cosh \chi - e^{-i\omega t} \sinh \chi}{\cosh \chi + e^{-i\omega t} \sinh \chi}. \tag{6.11}$$

For more details, significances of quantities found in (6.10)–(6.11) etc., see [13,14].

In Fig. 6.8, the real signals and the simulated ones are comparatively represented using the method in [15].

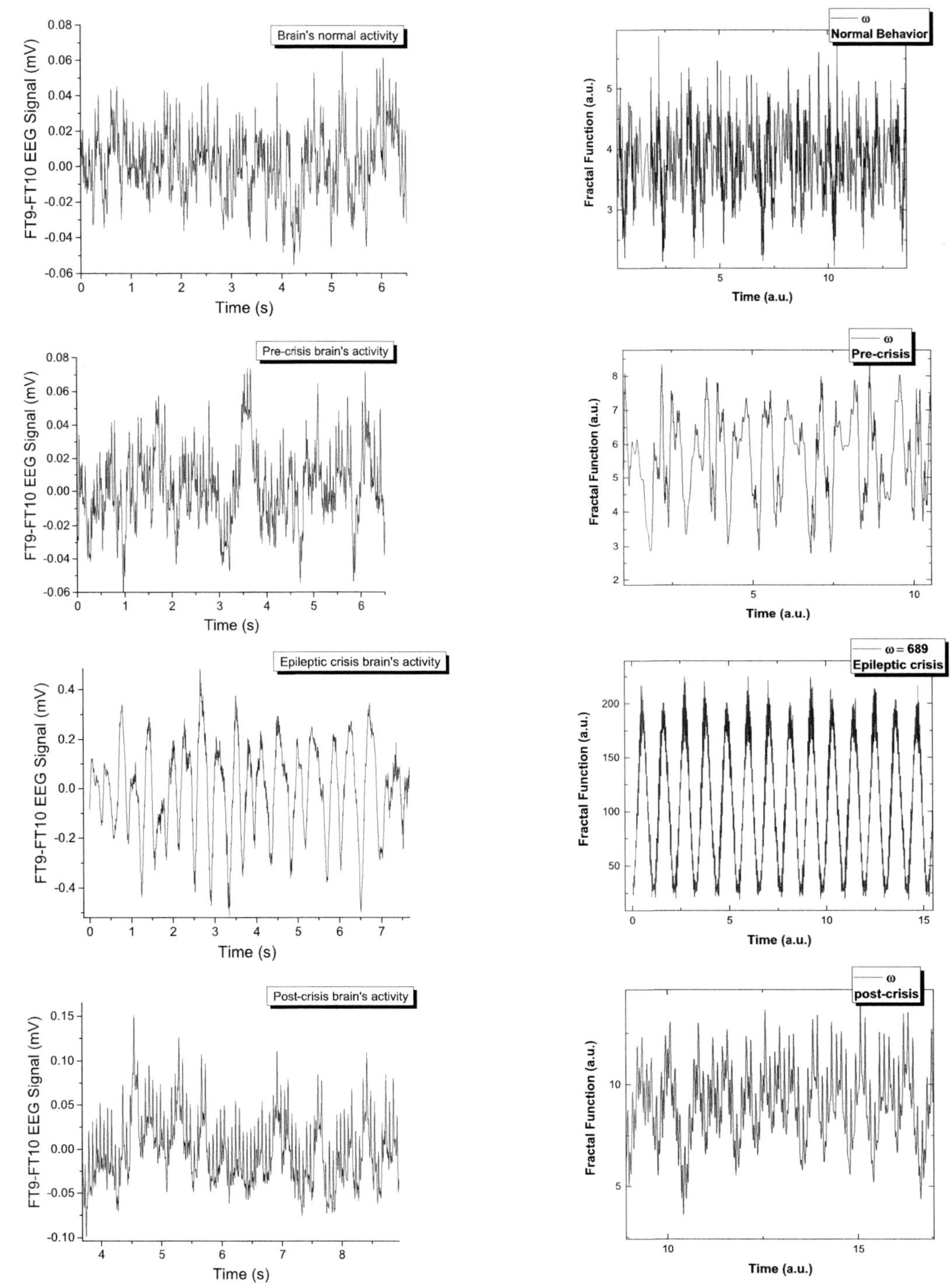


**FIGURE 6.8** Comparative representation of the empirical signal and the simulated multifractal one.

## 6.4 Conclusions

Two procedures have been proposed for evaluating and predicting the onset of epileptic and eclamptic seizures. The first procedure refers to the electrical activity of the brain (EEG signals) employing non-linear dynamic methods (time-dependent variation of the standard deviation, variance, skewness, kurtosis, time-dependent evolution of spatial-temporal entropy and largest Lyapunov exponent variation). The second procedure reconstructs any type of EEG signal, based on scale relativity theory (fundamented on two complementary scenarios). Using the non-linear dynamics method, the analysis performed on the signals corresponding to the electroencephalogram of an epileptic patient showed that some statistical parameters, such as standard deviation or variance, as well as the spatial–temporal entropy, can be used to predict in advance the onset of the epileptic seizure.

The novelty of the present work refers to the correlation between the experimental data (processed using statistical and non-linear procedures, such as standard deviation and variance, spatial–temporal entropy, Lyapunov exponents, etc.) obtained from a case of epilepsy concerning an underaged person, with a theoretical model. The model is based on the scale relativity theory, expressed through the concept of harmonic mappings, from the usual space to the hyperbolic one. Such an approach, concerning signal analysis, is singular in the framework of epileptic and eclamptic seizures [10–13]. Let it be noted that "strange" behaviors in epileptic and eclamptic seizures can be frequently contained in the hidden symmetries of the mathematical models, which can be used in the usual signal analyses (in particular, in the present case, the discussion concerns the involvement of SL(2R) symmetry).

## References

[1] C.E. Stafstrom, L. Carmant, Seizures and epilepsy: an overview for neuroscientists, Cold Spring Harb. Perspect. Med. 5 (6) (2015 Jun) a022426, https://doi.org/10.1101/cshperspect.a022426, https://www.ncbi.nlm.nih.gov/pmc/articles/PMC4448698/.

[2] A.T. Berg, S.F. Berkovic, M.J. Brodie, J. Buchhalter, J.H. Cross, W. van Emde Boas, J. Engel, T.A. Glauser, G.W. Mathern, S.L. Moshe, Revised terminology and concepts for organization of seizures and epilepsy: report of the ILAE commission on classification and terminology, 2005–2009, Epilepsia 51 (2010) 676–685.

[3] A.T. Berg, J.J. Millichap, The 2010 revised classification of seizures and epilepsy, Continuum (Minneap. Minn.) 19 (2013) 571–597.

[4] V.M. Muro, M.B. Connolly, Classifying epileptic seizures and the epilepsies, in: J.W. Miller, H.P. Goodkin (Eds.), Epilepsy, Wiley, Chichester, UK, 2014, pp. 10–14.

[5] C.E. Stafstrom, Pathophysiological mechanisms of seizures and epilepsy: a primer, in: J.M. Rho, R. Sankar, C.E. Stafstrom (Eds.), Epilepsy: Mechanisms, Models, and Translational Perspectives, CRC, Boca Raton, FL, 2010, pp. 3–19.

[6] S.F. Berkovic, Genetics of epilepsy in humans, Cold Spring Harb. Perspect. Med. (2015), https://doi.org/10.1101/cshperspect.a022400.

[7] Y. Ben-Ari, Excitatory actions of GABA during development: the nature of the nurture, Nat. Rev. Neurosci. 3 (2002) 728–739.

[8] A. Pitkänen, K. Lukasiuk, F.E. Dudek, K.J. Staley, Epileptogenesis, Cold Spring Harb. Perspect. Med. (2015), https://doi.org/10.1101/cshperspect.a022822.

[9] G. Holmes, Y. Ben-Ari, Seizures in the developing brain: perhaps not so benign after all, Neuron 21 (1998) 1231–1234.

[10] H. Kim, A.L. Paige, R.C. Knowlton, Advances in structural and functional neuroimaging: how are these guiding epilepsy surgery?, in: J.M. Rho, R. Sankar, C.E. Stafstrom (Eds.), Epilepsy: Mechanisms, Models, and Translational Perspectives, CRC, Boca Raton, FL, 2010, pp. 257–282.
[11] P.A. Caruso, J. Johnson, R. Thibert, O. Rapalino, S. Rincon, E.M. Ratai, The use of magnetic resonance spectroscopy in the evaluation of epilepsy, Neuroimaging Clin. N. Am. 23 (2013) 407–424.
[12] B. Kay, J.P. Szaflarski, EEG/fMRI contributions to our understanding of genetic generalized epilepsies, Epilepsy Behav. 34 (2014) 129–135.
[13] L. Nottale, Scale Relativity and Fractal Space-Time: A New Approach to Unifying Relativity and Quantum Mechanics, Imperial College Press, London, UK, 2011.
[14] I. Merches, M. Agop, Differentiability and Fractality in Dynamics of Physical Systems, World Scientific, Hackensack, NJ, USA, 2016.
[15] S.A. Irimiciuc, A. Zala, D. Dimitriu, L.M. Himiniuc, M. Agop, B.F. Toma, L.G. Gavril, D. Vasincu, L. Eva, Novel approach for EEG signal analysis in a multifractal paradigm of motions. Epileptic and eclamptic seizures as scale transitions, Symmetry 13 (2021) 1024, https://doi.org/10.3390/sym13061024.

# 7 Mathematical model of Alzheimer disease using the nonlocal and nonsingular fractional operators

Sonal Jain[a], Kolade M. Owolabi[b], and Edson Pindza[c]

[a]School of Technology, Woxsen University, Hyderabad, Telangana, India [b]Department of Mathematical Sciences, Federal University of Technology Akure, Akure, Ondo State, Nigeria [c]Department of Decision Sciences, College of Economic and Management Sciences, University of South Africa (UNISA), Pretoria, South Africa

## 7.1 Introduction

Alzheimer's disease (AD) is a progressive disorder represented by the degeneration and death of brain cells, leading to dementia, which manifests as a continuous decline in cognitive, behavioral, and social skills. This decline severely impairs an individual's ability to function independently [1,2]. The primary pathological hallmark of AD is the abnormal accumulation of proteins in and around brain cells. Amyloid proteins form plaques around brain cells, whereas tau proteins create tangles within the cells [3–6]. The increasing prevalence of AD among older adults poses significant challenges for global health. In 2006, there were approximately 26.12 million individuals worldwide with AD. By 2019, this number had risen to an estimated 46.8 million, and projections suggest that it will reach 74.7 million by 2030 [7]. By the year 2050, the prevalence is projected to rise by 1.2 percent [8]. The substantial costs associated with caring for individuals with AD underscore the urgent need for effective models to understand and manage the disease.

Developing accurate models to describe the transmission of AD, including the interactions of neurons, therapies, astrocytes, and other factors, is crucial. Recently, many researchers have utilized fractional derivatives and wavelets, among other systems, to create models that better reflect the complexities of diseases like AD compared to traditional ordinary differential equations.

Fractional differentiation and integration operators have emerged as powerful mathematical tools for addressing real-world problems. These operators, capable of modeling processes governed by power laws, exponential decay functions, and the novel Mittag-Leffler functions, offer a significant advantage in modeling complex phenomena [10,11,13]. Additionally, the concept of Brownian motion, or stochastic components, has been introduced to account for randomness and has found success in various fields of science, technology, and engineering [14,20–24]. However, neither fractional differential

Mathematical Modeling in Bioscience. https://doi.org/10.1016/B978-0-44-315445-4.00013-5

Copyright © 2025 Elsevier Inc. All rights are reserved, including those for text and data mining, AI training, and similar technologies.

operators nor stochastic approaches alone can fully capture the intricate nature of certain problems, particularly those exhibiting both deterministic and random processes [16,25].

To address this gap, researchers have proposed combining fractional differential operators with stochastic components, resulting in fractional stochastic differential and integral equations [9,12]. This novel approach is especially relevant for understanding the spread of infectious diseases, which depends on numerous factors and cannot be fully described by simple mathematical equations.

This chapter extends the AD model by using Atangana–Baleanu (AB) fractional derivatives, and Brownian motions. This approach aims to give a more accurate and realistic representation of the dynamics of AD.

## 7.2 Useful preliminaries

In this section, we review fundamental definitions and properties of fractional calculus theory that will be useful in the subsequent sections.

**Definition 7.2.1.** Assume that function $u$ is not necessarily differentiable, and let $\sigma$ be a positive real number ($\sigma > 0$). The Caputo fractional derivative of order $\sigma$ in the sense of a power-law representation is defined as [26]

$$ {}_0^C D_t^\sigma u(t) = \frac{1}{\Gamma(\sigma)} \times \int_0^t (t-s)^{\sigma-1} u(s) ds. \tag{7.1} $$

**Definition 7.2.2.** Let $u \in H^1(x, s)$, $s > x$, $0 < \sigma \leq 1$. Even though $u$ is not necessarily differentiable, the AB operator in the Riemann–Liouville (RL) sense is formulated as follows [15]:

$$ {}_a^{ABR} D_t^\omega [u(t)] = \frac{B(\sigma)}{1-\sigma} \frac{d}{dt} \int_a^t u(x) E_\sigma \left[ -\sigma \frac{(t-x)^\sigma}{1-\sigma} \right] dx. \tag{7.2} $$

**Definition 7.2.3.** The fractional integral corresponding to the new fractional operator with a nonlocal and nonsingular kernel, known as the AB fractional integral [17,18], is defined as follows:

$$ {}_a^{ABC} I_t^\sigma [u(t)] = \frac{1-\sigma}{B(\sigma)} u(t) + \frac{\sigma}{B(\sigma)\Gamma(\sigma)} \int_a^t u(j)(t-j)^{\sigma-1} dj. \tag{7.3} $$

Recall that whenever the value of parameter $\sigma \to 1$, one recovers the standard integral derivative.

## 7.3 Model formulation of AD with the ABC derivative

We examine a mathematical formulation of AD, as detailed in [19]. This model is composed of multi-component nonlinear PDEs:

$$ \frac{\partial \mathcal{A}_\beta^i}{\partial t} = \left( \lambda_\beta^i (1+R) - d_{\mathcal{A}_\beta^i} \mathcal{A}_\beta^i \right) N/N_0, \tag{7.4} $$

$$\frac{\partial \mathcal{A}_\beta^0}{\partial t} = \mathcal{A}_\beta^i \left|\frac{\partial N}{\partial t}\right| + \frac{\lambda_N N}{N_0} + \frac{\lambda_A A}{\mathcal{A}_0} - \left(d^0_{A_\beta \hat{M}} \left(\hat{\mathcal{M}}_1 + \hat{\mathcal{M}}_2\theta\right) + d^0_{A_\beta M}\left(\mathcal{M}_1 + \hat{\mathcal{M}}_2\theta\right)\right)\frac{A_\beta^0}{A_\beta^0 + \bar{K}_{A_\beta^0}}, \tag{7.5}$$

$$\frac{\partial \mu'}{\partial t} = \left(\lambda_{\mu'0} + \lambda_{\mu'} R - d_{\mu'}\mu'\right) N/N_0, \tag{7.6}$$

$$\frac{\partial \mathcal{F}_i}{\partial t} = \left(\lambda_{F\mu'} + d_{\mathcal{F}_i}\mathcal{F}_i\right) N/N_0, \tag{7.7}$$

$$\frac{\partial \mathcal{F}_0}{\partial t} = \mathcal{F}_i\left|\frac{\partial N}{\partial t}\right| - f_{\mathcal{F}_0}\mathcal{F}_0, \tag{7.8}$$

$$\frac{\partial N}{\partial t} = \frac{-d_{NF}\mathcal{F}_i N}{\mathcal{F}_i + K_{\mathcal{F}_i}} - \frac{d_{NT}\mathcal{T}_\alpha N}{(\mathcal{T}_\alpha + K_{\mathcal{T}_\alpha})(K_{\mathcal{I}_{10}} + \gamma \mathcal{I}_{10})}, \tag{7.9}$$

$$\frac{\partial A}{\partial t} = \lambda_{AA_0^\beta} A_0^\beta + \lambda_{A\mathcal{T}_\alpha}\mathcal{T}_\alpha - d_A A, \tag{7.10}$$

$$\frac{\partial \mathcal{N}_d}{\partial t} = \frac{d_{NF}\mathcal{F}_i N}{\mathcal{F}_i + K_{\mathcal{F}_i}} + \frac{d_{NT}\mathcal{T}_\alpha N K_{\mathcal{I}_{10}}}{(\mathcal{T}_\alpha + K_{\mathcal{T}_\alpha})(K_{\mathcal{I}_{10}} + \gamma \mathcal{I}_{10})} - d_{\mathcal{N}_d M}(\mathcal{M}_1 + \mathcal{M}_2)\frac{\mathcal{N}_d}{\mathcal{N}_d + \bar{K}\mathcal{N}_d} - d_{\mathcal{N}_d \hat{M}}\left(\hat{\mathcal{M}}_1 + \hat{\mathcal{M}}_2\right)\frac{\mathcal{N}_d}{\mathcal{N}_d + \bar{K}\mathcal{N}_d}, \tag{7.11}$$

$$\frac{\partial \mathcal{A}_0}{\partial t} = \lambda_{A_0} A_0^\beta - d_{\mathcal{A}_0}\mathcal{A}_0 + D_{\mathcal{A}_0} \triangle \mathcal{A}_0, \tag{7.12}$$

$$\frac{\partial H}{\partial t} = \lambda_H N_d - d_H H + D_H \triangle H, \tag{7.13}$$

$$\frac{\partial \mathcal{M}_1}{\partial t} = \left(\frac{M_G^0 \lambda_{MF}\mathcal{F}_0}{\mathcal{F}_0 + K_{\mathcal{F}_0}} + \frac{M_G^0 \lambda_{MA}\mathcal{A}_0}{\mathcal{A}_0 + K_{\mathcal{A}_0}}\right)\frac{\beta\epsilon_1}{\beta\epsilon_1 + \epsilon_2} - \nabla\cdot(\mathcal{M}_1 \nabla H) + \frac{\mathcal{T}_\beta \mathcal{M}_1 \lambda_{\mathcal{M}_1\mathcal{T}_\beta}}{\mathcal{T}_\beta + K_{\mathcal{T}_\beta}} - d_{\mathcal{M}_1}\mathcal{M}_1, \tag{7.14}$$

$$\frac{\partial \mathcal{M}_2}{\partial t} = \left(\frac{M_G^0 \lambda_{MF}\mathcal{F}_0}{\mathcal{F}_0 + K_{\mathcal{F}_0}} + \frac{M_G^0 \lambda_{MA}\mathcal{A}_0}{\mathcal{A}_0 + K_{\mathcal{A}_0}}\right)\frac{\beta\epsilon_2}{\beta\epsilon_1 + \epsilon_2} - \nabla\cdot(\mathcal{M}_2 \nabla H) + \frac{\mathcal{T}_\beta \mathcal{M}_2 \lambda_{\mathcal{M}_2\mathcal{T}_\beta}}{\mathcal{T}_\beta + K_{\mathcal{T}_\beta}} - d_{\mathcal{M}_2}\mathcal{M}_2, \tag{7.15}$$

$$\frac{\partial \hat{\mathcal{M}}_1}{\partial t} = \alpha(P)(M_0 - \hat{M})\frac{\beta\epsilon_1}{\beta\epsilon_1 + \epsilon_2} - \nabla\cdot(\hat{\mathcal{M}}_1 \nabla \mathcal{A}_0) + \frac{\mathcal{T}_\beta \hat{\mathcal{M}}_1 \lambda_{\hat{\mathcal{M}}_1\mathcal{T}_\beta}}{\mathcal{T}_\beta + K_{\mathcal{T}_\beta}} d_{\hat{\mathcal{M}}_1}\hat{\mathcal{M}}_1, \tag{7.16}$$

$$\frac{\partial \hat{\mathcal{M}}_2}{\partial t} = \alpha(P)(M_0 - \hat{M})\frac{\beta\epsilon_2}{\beta\epsilon_1 + \epsilon_2} - \nabla\cdot(\hat{\mathcal{M}}_2 \nabla \mathcal{A}_0) + \frac{\mathcal{T}_\beta \hat{\mathcal{M}}_2 \lambda_{\hat{\mathcal{M}}_2\mathcal{T}_\beta}}{\mathcal{T}_\beta + K_{\mathcal{T}_\beta}} d_{\hat{\mathcal{M}}_2}\hat{\mathcal{M}}_2, \tag{7.17}$$

$$\frac{\partial \mathcal{T}_\beta}{\partial t} = \lambda_{\mathcal{T}_\beta M}\mathcal{M}_2 + D_{\mathcal{T}_\beta} \triangle \mathcal{T}_\beta + \lambda_{\mathcal{T}_\beta \hat{M}}\hat{\mathcal{M}}_2 - d_{\mathcal{T}_\beta}\mathcal{T}_\beta, \tag{7.18}$$

$$\frac{\partial \mathcal{I}_{10}}{\partial t} = \lambda_{\mathcal{I}_{10}\mathcal{M}_1}\mathcal{M}_2 + D_{\mathcal{I}_{10}} \triangle \mathcal{T}_\beta + \lambda_{\mathcal{I}_{10}\hat{\mathcal{M}}}\hat{\mathcal{M}}_2 - d_{\mathcal{I}_{10}}\mathcal{I}_{10}, \tag{7.19}$$

$$\frac{\partial \mathcal{T}_\alpha}{\partial t} = \lambda_{\mathcal{T}_\alpha\mathcal{M}_1}\mathcal{M}_1 + D_{\mathcal{T}_\alpha} \triangle \mathcal{T}_\alpha + \lambda_{\mathcal{T}_\alpha\hat{\mathcal{M}}_1}\hat{\mathcal{M}}_1 - d_{\mathcal{T}_\alpha}\mathcal{T}_\alpha, \tag{7.20}$$

$$\frac{\partial P}{\partial t} = \lambda_{PA}A + D_P \triangle P_{\lambda P\mathcal{M}_2}\mathcal{M}_2 - d_p P, \tag{7.21}$$

where

$$\epsilon_1 = \frac{\mathcal{T}_\alpha}{\mathcal{T}_\alpha + K_\alpha},$$
$$\epsilon_2 = \frac{\mathcal{I}_{10}}{\mathcal{I}_{10} + K_{\mathcal{I}_{10}}},$$
$$\hat{M} = \hat{\mathcal{M}}_1 + \hat{\mathcal{M}}_2,$$
$$\alpha(P) = \frac{\alpha P}{P + K_P},$$

and

$$R(t) = \begin{cases} 0.01R_0 t, & 0 \le t \le 100 \\ R_0, & t > 100 \end{cases}. \tag{7.22}$$

In Table 7.1, variables description are given. The proposed AD model is subject to the following conditions:

$$A^0_\beta(0,q,t) = A^0_\beta(1,q,t);\ A^0_\beta(p,0,t) = A^0_\beta(p,1,t);$$
$$h(0,q,t) = h(1,q,t);\ h(p,0,t) = h(p,1,t);$$
$$\mathcal{T}_\beta(0,q,t) = \mathcal{T}_\beta(1,q,t);\ \mathcal{T}_\beta(p,0,t) = \mathcal{T}_\beta(p,1,t);$$
$$\mathcal{I}_{10}(0,q,t) = \mathcal{I}_{10}(1,q,t);\ \mathcal{I}_{10}(p,0,t) = \mathcal{I}_{10}(p,1,t);$$
$$\mathcal{T}_\alpha(0,q,t) = \mathcal{T}_\alpha(1,q,t);\ \mathcal{T}_\alpha(p,0,t) = \mathcal{T}_\alpha(p,1,t);$$
$$P(0,q,t) = P(1,q,t);\ P(p,0,t) = P(p,1,t);$$

with initial conditions:

$$A_\beta(p,q,0) = \frac{1}{10^6};\ A^0_\beta(p,q,0) = \frac{1}{10^8};\ \mu'(p,q,0) = \frac{1.37}{10^{10}};\ \mathcal{F}_0(p,q,0) = \frac{3.36}{10^{11}};$$
$$\mathcal{M}_1(p,q,0) = 0.02;\ N(p,q,0) = 0.14;\ A(p,q,0) = 0.14;\ \mathcal{M}_2(p,q,0) = 0.02;\ \hat{\mathcal{M}}_1 = (p,q,0) = 0;$$
$$\hat{\mathcal{M}}_2(p,q,0) = 0,\ \mathcal{N}_d(p,q,0) = 0;\ H(p,q,0) = \frac{1.3}{10^{11}};\ \mathcal{T}_\beta(p,q,0) = \frac{1}{10^6};\ \mathcal{T}_\alpha(p,q,0) = \frac{2}{10^5};$$
$$\mathcal{I}_{10}(p,q,0) = \frac{1}{10^5};\ P(p,q,0) = \frac{5}{10^9};\ \mathcal{A}_0(p,q,0) = 0. \tag{7.23}$$

**Table 7.1** Model parameters with their descriptions.

| Symbols | Description |
|---|---|
| $A^i_\beta$ | The amyloid-$\beta$ present within neurons |
| $A^0_\beta$ | The amyloid-$\beta$ peptides located outside neurons |
| $\mu'$ | Protein with excessive phosphorylation |
| $NFT(\mathcal{F}_i)$ | The intracellular neuron fibrillary tangle |
| $NFT(\mathcal{F}_0)$ | The tangled fibrils found within neurons |
| $N$ | Live neurons |
| $A$ | Astrocytes |
| $ROS(R)$ | Reactive species of oxygen |
| $APP(A_P)$ | The precursor protein of amyloid |
| $TNf-\alpha(\mathcal{T}_\alpha)$ | Tumor necrosis due to factor alpha |
| $IL-10(I_10)$ | Interleukin 10 |
| $MG(M_G)$ | Microglias |
| $GSK-3G$ | Glycogen synthase (of kinase-type 3) |
| $A\beta0(\mathcal{A}_0)$ | The soluble Amyloid $\beta$ oligomer |
| $TGF-\beta(\mathcal{T}_\beta)$ | The growth factor known as transforming growth factor beta |
| $P$ | MCP-1 |
| $\mathcal{M}_1$ | Proinammatory microglias |
| $\hat{\mathcal{M}}_1$ | Proinflammatory macrophages in the periphery |
| $\mathcal{M}_2$ | Microglia with anti-inflammatory properties |
| $\hat{\mathcal{M}}_2$ | Macrophages with anti-inflammatory properties in the periphery |
| $\mathcal{N}_d$ | Dead neurons |
| $H$ | High mobility or high-mobility-group-box-1 (HMGB1) |

## 7.4 A model integrating fractional calculus and stochastic processes to depict the propagation of Alzheimer's disease

In this study, we transform the previously mentioned deterministic model into a stochastic model to incorporate randomness associated with the spread of infectious diseases. The stochastic model allows us to account for the inherent variability and unpredictability in the transmission dynamics, which is crucial for accurately capturing real-world scenarios. By introducing stochasticity, we aim to enhance the model's ability to simulate and analyze the complex dynamics of disease spread, particularly in heterogeneous populations and under varying conditions. This approach provides a more comprehensive understanding of the epidemic dynamics and can lead to improved strategies for disease control and prevention. To accomplish this, we incorporate a stochastic element into the existing model, as well as introducing the fractional derivative:

$$A_{\beta}^{i}(t) - A_{\beta}^{i}(0) = \frac{1}{\gamma(\sigma)} \int_{0}^{t} G_{1}(A_{\beta}^{i}, \mu)(t-\mu)^{\sigma-1} dB_{1}(\mu)$$
$$+ \frac{1}{\Gamma(\sigma)} \int_{0}^{t} (t-\mu)^{\sigma-1} f_{1}\left(A_{\beta}^{i}, A_{\beta}^{0}, \mu', \mathcal{F}_{i}, \mathcal{F}_{0}, N, A, N_{d}, \mathcal{A}_{0}, M, \mathcal{M}_{1}, \mathcal{M}_{2}, \hat{\mathcal{M}}_{1}, \hat{\mathcal{M}}_{2}, \mathcal{T}_{\beta}, \mathcal{I}_{10}, \mathcal{T}_{\alpha}, P\right) d\mu, \tag{7.24}$$

$$A_{\beta}^{0}(t) - A_{\beta}^{0}(0) = \frac{1}{\gamma(\sigma)} \int_{0}^{t} G_{2}(A_{\beta}^{0}, \mu)(t-\mu)^{\sigma-1} dB_{2}(\mu)$$
$$+ \frac{1}{\Gamma(\sigma)} \int_{0}^{t} (t-\mu)^{\sigma-1} f_{2}\left(A_{\beta}^{i}, A_{\beta}^{0}, \mu', \mathcal{F}_{i}, \mathcal{F}_{0}, N, A, N_{d}, \mathcal{A}_{0}, M, \mathcal{M}_{1}, \mathcal{M}_{2}, \hat{\mathcal{M}}_{1}, \hat{\mathcal{M}}_{2}, \mathcal{T}_{\beta}, \mathcal{I}_{10}, \mathcal{T}_{\alpha}, P\right) d\mu, \tag{7.25}$$

$$\mu'(t) - \mu'(0) = \frac{1}{\gamma(\sigma)} \int_{0}^{t} G_{3}(\mu', \mu)(t-\mu)^{\sigma-1} dB_{3}(\mu)$$
$$+ \frac{1}{\Gamma(\sigma)} \int_{0}^{t} (t-\mu)^{\sigma-1} f_{3}\left(A_{\beta}^{i}, A_{\beta}^{0}, \mu', \mathcal{F}_{i}, \mathcal{F}_{0}, N, A, N_{d}, \mathcal{A}_{0}, M, \mathcal{M}_{1}, \mathcal{M}_{2}, \hat{\mathcal{M}}_{1}, \hat{\mathcal{M}}_{2}, \mathcal{T}_{\beta}, \mathcal{I}_{10}, \mathcal{T}_{\alpha}, P\right) d\mu, \tag{7.26}$$

$$\mathcal{F}_{i}(t) - \mathcal{F}_{i}(0) = \frac{1}{\gamma(\sigma)} \int_{0}^{t} G_{4}(\mathcal{F}_{i}, \mu)(t-\mu)^{\sigma-1} dB_{4}(\mu)$$
$$+ \frac{1}{\Gamma(\sigma)} \int_{0}^{t} (t-\mu)^{\sigma-1} f_{4}\left(A_{\beta}^{i}, A_{\beta}^{0}, \mu', \mathcal{F}_{i}, \mathcal{F}_{0}, N, A, N_{d}, \mathcal{A}_{0}, M, \mathcal{M}_{1}, \mathcal{M}_{2}, \hat{\mathcal{M}}_{1}, \hat{\mathcal{M}}_{2}, \mathcal{T}_{\beta}, \mathcal{I}_{10}, \mathcal{T}_{\alpha}, P\right) d\mu, \tag{7.27}$$

$$\mathcal{F}_{0}(t) - \mathcal{F}_{0}(0) = \frac{1}{\gamma(\sigma)} \int_{0}^{t} G_{5}(\mathcal{F}_{0}, \mu)(t-\mu)^{\sigma-1} dB_{5}(\mu)$$
$$+ \frac{1}{\Gamma(\sigma)} \int_{0}^{t} (t-\mu)^{\sigma-1} f_{5}\left(A_{\beta}^{i}, A_{\beta}^{0}, \mu', \mathcal{F}_{i}, \mathcal{F}_{0}, N, A, N_{d}, \mathcal{A}_{0}, M, \mathcal{M}_{1}, \mathcal{M}_{2}, \hat{\mathcal{M}}_{1}, \hat{\mathcal{M}}_{2}, \mathcal{T}_{\beta}, \mathcal{I}_{10}, \mathcal{T}_{\alpha}, P\right) d\mu, \tag{7.28}$$

$$N(t) - N(0) = \frac{1}{\gamma(\sigma)} \int_{0}^{t} G_{6}(N, \mu)(t-\mu)^{\sigma-1} dB_{6}(\mu)$$
$$+ \frac{1}{\Gamma(\sigma)} \int_{0}^{t} (t-\mu)^{\sigma-1} f_{6}\left(A_{\beta}^{i}, A_{\beta}^{0}, \mu', \mathcal{F}_{i}, \mathcal{F}_{0}, N, A, N_{d}, \mathcal{A}_{0}, M, \mathcal{M}_{1}, \mathcal{M}_{2}, \hat{\mathcal{M}}_{1}, \hat{\mathcal{M}}_{2}, \mathcal{T}_{\beta}, \mathcal{I}_{10}, \mathcal{T}_{\alpha}, P\right) d\mu, \tag{7.29}$$

$$A(t) - A(0) = \frac{1}{\gamma(\sigma)} \int_{0}^{t} G_{7}(A, \mu)(t-\mu)^{\sigma-1} dB_{7}(\mu)$$
$$+ \frac{1}{\Gamma(\sigma)} \int_{0}^{t} (t-\mu)^{\sigma-1} f_{7}\left(A_{\beta}^{i}, A_{\beta}^{0}, \mu', \mathcal{F}_{i}, \mathcal{F}_{0}, N, A, N_{d}, \mathcal{A}_{0}, M, \mathcal{M}_{1}, \mathcal{M}_{2}, \hat{\mathcal{M}}_{1}, \hat{\mathcal{M}}_{2}, \mathcal{T}_{\beta}, \mathcal{I}_{10}, \mathcal{T}_{\alpha}, P\right) d\mu, \tag{7.30}$$

$$N_{d}(t) - N_{d}(0) = \frac{1}{\gamma(\sigma)} \int_{0}^{t} G_{8}(N_{d}, \mu)(t-\mu)^{\sigma-1} dB_{8}(\mu)$$
$$+ \frac{1}{\Gamma(\sigma)} \int_{0}^{t} (t-\mu)^{\sigma-1} f_{8}\left(A_{\beta}^{i}, A_{\beta}^{0}, \mu', \mathcal{F}_{i}, \mathcal{F}_{0}, N, A, N_{d}, \mathcal{A}_{0}, M, \mathcal{M}_{1}, \mathcal{M}_{2}, \hat{\mathcal{M}}_{1}, \hat{\mathcal{M}}_{2}, \mathcal{T}_{\beta}, \mathcal{I}_{10}, \mathcal{T}_{\alpha}, P\right) d\mu, \tag{7.31}$$

$$\mathcal{A}_0(t)-\mathcal{A}_0(0)=\frac{1}{\gamma(\sigma)}\int_0^t G_9(\mathcal{A}_0,\mu)(t-\mu)^{\sigma-1}dB_9(\mu)$$
$$+\frac{1}{\Gamma(\sigma)}\int_0^t(t-\mu)^{\sigma-1}f_9\left(A^i_\beta,A^0_\beta,\mu',\mathcal{F}_i,\mathcal{F}_0,N,A,\mathcal{N}_d,\mathcal{A}_0,M,\mathcal{M}_1,\mathcal{M}_2,\hat{\mathcal{M}}_1,\hat{\mathcal{M}}_2,\mathcal{T}_\beta,\mathcal{I}_{10},\mathcal{T}_\alpha,P\right)d\mu, \tag{7.32}$$

$$H(t)-H(0)=\frac{1}{\gamma(\sigma)}\int_0^t G_{10}(H,\mu)(t-\mu)^{\sigma-1}dB_{10}(\mu)$$
$$+\frac{1}{\Gamma(\sigma)}\int_0^t(t-\mu)^{\sigma-1}f_{10}\left(A^i_\beta,A^0_\beta,\mu',\mathcal{F}_i,\mathcal{F}_0,N,A,\mathcal{N}_d,\mathcal{A}_0,H,\mathcal{M}_1,\mathcal{M}_2,\hat{\mathcal{M}}_1,\hat{\mathcal{M}}_2,\mathcal{T}_\beta,\mathcal{I}_{10},\mathcal{T}_\alpha,P\right)d\mu, \tag{7.33}$$

$$\mathcal{M}_1(t)-\mathcal{M}_1(0)=\frac{1}{\gamma(\sigma)}\int_0^t G_{11}(\mathcal{M}_1,\mu)(t-\mu)^{\sigma-1}dB_{11}(\mu)$$
$$+\frac{1}{\Gamma(\sigma)}\int_0^t(t-\mu)^{\sigma-1}f_{11}\left(A^i_\beta,A^0_\beta,\mu',\mathcal{F}_i,\mathcal{F}_0,N,A,\mathcal{N}_d,\mathcal{A}_0,M,\mathcal{M}_1,\mathcal{M}_2,\hat{\mathcal{M}}_1,\hat{\mathcal{M}}_2,\mathcal{T}_\beta,\mathcal{I}_{10},\mathcal{T}_\alpha,P\right)d\mu, \tag{7.34}$$

$$\mathcal{M}_2(t)-\mathcal{M}_2(0)=\frac{1}{\gamma(\sigma)}\int_0^t G_{12}(\mathcal{M}_2,\mu)(t-\mu)^{\sigma-1}dB_{12}(\mu)$$
$$+\frac{1}{\Gamma(\sigma)}\int_0^t(t-\mu)^{\sigma-1}f_{12}\left(A^i_\beta,A^0_\beta,\mu',\mathcal{F}_i,\mathcal{F}_0,N,A,\mathcal{N}_d,\mathcal{A}_0,M,\mathcal{M}_1,\mathcal{M}_2,\hat{\mathcal{M}}_1,\hat{\mathcal{M}}_2,\mathcal{T}_\beta,\mathcal{I}_{10},\mathcal{T}_\alpha,P\right)d\mu, \tag{7.35}$$

$$\hat{\mathcal{M}}_1(t)-\hat{\mathcal{M}}_1(0)=\frac{1}{\gamma(\sigma)}\int_0^t G_{13}(\hat{\mathcal{M}}_1,\mu)(t-\mu)^{\sigma-1}dB_{13}(\mu)$$
$$+\frac{1}{\Gamma(\sigma)}\int_0^t(t-\mu)^{\sigma-1}f_{13}\left(A^i_\beta,A^0_\beta,\mu',\mathcal{F}_i,\mathcal{F}_0,N,A,\mathcal{N}_d,\mathcal{A}_0,M,\mathcal{M}_1,\mathcal{M}_2,\hat{\mathcal{M}}_1,\hat{\mathcal{M}}_2,\mathcal{T}_\beta,\mathcal{I}_{10},\mathcal{T}_\alpha,P\right)d\mu, \tag{7.36}$$

$$\hat{\mathcal{M}}_2(t)-\hat{\mathcal{M}}_2(0)=\frac{1}{\gamma(\sigma)}\int_0^t G_{14}(\hat{\mathcal{M}}_2,\mu)(t-\mu)^{\sigma-1}dB_{14}(\mu)$$
$$+\frac{1}{\Gamma(\sigma)}\int_0^t(t-\mu)^{\sigma-1}f_{14}\left(A^i_\beta,A^0_\beta,\mu',\mathcal{F}_i,\mathcal{F}_0,N,A,\mathcal{N}_d,\mathcal{A}_0,M,\mathcal{M}_1,\mathcal{M}_2,\hat{\mathcal{M}}_1,\hat{\mathcal{M}}_2,\mathcal{T}_\beta,\mathcal{I}_{10},\mathcal{T}_\alpha,P\right)d\mu, \tag{7.37}$$

$$\mathcal{T}_\beta(t)-\mathcal{T}_\beta(0)=\frac{1}{\gamma(\sigma)}\int_0^t G_{15}(\mathcal{T}_\beta,\mu)(t-\mu)^{\sigma-1}dB_{15}(\mu)$$
$$+\frac{1}{\Gamma(\sigma)}\int_0^t(t-\mu)^{\sigma-1}f_{15}\left(A^i_\beta,A^0_\beta,\mu',\mathcal{F}_i,\mathcal{F}_0,N,A,\mathcal{N}_d,\mathcal{A}_0,M,\mathcal{M}_1,\mathcal{M}_2,\hat{\mathcal{M}}_1,\hat{\mathcal{M}}_2,\mathcal{T}_\beta,\mathcal{I}_{10},\mathcal{T}_\alpha,P\right)d\mu, \tag{7.38}$$

$$\mathcal{I}_{10}(t)-\mathcal{I}_{10}(0)=\frac{1}{\gamma(\sigma)}\int_0^t G_{16}(\mathcal{I}_{10},\mu)(t-\mu)^{\sigma-1}dB_{16}(\mu)$$
$$+\frac{1}{\Gamma(\sigma)}\int_0^t(t-\mu)^{\sigma-1}f_{16}\left(A^i_\beta,A^0_\beta,\mu',\mathcal{F}_i,\mathcal{F}_0,N,A,\mathcal{N}_d,\mathcal{A}_0,M,\mathcal{M}_1,\mathcal{M}_2,\hat{\mathcal{M}}_1,\hat{\mathcal{M}}_2,\mathcal{T}_\beta,\mathcal{I}_{10},\mathcal{T}_\alpha,P\right)d\mu, \tag{7.39}$$

$$\mathcal{T}_\alpha(t)-\mathcal{T}_\alpha(0)=\frac{1}{\gamma(\sigma)}\int_0^t G_{17}(\mathcal{T}_\alpha,\mu)(t-\mu)^{\sigma-1}dB_{17}(\mu)$$
$$+\frac{1}{\Gamma(\sigma)}\int_0^t(t-\mu)^{\sigma-1}f_{17}\left(A^i_\beta,A^0_\beta,\mu',\mathcal{F}_i,\mathcal{F}_0,N,A,N_d,\mathcal{A}_0,M,\mathcal{M}_1,\mathcal{M}_2,\hat{\mathcal{M}}_1,\hat{\mathcal{M}}_2,\mathcal{T}_\beta,\mathcal{I}_{10},\mathcal{T}_\alpha,P\right)d\mu, \tag{7.40}$$

$$P(t)-P(0)=\frac{1}{\gamma(\sigma)}\int_0^t G_{18}(P,\mu)(t-\mu)^{\sigma-1}dB_{18}(\mu)$$
$$+\frac{1}{\Gamma(\sigma)}\int_0^t(t-\mu)^{\sigma-1}f_{18}\left(A^i_\beta,A^0_\beta,\mu',\mathcal{F}_i,\mathcal{F}_0,N,A,N_d,\mathcal{A}_0,M,\mathcal{M}_1,\mathcal{M}_2,\hat{\mathcal{M}}_1,\hat{\mathcal{M}}_2,\mathcal{T}_\beta,\mathcal{I}_{10},\mathcal{T}_\alpha,P\right)d\mu. \tag{7.41}$$

## 7.5 Numerical schemes

This section introduces numerical solutions utilizing the Newton polynomial scheme. We begin by applying classical differentiation. We make the assumption that the stochastic functions are differentiable, that is,

$$dB_i'(t)=B_i'(t)dt. \tag{7.42}$$

### 7.5.1 The ABC operator

Now, we explore the scenario where the differential operator is the ABC derivative:

$$\begin{aligned}\mathcal{A}^i_\beta(t)-\mathcal{A}^i_\beta(0)&=\frac{1-\sigma}{Q}f_1(\mathcal{A}^i_\beta,z,t)+\frac{\sigma}{Q\Gamma(\sigma)}\int_0^t f_1(\mathcal{A}^i_\beta,z,\mu)(t-\mu)^{\sigma-1}d\mu\\&+\frac{1-\sigma}{Q}G_1(\mathcal{A}^i_\beta,t)B_1'(t)+\frac{\sigma}{Q\Gamma(\sigma)}\int_0^t G_1(\mathcal{A}^i_\beta,\mu)B_1'(\mathcal{A}^i_\beta,\mu)(t-\mu)^{\sigma-1}d\mu;\end{aligned} \tag{7.43}$$

$$\begin{aligned}A^0_\beta(t)-\mathcal{A}^0_\beta(0)&=\frac{1-\sigma}{Q}f_2(\mathcal{A}^0_\beta,\mathcal{A}^i_\beta,t)+\frac{\sigma}{Q\Gamma(\sigma)}\int_0^t f_2(\mathcal{A}^0_\beta,\mathcal{A}^i_\beta,\mu)(t-\mu)^{\sigma-1}d\mu\\&+\frac{1-\sigma}{Q}G_2(\mathcal{A}^0_\beta,t)B_2'(t)+\frac{\sigma}{Q\Gamma(\sigma)}\int_0^t G_2(\mathcal{A}^0_\beta,\mu)B_2'(\mathcal{A}^0_\beta,\mu)(t-\mu)^{\sigma-1}d\mu;\end{aligned} \tag{7.44}$$

$$\begin{aligned}\mu'(t)-\mu'(0)&=\frac{1-\sigma}{Q}f_3(\mu',\mathcal{A}^0_\beta,t)+\frac{\sigma}{Q\Gamma(\sigma)}\int_0^t f_3(\mu',\mathcal{A}^0_\beta,\mu)(t-\mu)^{\sigma-1}d\mu\\&+\frac{1-\sigma}{Q}G_3(\mu',\mathcal{A}^0_\beta,t)B_3'(t)+\frac{\sigma}{Q\Gamma(\sigma)}\int_0^t G_3(\mu',\mu)B_3'(\mu',\mu)(t-\mu)^{\sigma-1}d\mu;\end{aligned} \tag{7.45}$$

$$\begin{aligned}\mathcal{F}_i(t)-\mathcal{F}_i(0)&=\frac{1-\sigma}{Q}f_4(\mathcal{F}_i,\mu',t)+\frac{\sigma}{Q\Gamma(\sigma)}\int_0^t f_4(\mathcal{F}_i,\mu',\mu)(t-\mu)^{\sigma-1}d\mu\\&+\frac{1-\sigma}{Q}G_4(\mathcal{F}_i,\mu',t)B_4'(t)+\frac{\sigma}{Q\Gamma(\sigma)}\int_0^t G_4(\mathcal{F}_i,\mu)B_4'(\mathcal{F}_i,\mu)(t-\mu)^{\sigma-1}d\mu;\end{aligned} \tag{7.46}$$

$$\begin{aligned}\mathcal{F}_0(t)-\mathcal{F}_0(0)&=\frac{1-\sigma}{Q}f_5(\mathcal{F}_0,\mathcal{F}_i,t)+\frac{\sigma}{Q\Gamma(\sigma)}\int_0^t f_5(\mathcal{F}_0,\mathcal{F}_i,\mu)(t-\mu)^{\sigma-1}d\mu\\&+\frac{1-\sigma}{Q}G_5(\mathcal{F}_0,\mathcal{F}_i,t)B_5'(t)+\frac{\sigma}{Q\Gamma(\sigma)}\int_0^t G_5(\mathcal{F}_0,\mu)B_5'(\mathcal{F}_0,\mu)(t-\mu)^{\sigma-1}d\mu;\end{aligned}\tag{7.47}$$

$$\begin{aligned}N(t)-N(0)&=\frac{1-\sigma}{Q}f_6(N,\mathcal{F}_0,t)+\frac{\sigma}{Q\Gamma(\sigma)}\int_0^t f_6(N,\mathcal{F}_0,\mu)(t-\mu)^{\sigma-1}d\mu\\&+\frac{1-\sigma}{Q}G_6(N,\mathcal{F}_0,t)B_6'(t)+\frac{\sigma}{Q\Gamma(\sigma)}\int_0^t G_6(N,\mu)B_6'(N,\mu)(t-\mu)^{\sigma-1}d\mu;\end{aligned}\tag{7.48}$$

$$\begin{aligned}A(t)-A(0)&=\frac{1-\sigma}{Q}f_7(A,N,t)+\frac{\sigma}{Q\Gamma(\sigma)}\int_0^t f_7(A,N,\mu)(t-\mu)^{\sigma-1}d\mu\\&+\frac{1-\sigma}{Q}G_7(A,N,t)B_7'(t)+\frac{\sigma}{Q\Gamma(\sigma)}\int_0^t G_7(A,\mu)B_7'(A,\mu)(t-\mu)^{\sigma-1}d\mu;\end{aligned}\tag{7.49}$$

$$\begin{aligned}\mathcal{N}_d(t)-\mathcal{N}_d(0)&=\frac{1-\sigma}{Q}f_8(\mathcal{N}_d,A,t)+\frac{\sigma}{Q\Gamma(\sigma)}\int_0^t f_8(\mathcal{N}_d,A,\mu)(t-\mu)^{\sigma-1}d\mu\\&+\frac{1-\sigma}{Q}G_8(\mathcal{N}_d,A,t)B_8'(t)+\frac{\sigma}{Q\Gamma(\sigma)}\int_0^t G_8(\mathcal{N}_d,\mu)B_8'(\mathcal{N}_d,\mu)(t-\mu)^{\sigma-1}d\mu;\end{aligned}\tag{7.50}$$

$$\begin{aligned}\mathcal{A}_0(t)-\mathcal{A}_0(0)&=\frac{1-\sigma}{Q}f_9(\mathcal{A}_0,\mathcal{N}_d,t)+\frac{\sigma}{Q\Gamma(\sigma)}\int_0^t f_9(\mathcal{A}_0,\mathcal{N}_d,\mu)(t-\mu)^{\sigma-1}d\mu\\&+\frac{1-\sigma}{Q}G_9(\mathcal{A}_0,\mathcal{N}_d,t)B_9'(t)+\frac{\sigma}{Q\Gamma(\sigma)}\int_0^t G_9(\mathcal{A}_0,\mu)B_9'(\mathcal{A}_0,\mu)(t-\mu)^{\sigma-1}d\mu;\end{aligned}\tag{7.51}$$

$$\begin{aligned}H(t)-H(0)&=\frac{1-\sigma}{Q}f_{10}(H,\mathcal{A}_0,t)+\frac{\sigma}{Q\Gamma(\sigma)}\int_0^t f_{10}(H,\mathcal{A}_0,\mu)(t-\mu)^{\sigma-1}d\mu\\&+\frac{1-\sigma}{Q}G_{10}(H,\mathcal{A}_0,t)B_{10}'(t)+\frac{\sigma}{Q\Gamma(\sigma)}\int_0^t G_{10}(H,\mu)B_{10}'(H,\mu)(t-\mu)^{\sigma-1}d\mu;\end{aligned}\tag{7.52}$$

$$\begin{aligned}\mathcal{M}_1(t)-\mathcal{M}_1(0)&=\frac{1-\sigma}{Q}f_{11}(\mathcal{M}_1,H,t)+\frac{\sigma}{Q\Gamma(\sigma)}\int_0^t f_{11}(\mathcal{M}_1,H,\mu)(t-\mu)^{\sigma-1}d\mu\\&+\frac{1-\sigma}{Q}G_{11}(\mathcal{M}_1,H,t)B_{11}'(t)+\frac{\sigma}{Q\Gamma(\sigma)}\int_0^t G_{11}(\mathcal{M}_1,\mu)B_{11}'(\mathcal{M}_1,\mu)(t-\mu)^{\sigma-1}d\mu;\end{aligned}\tag{7.53}$$

$$\begin{aligned}\mathcal{M}_2(t)-\mathcal{M}_2(0)&=\frac{1-\sigma}{Q}f_{12}(\mathcal{M}_2,\mathcal{M}_1,t)+\frac{\sigma}{Q\Gamma(\sigma)}\int_0^t f_{12}(\mathcal{M}_2,\mathcal{M}_1,\mu)(t-\mu)^{\sigma-1}d\mu\\&+\frac{1-\sigma}{Q}G_{12}(\mathcal{M}_2,\mathcal{M}_1,t)B_{12}'(t)+\frac{\sigma}{Q\Gamma(\sigma)}\int_0^t G_{12}(\mathcal{M}_2,\mu)B_{12}'(\mathcal{M}_2,\mu)(t-\mu)^{\sigma-1}d\mu;\end{aligned}\tag{7.54}$$

$$\begin{aligned}\hat{\mathcal{M}}_1(t)-\hat{\mathcal{M}}_1(0)&=\frac{1-\sigma}{Q}f_{13}(\hat{\mathcal{M}}_1,\mathcal{M}_2,t)+\frac{\sigma}{Q\Gamma(\sigma)}\int_0^t f_{13}(\hat{\mathcal{M}}_1,\mathcal{M}_2,\mu)(t-\mu)^{\sigma-1}d\mu\\&+\frac{1-\sigma}{Q}G_{13}(\hat{\mathcal{M}}_1,\mathcal{M}_2,t)B'_{13}(t)+\frac{\sigma}{Q\Gamma(\sigma)}\int_0^t G_{13}(\hat{\mathcal{M}}_1,\mu)B'_{13}(\hat{\mathcal{M}}_1,\mu)(t-\mu)^{\sigma-1}d\mu;\end{aligned}\tag{7.55}$$

$$\begin{aligned}\hat{\mathcal{M}}_2(t)-\hat{\mathcal{M}}_2(0)&=\frac{1-\sigma}{Q}f_{14}(\hat{\mathcal{M}}_2,\hat{\mathcal{M}}_1,t)+\frac{\sigma}{Q\Gamma(\sigma)}\int_0^t f_{14}(\hat{\mathcal{M}}_2,\hat{\mathcal{M}}_1,\mu)(t-\mu)^{\sigma-1}d\mu\\&+\frac{1-\sigma}{Q}G_{14}(\hat{\mathcal{M}}_2,\hat{\mathcal{M}}_1,t)B'_{14}(t)+\frac{\sigma}{Q\Gamma(\sigma)}\int_0^t G_{14}(\hat{\mathcal{M}}_2,\mu)B'_{14}(\hat{\mathcal{M}}_2,\mu)(t-\mu)^{\sigma-1}d\mu;\end{aligned}\tag{7.56}$$

$$\begin{aligned}\mathcal{T}_\beta(t)-\mathcal{T}_\beta(0)&=\frac{1-\sigma}{Q}f_{15}(\mathcal{T}_\beta,\hat{\mathcal{M}}_1,t)+\frac{\sigma}{Q\Gamma(\sigma)}\int_0^t f_{15}(\mathcal{T}_\beta,\hat{\mathcal{M}}_1,\mu)(t-\mu)^{\sigma-1}d\mu\\&+\frac{1-\sigma}{Q}G_{15}(\mathcal{T}_\beta,\hat{\mathcal{M}}_1,t)B'_{15}(t)+\frac{\sigma}{Q\Gamma(\sigma)}\int_0^t G_{15}(\mathcal{T}_\beta,\mu)B'_{15}(\mathcal{T}_\beta,\mu)(t-\mu)^{\sigma-1}d\mu;\end{aligned}\tag{7.57}$$

$$\begin{aligned}\mathcal{I}_{10}(t)-\mathcal{I}_{10}(0)&=\frac{1-\sigma}{Q}f_{16}(\mathcal{I}_{10},\mathcal{T}_\beta,t)+\frac{\sigma}{Q\Gamma(\sigma)}\int_0^t f_{16}(\mathcal{I}_{10},\mathcal{T}_\beta,\mu)(t-\mu)^{\sigma-1}d\mu\\&+\frac{1-\sigma}{Q}G_{16}(\mathcal{I}_{10},\mathcal{T}_\beta,t)B'_{16}(t)+\frac{\sigma}{Q\Gamma(\sigma)}\int_0^t G_{16}(\mathcal{I}_{10},\mu)B'_{16}(\mathcal{I}_{10},\mu)(t-\mu)^{\sigma-1}d\mu;\end{aligned}\tag{7.58}$$

$$\begin{aligned}\mathcal{T}_\alpha(t)-\mathcal{T}_\alpha(0)&=\frac{1-\sigma}{Q}f_{17}(\mathcal{T}_\alpha,\mathcal{I}_{10},t)+\frac{\sigma}{Q\Gamma(\sigma)}\int_0^t f_{17}(\mathcal{T}_\alpha,\mathcal{I}_{10},\mu)(t-\mu)^{\sigma-1}d\mu\\&+\frac{1-\sigma}{Q}G_{17}(\mathcal{T}_\alpha,\mathcal{I}_{10},t)B'_{17}(t)+\frac{\sigma}{Q\Gamma(\sigma)}\int_0^t G_{17}(\mathcal{T}_\alpha,\mu)B'_{17}(\mathcal{T}_\alpha,\mu)(t-\mu)^{\sigma-1}d\mu;\end{aligned}\tag{7.59}$$

$$\begin{aligned}P(t)-P(0)&=\frac{1-\sigma}{Q}f_{18}(P,\mathcal{I}_{10},t)+\frac{\sigma}{Q\Gamma(\sigma)}\int_0^t f_{18}(P,\mathcal{I}_{10},\mu)(t-\mu)^{\sigma-1}d\mu\\&+\frac{1-\sigma}{Q}G_{18}(P,\mathcal{I}_{10},t)B'_{18}(t)+\frac{\sigma}{Q\Gamma(\sigma)}\int_0^t G_{18}(P,\mu)B'_{18}(P,\mu)(t-\mu)^{\sigma-1}d\mu,\end{aligned}\tag{7.60}$$

where $Q = AB(\sigma)$. Here the function $f_1$ and $G_1$ are approximated via the Lagrange polynomials, as suggested in [18], to get

$$\begin{aligned}\mathcal{A}^i_\beta(t_{k+1})&=\mathcal{A}^i_\beta(0)+\frac{1-\sigma}{Q}f_1\left(\mathcal{A}^i_\beta(t_k),z(t_k),t_k\right)\\&+\frac{\sigma(\Delta t)^\sigma}{Q\Gamma(\sigma+1)}\sum_{a=2}^{k}f_1\left(\mathcal{A}^i_{\beta a-2},z_{a-2},t_{a-2}\right)[(k-a+1)^\sigma-(k-a)^\sigma]\\&+\frac{\sigma(\Delta t)^\sigma}{Q\Gamma(\sigma+2)}\\&\times\sum_{a=2}^{k}\left(f_1(\mathcal{A}^i_{\beta a-1},z_{a-1},t_{a-1})-f_1(\mathcal{A}^i_{\beta a-2},z_{a-2},t_{a-2})\right)\end{aligned}$$

$$\times\left[\begin{array}{r}(k-a+1)^{\sigma}(k-a+3+2\sigma)\\-(k-a)^{\sigma}(k-a+3+3\sigma)\end{array}\right]+\frac{\sigma(\Delta t)^{\sigma}}{2Q\Gamma(\sigma+3)}$$

$$\times\sum_{a=2}^{k}\left[f_1(\mathcal{A}_{\beta a}^{i},z_a,t_a)-2f_1(\mathcal{A}_{\beta a-1}^{i},z_{a-1},t_{a-1})+f_1\left(\mathcal{A}_{\beta a-2}^{i},z_{a-2},t_{a-2}\right)\right]$$

$$\times\left[\begin{array}{c}(k-a+1)^{\sigma}\left[\begin{array}{r}2(k-a)^2+(3\sigma+10)(k-a)\\+2\sigma^2+9\sigma+12\end{array}\right]\\-(k-a)^{\sigma}\left[\begin{array}{r}2(k-a)^2+(5\sigma+10)(k-a)\\6\sigma^2+18\sigma+12\end{array}\right]\end{array}\right]$$

$$+\frac{1-\sigma}{Q}G_1\left(\mathcal{A}_{\beta}^{i}(t_k),z(t_k),t_k\right)\left\{B_1'(t_{k+1})-B_1'(t_k)\right\}$$

$$+\frac{\sigma(\Delta t)^{\sigma-1}}{Q\Gamma(\sigma+1)}\sum_{a=2}^{k}G_1\left(\mathcal{A}_{\beta a-2}^{i},z_{a-2},t_{a-2}\right)\left\{B_1'(t_{a-1})-B_1'(t_{a-2})\right\}\left[(k-a+1)^{\sigma}-(k-a)^{\sigma}\right]$$

$$+\frac{\sigma(\Delta t)^{\sigma-1}}{Q\Gamma(\sigma+2)}$$

$$\times\sum_{a=2}^{k}\Big(G_1(\mathcal{A}_{\beta a-1}^{i},z_{a-1},t_{a-1})\left\{B_1'(t_a)-B_1'(t_{a-1})\right\}-G_1(\mathcal{A}_{\beta a-2}^{i},z_{a-2},t_{a-2})\Big)$$

$$\times\left\{B_1'(t_{a-1})-B_1'(t_{a-2})\right\}$$

$$\times\left[\begin{array}{r}(k-a+1)^{\sigma}(k-a+3+2\sigma)\\-(k-a)^{\sigma}(k-a+3+3\sigma)\end{array}\right]+\frac{\sigma(\Delta t)^{\sigma-1}}{2Q\Gamma(\sigma+3)}$$

$$\times\sum_{a=2}^{k}\Big[G_1(\mathcal{A}_{\beta a}^{i},z_a,t_a)\left\{B_1'(t_{a+1})-B_1'(t_a)\right\}-2G_1(\mathcal{A}_{\beta a-1}^{i},z_{a-1},t_{a-1})\left\{B_1'(t_a)-B_1'(t_{a-1})\right\}$$

$$+G_1\left(\mathcal{A}_{\beta a-2}^{i},z_{a-2},t_{a-2}\right)\left\{B_1'(t_{a-1})-B_1'(t_{a-2})\right\}\Big]$$

$$\times\left[\begin{array}{c}(k-a+1)^{\sigma}\left[\begin{array}{r}2(k-a)^2+(3\sigma+10)(k-a)\\+2\sigma^2+9\sigma+12\end{array}\right]\\-(k-a)^{\sigma}\left[\begin{array}{r}2(k-a)^2+(5\sigma+10)(k-a)\\6\sigma^2+18\sigma+12\end{array}\right]\end{array}\right];\tag{7.61}$$

$$\mathcal{A}_{\beta}^{0}(t_{k+1})=\mathcal{A}_{\beta}^{0}(0)+\frac{1-\sigma}{Q}f_2\left(\mathcal{A}_{\beta}^{0}(t_k),\mathcal{A}_{\beta}^{i}(t_k),t_k\right)$$

$$+\frac{\sigma(\Delta t)^{\sigma}}{Q\Gamma(\sigma+1)}\sum_{a=2}^{k}f_2\left(\mathcal{A}_{\beta a-2}^{0},\mathcal{A}_{\beta a-2}^{i},t_{a-2}\right)\left[(k-a+1)^{\sigma}-(k-a)^{\sigma}\right]$$

$$+\frac{\sigma(\Delta t)^{\sigma}}{Q\Gamma(\sigma+2)}$$

$$\times\sum_{a=2}^{k}\left(f_2(\mathcal{A}_{\beta a-1}^{0},\mathcal{A}_{\beta a-1}^{i},t_{a-1})-f_2(\mathcal{A}_{\beta a-2}^{0},\mathcal{A}_{\beta a-2}^{i},t_{a-2})\right)$$

$$
\times\begin{bmatrix} (k-a+1)^{\sigma}(k-a+3+2\sigma) \\ -(k-a)^{\sigma}(k-a+3+3\sigma) \end{bmatrix}+\frac{\sigma(\Delta t)^{\sigma}}{2Q\Gamma(\sigma+3)}
$$

$$
\times\sum_{a=2}^{k}\left[f_2(\mathcal{A}^0_{\beta a},\mathcal{A}^i_{\beta a},t_a)-2f_2(\mathcal{A}^0_{\beta a-1},\mathcal{A}^i_{\beta a-1},t_{a-1})+f_2\left(\mathcal{A}^0_{\beta a-2},\mathcal{A}^i_{\beta a-2},t_{a-2}\right)\right]
$$

$$
\times\begin{bmatrix} (k-a+1)^{\sigma}\begin{bmatrix} 2(k-a)^2+(3\sigma+10)(k-a) \\ +2\sigma^2+9\sigma+12 \end{bmatrix} \\ -(k-a)^{\sigma}\begin{bmatrix} 2(k-a)^2+(5\sigma+10)(k-a) \\ 6\sigma^2+18\sigma+12 \end{bmatrix} \end{bmatrix}
$$

$$
+\frac{1-\sigma}{Q}G_2\left(\mathcal{A}^0_{\beta}(t_k),\mathcal{A}^i_{\beta}(t_k),t_k\right)\left\{B_2'(t_{k+1})-B_2'(t_k)\right\}
$$

$$
+\frac{\sigma(\Delta t)^{\sigma-1}}{Q\Gamma(\sigma+1)}\sum_{a=2}^{k}G_2\left(\mathcal{A}^0_{\beta a-2},\mathcal{A}^i_{\beta a-2},t_{a-2}\right)\left\{B_2'(t_{a-1})-B_2'(t_{a-2})\right\}\left[(k-a+1)^{\sigma}-(k-a)^{\sigma}\right]
$$

$$
+\frac{\sigma(\Delta t)^{\sigma-1}}{Q\Gamma(\sigma+2)}
$$

$$
\times\sum_{a=2}^{k}\left(G_2(\mathcal{A}^0_{\beta a-1},\mathcal{A}^i_{\beta a-1},t_{a-1})\left\{B_2'(t_a)-B_2'(t_{a-1})\right\}-G_2(\mathcal{A}^0_{\beta a-2},\mathcal{A}^i_{\beta a-2},t_{a-2})\right)
$$

$$
\times\left\{B_2'(t_{a-1})-B_2'(t_{a-2})\right\}
$$

$$
\times\begin{bmatrix} (k-a+1)^{\sigma}(k-a+3+2\sigma) \\ -(k-a)^{\sigma}(k-a+3+3\sigma) \end{bmatrix}+\frac{\sigma(\Delta t)^{\sigma-1}}{2Q\Gamma(\sigma+3)}
$$

$$
\times\sum_{a=2}^{k}\left[G_2(\mathcal{A}^0_{\beta a},\mathcal{A}^i_{\beta a},t_a)\left\{B_2'(t_{a+1})-B_2'(t_a)\right\}-2G_2(\mathcal{A}^0_{\beta a-1},\mathcal{A}^i_{\beta a-1},t_{a-1})\left\{B_2'(t_a)-B_2'(t_{a-1})\right\}\right.
$$

$$
\left.+G_2\left(\mathcal{A}^0_{\beta a-2},\mathcal{A}^i_{\beta a-2},t_{a-2}\right)\left\{B_2'(t_{a-1})-B_2'(t_{a-2})\right\}\right]
$$

$$
\times\begin{bmatrix} (k-a+1)^{\sigma}\begin{bmatrix} 2(k-a)^2+(3\sigma+10)(k-a) \\ +2\sigma^2+9\sigma+12 \end{bmatrix} \\ -(k-a)^{\sigma}\begin{bmatrix} 2(k-a)^2+(5\sigma+10)(k-a) \\ 6\sigma^2+18\sigma+12 \end{bmatrix} \end{bmatrix}; \tag{7.62}
$$

$$
\mu'(t_{k+1})=\mu'(0)+\frac{1-\sigma}{Q}f_3\left(\mu'(t_k),\mathcal{A}^0_{\beta}(t_k),t_k\right)
$$

$$
+\frac{\sigma(\Delta t)^{\sigma}}{Q\Gamma(\sigma+1)}\sum_{a=2}^{k}f_3\left(\mu'_{a-2},\mathcal{A}^0_{\beta a-2},t_{a-2}\right)\left[(k-a+1)^{\sigma}-(k-a)^{\sigma}\right]
$$

$$
+\frac{\sigma(\Delta t)^{\sigma}}{Q\Gamma(\sigma+2)}
$$

$$
\times\sum_{a=2}^{k}\left(f_3(\mu'_{a-1},\mathcal{A}^0_{\beta a-1},t_{a-1})-f_3(\mu'_{a-2},\mathcal{A}^0_{\beta a-2},t_{a-2})\right)
$$

$$\times\left[\begin{array}{r}(k-a+1)^{\sigma}(k-a+3+2\sigma)\\-(k-a)^{\sigma}(k-a+3+3\sigma)\end{array}\right]+\frac{\sigma(\Delta t)^{\sigma}}{2Q\Gamma(\sigma+3)}$$

$$\times\sum_{a=2}^{k}\left[f_3(\mu'_a,\mathcal{A}^0_{\beta a},t_a)-2f_3(\mu'_{a-1},\mathcal{A}^0_{\beta a-1},t_{a-1})+f_3\left(\mu'_{a-2},\mathcal{A}^0_{\beta a-2},t_{a-2}\right)\right]$$

$$\times\left[\begin{array}{c}(k-a+1)^{\sigma}\left[\begin{array}{r}2(k-a)^2+(3\sigma+10)(n-a)\\+2\sigma^2+9\sigma+12\end{array}\right]\\-(k-a)^{\sigma}\left[\begin{array}{r}2(k-a)^2+(5\sigma+10)(k-a)\\6\sigma^2+18\sigma+12\end{array}\right]\end{array}\right]$$

$$+\frac{1-\sigma}{Q}G_3\left(\mu'(t_k),\mathcal{A}^0_{\beta}(t_k),t_k\right)\left\{B'_3(t_{k+1})-B'_3(t_k)\right\}$$

$$+\frac{\sigma(\Delta t)^{\sigma-1}}{Q\Gamma(\sigma+1)}\sum_{a=2}^{k}G_3\left(\mu'_{a-2},\mathcal{A}^0_{\beta a-2},t_{a-2}\right)\left\{B'_3(t_{a-1})-B'_3(t_{a-2})\right\}\left[(k-a+1)^{\sigma}-(k-a)^{\sigma}\right]$$

$$+\frac{\sigma(\Delta t)^{\sigma-1}}{Q\Gamma(\sigma+2)}$$

$$\times\sum_{a=2}^{k}\left(G_3(\mu'_{a-1},\mathcal{A}^0_{\beta a-1},t_{a-1})\left\{B'_3(t_a)-B'_3(t_{a-1})\right\}-G_3(\mu'_{a-2},\mathcal{A}^0_{\beta a-2},t_{a-2})\right)$$

$$\times\left\{B'_3(t_{a-1})-B'_3(t_{a-2})\right\}$$

$$\times\left[\begin{array}{r}(k-a+1)^{\sigma}(k-a+3+2\sigma)\\-(k-a)^{\sigma}(k-a+3+3\sigma)\end{array}\right]+\frac{\sigma(\Delta t)^{\sigma-1}}{2Q\Gamma(\sigma+3)}$$

$$\times\sum_{a=2}^{k}\left[G_3(\mu'_a,\mathcal{A}^0_{\beta a},t_a)\left\{B'_3(t_{a+1})-B'_3(t_a)\right\}-2G_3(\mu'_{a-1},\mathcal{A}^0_{\beta a-1},t_{a-1})\left\{B'_3(t_a)-B'_3(t_{a-1})\right\}\right.$$

$$\left.+G_3\left(\mu'_{a-2},\mathcal{A}^0_{\beta a-2},t_{a-2}\right)\left\{B'_3(t_{a-1})-B'_3(t_{a-2})\right\}\right]$$

$$\times\left[\begin{array}{c}(k-a+1)^{\sigma}\left[\begin{array}{r}2(k-a)^2+(3\sigma+10)(k-a)\\+2\sigma^2+9\sigma+12\end{array}\right]\\-(k-a)^{\sigma}\left[\begin{array}{r}2(k-a)^2+(5\sigma+10)(k-a)\\6\sigma^2+18\sigma+12\end{array}\right]\end{array}\right];\tag{7.63}$$

$$\mathcal{F}_i(t_{k+1})=\mathcal{F}_i(0)+\frac{1-\sigma}{Q}f_4\left(\mathcal{F}_i(t_k),\mu'(t_k),t_k\right)$$

$$+\frac{\sigma(\Delta t)^{\sigma}}{Q\Gamma(\sigma+1)}\sum_{a=2}^{k}f_4\left(F_{i(a-2)},\mu'_{a-2},t_{a-2}\right)\left[(k-a+1)^{\sigma}-(k-a)^{\sigma}\right]$$

$$+\frac{\sigma(\Delta t)^{\sigma}}{Q\Gamma(\sigma+2)}$$

$$\times\sum_{a=2}^{k}\left(f_4(F_{i(a-1)},\mu'_{a-1},t_{a-1})-f_4(F_{i(a-2)},\mu'_{a-2},t_{a-2})\right)$$

$$\times\begin{bmatrix} (k-a+1)^{\sigma}(k-a+3+2\sigma) \\ -(k-a)^{\sigma}(k-a+3+3\sigma) \end{bmatrix}+\frac{\sigma(\Delta t)^{\sigma}}{2Q\Gamma(\sigma+3)}$$

$$\times\sum_{a=2}^{k}\left[f_4(F_{ia},\mu'_a,t_a)-2f_4(F_{i(a-1)},\mu'_{a-1},t_{a-1})+f_4\left(F_{i(a-2)},\mu'_{a-2},t_{a-2}\right)\right]$$

$$\times\begin{bmatrix} (k-a+1)^{\sigma}\begin{bmatrix} 2(k-a)^2+(3\sigma+10)(k-a) \\ +2\sigma^2+9\sigma+12 \end{bmatrix} \\ -(k-a)^{\sigma}\begin{bmatrix} 2(k-a)^2+(5\sigma+10)(k-a) \\ 6\sigma^2+18\sigma+12 \end{bmatrix} \end{bmatrix}$$

$$+\frac{1-\sigma}{Q}G_4\left(\mathcal{F}_i(t_k),\mu'(t_k),t_k\right)\left\{B'_4(t_{k+1})-B'_4(t_k)\right\}$$

$$+\frac{\sigma(\Delta t)^{\sigma-1}}{Q\Gamma(\sigma+1)}\sum_{a=2}^{k}G_4\left(F_{i(a-2)},\mu'_{a-2},t_{a-2}\right)\left\{B'_4(t_{a-1})-B'_4(t_{a-2})\right\}\left[(k-a+1)^{\sigma}-(k-a)^{\sigma}\right]$$

$$+\frac{\sigma(\Delta t)^{\sigma-1}}{Q\Gamma(\sigma+2)}$$

$$\times\sum_{a=2}^{k}\left(G_4(F_{i(a-1)},\mu'_{a-1},t_{a-1})\left\{B'_4(t_a)-B'_4(t_{a-1})\right\}-G_4(F_{i(a-2)},\mu'_{a-2},t_{a-2})\right)$$

$$\times\left\{B'_4(t_{a-1})-B'_4(t_{a-2})\right\}$$

$$\times\begin{bmatrix} (k-a+1)^{\sigma}(k-a+3+2\sigma) \\ -(k-a)^{\sigma}(k-a+3+3\sigma) \end{bmatrix}+\frac{\sigma(\Delta t)^{\sigma-1}}{2Q\Gamma(\sigma+3)}$$

$$\times\sum_{a=2}^{k}\left[G_4(F_{ia},\mu'_a,t_a)\left\{B'_4(t_{a+1})-B'_4(t_a)\right\}-2G_4(F_{i(a-1)},\mu'_{a-1},t_{a-1})\left\{B'_4(t_a)-B'_4(t_{a-1})\right\}\right.$$

$$\left.+G_4\left(F_{i(a-2)},\mu'_{a-2},t_{a-2}\right)\left\{B'_4(t_{a-1})-B'_4(t_{a-2})\right\}\right]$$

$$\times\begin{bmatrix} (n-a+1)^{\sigma}\begin{bmatrix} 2(k-a)^2+(3\sigma+10)(k-a) \\ +2\sigma^2+9\sigma+12 \end{bmatrix} \\ -(k-a)^{\sigma}\begin{bmatrix} 2(k-a)^2+(5\sigma+10)(k-a) \\ 6\sigma^2+18\sigma+12 \end{bmatrix} \end{bmatrix}; \tag{7.64}$$

$$\mathcal{F}_0(t_{k+1})=\mathcal{F}_0(0)+\frac{1-\sigma}{Q}f_5\left(\mathcal{F}_0(t_k),\mathcal{F}_i(t_k),t_k\right)$$

$$+\frac{\sigma(\Delta t)^{\sigma}}{Q\Gamma(\sigma+1)}\sum_{a=2}^{k}f_5\left(\mathcal{F}_{0a-2},F_{i(a-2)},t_{a-2}\right)\left[(k-a+1)^{\sigma}-(k-a)^{\sigma}\right]$$

$$+\frac{\sigma(\Delta t)^{\sigma}}{Q\Gamma(\sigma+2)}$$

$$\times\sum_{a=2}^{k}\left(f_5(\mathcal{F}_{0a-1},F_{i(a-1)},t_{a-1})-f_5(\mathcal{F}_{0a-2},F_{i(a-2)},t_{a-2})\right)$$

$$\times\left[\begin{array}{r}(k-a+1)^{\sigma}(k-a+3+2\sigma)\\-(k-a)^{\sigma}(k-a+3+3\sigma)\end{array}\right]+\frac{\sigma(\Delta t)^{\sigma}}{2Q\Gamma(\sigma+3)}$$

$$\times\sum_{a=2}^{k}\left[f_5(\mathcal{F}_{0a},F_{ia},t_a)-2f_5(\mathcal{F}_{0a-1},F_{i(a-1)},t_{a-1})+f_5\left(\mathcal{F}_{0a-2},F_{i(a-2)},t_{a-2}\right)\right]$$

$$\times\left[\begin{array}{c}(k-a+1)^{\sigma}\left[\begin{array}{r}2(k-a)^2+(3\sigma+10)(k-a)\\+2\sigma^2+9\sigma+12\end{array}\right]\\-(k-a)^{\sigma}\left[\begin{array}{r}2(k-a)^2+(5\sigma+10)(k-a)\\6\sigma^2+18\sigma+12\end{array}\right]\end{array}\right]$$

$$+\frac{1-\sigma}{Q}G_5\left(\mathcal{F}_{0k},\mathcal{F}_i(t_k),t_k\right)\left\{B_5'(t_{k+1})-B_5'(t_k)\right\}$$

$$+\frac{\sigma(\Delta t)^{\sigma-1}}{Q\Gamma(\sigma+1)}\sum_{a=2}^{k}G_5\left(\mathcal{F}_{0a-2},F_{i(a-2)},t_{a-2}\right)\left\{B_5'(t_{a-1})-B_5'(t_{a-2})\right\}\left[(k-a+1)^{\sigma}-(k-a)^{\sigma}\right]$$

$$+\frac{\sigma(\Delta t)^{\sigma-1}}{Q\Gamma(\sigma+2)}$$

$$\times\sum_{a=2}^{k}\left(G_5(\mathcal{F}_{0a-1},F_{i(a-1)},t_{a-1})\left\{B_5'(t_a)-B_5'(t_{a-1})\right\}-G_5(\mathcal{F}_{0a-2},F_{i(a-2)},t_{a-2})\right)$$

$$\times\left\{B_5'(t_{a-1})-B_5'(t_{a-2})\right\}$$

$$\times\left[\begin{array}{r}(k-a+1)^{\sigma}(k-a+3+2\sigma)\\-(k-a)^{\sigma}(k-a+3+3\sigma)\end{array}\right]+\frac{\sigma(\Delta t)^{\sigma-1}}{2Q\Gamma(\sigma+3)}$$

$$\times\sum_{a=2}^{k}\left[G_5(\mathcal{F}_{0a},F_{ia},t_a)\left\{B_5'(t_{a+1})-B_5'(t_a)\right\}-2G_5(\mathcal{F}_{0a-1},F_{i(a-1)},t_{a-1})\left\{B_5'(t_a)-B_5'(t_{a-1})\right\}\right.$$

$$\left.+G_5\left(\mathcal{F}_{0a-2},F_{i(a-2)},t_{a-2}\right)\left\{B_5'(t_{a-1})-B_5'(t_{a-2})\right\}\right]$$

$$\times\left[\begin{array}{c}(k-a+1)^{\sigma}\left[\begin{array}{r}2(k-a)^2+(3\sigma+10)(k-a)\\+2\sigma^2+9\sigma+12\end{array}\right]\\-(k-a)^{\sigma}\left[\begin{array}{r}2(k-a)^2+(5\sigma+10)(k-a)\\6\sigma^2+18\sigma+12\end{array}\right]\end{array}\right]; \tag{7.65}$$

$$N(t_{k+1})=N(0)+\frac{1-\sigma}{Q}f_6\left(N(t_k),\mathcal{F}_0(t_k),t_k\right)$$

$$+\frac{\sigma(\Delta t)^{\sigma}}{Q\Gamma(\sigma+1)}\sum_{a=2}^{k}f_6\left(N_{a-2},\mathcal{F}_{0a-2},t_{a-2}\right)\left[(k-a+1)^{\sigma}-(k-a)^{\sigma}\right]$$

$$+\frac{\sigma(\Delta t)^{\sigma}}{Q\Gamma(\sigma+2)}$$

$$\times\sum_{a=2}^{k}\left(f_6(N_{a-1},\mathcal{F}_{0a-1},t_{a-1})-f_6(N_{a-2},\mathcal{F}_{0a-2},t_{a-2})\right)$$

$$\times\left[\begin{array}{c}(k-a+1)^{\sigma}(k-a+3+2\sigma)\\-(k-a)^{\sigma}(k-a+3+3\sigma)\end{array}\right]+\frac{\sigma(\Delta t)^{\sigma}}{2Q\Gamma(\sigma+3)}$$

$$\times\sum_{a=2}^{k}[f_6(N_a,\mathcal{F}_{0a},t_a)-2f_6(N_{a-1},\mathcal{F}_{0a-1},t_{a-1})+f_6(N_{a-2},\mathcal{F}_{0a-2},t_{a-2})]$$

$$\times\left[\begin{array}{c}(k-a+1)^{\sigma}\left[\begin{array}{r}2(k-a)^2+(3\sigma+10)(k-a)\\+2\sigma^2+9\sigma+12\end{array}\right]\\-(k-a)^{\sigma}\left[\begin{array}{r}2(k-a)^2+(5\sigma+10)(k-a)\\6\sigma^2+18\sigma+12\end{array}\right]\end{array}\right]$$

$$+\frac{1-\sigma}{Q}G_6(N_k,\mathcal{F}_{0k},t_k)\left\{B_6'(t_{k+1})-B_6'(t_k)\right\}$$

$$+\frac{\sigma(\Delta t)^{\sigma-1}}{Q\Gamma(\sigma+1)}\sum_{a=2}^{k}G_6(N_{a-2},\mathcal{F}_{0a-2},t_{a-2})\left\{B_6'(t_{a-1})-B_6'(t_{a-2})\right\}[(k-a+1)^{\sigma}-(k-a)^{\sigma}]$$

$$+\frac{\sigma(\Delta t)^{\sigma-1}}{Q\Gamma(\sigma+2)}$$

$$\times\sum_{a=2}^{k}\left(G_6(N_{a-1},\mathcal{F}_{0a-1},t_{a-1})\left\{B_6'(t_a)-B_6'(t_{a-1})\right\}-G_6(N_{a-2},\mathcal{F}_{0a-2},t_{a-2})\right)$$

$$\times\left\{B_6'(t_{a-1})-B_6'(t_{a-2})\right\}$$

$$\times\left[\begin{array}{c}(k-a+1)^{\sigma}(k-a+3+2\sigma)\\-(k-a)^{\sigma}(k-a+3+3\sigma)\end{array}\right]+\frac{\sigma(\Delta t)^{\sigma-1}}{2Q\Gamma(\sigma+3)}$$

$$\times\sum_{a=2}^{k}\left[G_6(N_a,\mathcal{F}_{0a},t_a)\left\{B_6'(t_{a+1})-B_6'(t_a)\right\}-2G_6(N_{a-1},\mathcal{F}_{0a-1},t_{a-1})\left\{B_6'(t_a)-B_6'(t_{a-1})\right\}\right.$$

$$\left.+G_6(N_{a-2},\mathcal{F}_{0a-2},t_{a-2})\left\{B_6'(t_{a-1})-B_6'(t_{a-2})\right\}\right]$$

$$\times\left[\begin{array}{c}(k-a+1)^{\sigma}\left[\begin{array}{r}2(k-a)^2+(3\sigma+10)(k-a)\\+2\sigma^2+9\sigma+12\end{array}\right]\\-(k-a)^{\sigma}\left[\begin{array}{r}2(k-a)^2+(5\sigma+10)(k-a)\\6\sigma^2+18\sigma+12\end{array}\right]\end{array}\right];\tag{7.66}$$

$$A(t_{k+1})=A(0)+\frac{1-\sigma}{Q}f_7(A(t_k),N(t_k),t_k)$$

$$+\frac{\sigma(\Delta t)^{\sigma}}{Q\Gamma(\sigma+1)}\sum_{a=2}^{k}f_7(A_{a-2},N_{a-2},t_{a-2})[(k-a+1)^{\sigma}-(k-a)^{\sigma}]$$

$$+\frac{\sigma(\Delta t)^{\sigma}}{Q\Gamma(\sigma+2)}$$

$$\times\sum_{a=2}^{k}(f_7(A_{a-1},N_{a-1},t_{a-1})-f_7(A_{a-2},N_{a-2},t_{a-2}))$$

$$\times\left[\begin{array}{r}(k-a+1)^{\sigma}(k-a+3+2\sigma)\\-(k-a)^{\sigma}(k-a+3+3\sigma)\end{array}\right]+\frac{\sigma(\Delta t)^{\sigma}}{2Q\Gamma(\sigma+3)}$$

$$\times\sum_{a=2}^{k}[f_7(A_a,N_a,t_a)-2f_7(A_{a-1},N_{a-1},t_{a-1})+f_7(A_{a-2},N_{a-2},t_{a-2})]$$

$$\times\left[\begin{array}{c}(n-a+1)^{\sigma}\left[\begin{array}{r}2(k-a)^2+(3\sigma+10)(k-a)\\+2\sigma^2+9\sigma+12\end{array}\right]\\-(k-a)^{\sigma}\left[\begin{array}{r}2(k-a)^2+(5\sigma+10)(k-a)\\6\sigma^2+18\sigma+12\end{array}\right]\end{array}\right]$$

$$+\frac{1-\sigma}{Q}G_7(A_k,N_k,t_k)\left\{B_7'(t_{k+1})-B_7'(t_k)\right\}$$

$$+\frac{\sigma(\Delta t)^{\sigma-1}}{Q\Gamma(\sigma+1)}\sum_{a=2}^{k}G_7(A_{a-2},N_{a-2},t_{a-2})\left\{B_7'(t_{a-1})-B_7'(t_{a-2})\right\}[(k-a+1)^{\sigma}-(k-a)^{\sigma}]$$

$$+\frac{\sigma(\Delta t)^{\sigma-1}}{Q\Gamma(\sigma+2)}$$

$$\times\sum_{a=2}^{k}\left(G_7(A_{a-1},N_{a-1},t_{a-1})\left\{B_7'(t_a)-B_7'(t_{a-1})\right\}-G_7(A_{a-2},N_{a-2},t_{a-2})\right)$$

$$\times\left\{B_7'(t_{a-1})-B_7'(t_{a-2})\right\}$$

$$\times\left[\begin{array}{r}(k-a+1)^{\sigma}(k-a+3+2\sigma)\\-(k-a)^{\sigma}(k-a+3+3\sigma)\end{array}\right]+\frac{\sigma(\Delta t)^{\sigma-1}}{2Q\Gamma(\sigma+3)}$$

$$\times\sum_{a=2}^{k}\left[G_7(A_a,N_a,t_a)\left\{B_7'(t_{a+1})-B_7'(t_a)\right\}-2G_7(A_{a-1},N_{a-1},t_{a-1})\left\{B_7'(t_a)-B_7'(t_{a-1})\right\}\right.$$

$$\left.+G_7(A_{a-2},N_{a-2},t_{a-2})\left\{B_7'(t_{a-1})-B_7'(t_{a-2})\right\}\right]$$

$$\times\left[\begin{array}{c}(k-a+1)^{\sigma}\left[\begin{array}{r}2(k-a)^2+(3\sigma+10)(k-a)\\+2\sigma^2+9\sigma+12\end{array}\right]\\-(k-a)^{\sigma}\left[\begin{array}{r}2(k-a)^2+(5\sigma+10)(k-a)\\6\sigma^2+18\sigma+12\end{array}\right]\end{array}\right]; \qquad (7.67)$$

$$\mathcal{N}_d(t_{k+1})=\mathcal{N}_d(0)+\frac{1-\sigma}{Q}f_8(\mathcal{N}_d(t_k),A(t_k),t_k)$$

$$+\frac{\sigma(\Delta t)^{\sigma}}{Q\Gamma(\sigma+1)}\sum_{a=2}^{k}f_8(\mathcal{N}_{da-2},A_{a-2},t_{a-2})[(k-a+1)^{\sigma}-(k-a)^{\sigma}]$$

$$+\frac{\sigma(\Delta t)^{\sigma}}{Q\Gamma(\sigma+2)}$$

$$\times\sum_{a=2}^{k}(f_8(\mathcal{N}_{da-1},A_{a-1},t_{a-1})-f_8(\mathcal{N}_{da-2},A_{a-2},t_{a-2}))$$

$$
\begin{aligned}
&\times\begin{bmatrix} (k-a+1)^{\sigma}(k-a+3+2\sigma) \\ -(k-a)^{\sigma}(k-a+3+3\sigma) \end{bmatrix}+\frac{\sigma(\Delta t)^{\sigma}}{2Q\Gamma(\sigma+3)} \\
&\times\sum_{a=2}^{k}[f_8(\mathcal{N}_{da},A_a,t_a)-2f_8(\mathcal{N}_{da-1},A_{a-1},t_{a-1})+f_8(\mathcal{N}_{da-2},A_{a-2},t_{a-2})] \\
&\times\begin{bmatrix} (k-a+1)^{\sigma}\begin{bmatrix} 2(k-a)^2+(3\sigma+10)(k-a) \\ +2\sigma^2+9\sigma+12 \end{bmatrix} \\ -(k-a)^{\sigma}\begin{bmatrix} 2(k-a)^2+(5\sigma+10)(k-a) \\ 6\sigma^2+18\sigma+12 \end{bmatrix} \end{bmatrix} \\
&+\frac{1-\sigma}{Q}G_8(\mathcal{N}_{dn},A_n,t_n)\left\{B_8'(t_{k+1})-B_8'(t_k)\right\} \\
&+\frac{\sigma(\Delta t)^{\sigma-1}}{Q\Gamma(\sigma+1)}\sum_{a=2}^{k}G_8(\mathcal{N}_{da-2},A_{a-2},t_{a-2})\left\{B_8'(t_{a-1})-B_8'(t_{a-2})\right\}[(k-a+1)^{\sigma}-(k-a)^{\sigma}] \\
&+\frac{\sigma(\Delta t)^{\sigma-1}}{Q\Gamma(\sigma+2)} \\
&\times\sum_{a=2}^{k}\left(G_8(\mathcal{N}_{da-1},A_{a-1},t_{a-1})\left\{B_8'(t_a)-B_8'(t_{a-1})\right\}-G_8(Z_{a-2},A_{a-2},t_{a-2})\right) \\
&\times\left\{B_8'(t_{a-1})-B_8'(t_{a-2})\right\} \\
&\times\begin{bmatrix} (k-a+1)^{\sigma}(k-a+3+2\sigma) \\ -(k-a)^{\sigma}(k-a+3+3\sigma) \end{bmatrix}+\frac{\sigma(\Delta t)^{\sigma-1}}{2Q\Gamma(\sigma+3)} \\
&\times\sum_{a=2}^{k}\Big[G_8(Z_a,A_a,t_a)\left\{B_8'(t_{a+1})-B_8'(t_a)\right\}-2G_8(Z_{a-1},A_{a-1},t_{a-1})\left\{B_8'(t_a)-B_8'(t_{a-1})\right\} \\
&+G_8(Z_{a-2},A_{a-2},t_{a-2})\left\{B_8'(t_{a-1})-B_8'(t_{a-2})\right\}\Big] \\
&\times\begin{bmatrix} (k-a+1)^{\sigma}\begin{bmatrix} 2(k-a)^2+(3\sigma+10)(k-a) \\ +2\sigma^2+9\sigma+12 \end{bmatrix} \\ -(k-a)^{\sigma}\begin{bmatrix} 2(k-a)^2+(5\sigma+10)(k-a) \\ 6\sigma^2+18\sigma+12 \end{bmatrix} \end{bmatrix}; \qquad (7.68)
\end{aligned}
$$

$$
\begin{aligned}
&\mathcal{A}_0(t_{k+1})=\mathcal{A}_0(0)+\frac{1-\sigma}{Q}f_9(\mathcal{A}_0(t_k),\mathcal{N}_d(t_k),t_k) \\
&+\frac{\sigma(\Delta t)^{\sigma}}{Q\Gamma(\sigma+1)}\sum_{a=2}^{k}f_9(\mathcal{A}_{0a-2},\mathcal{N}_{da-2},t_{a-2})[(k-a+1)^{\sigma}-(k-a)^{\sigma}] \\
&+\frac{\sigma(\Delta t)^{\sigma}}{Q\Gamma(\sigma+2)} \\
&\times\sum_{a=2}^{k}(f_9(\mathcal{A}_{0a-1},\mathcal{N}_{da-1},t_{a-1})-f_9(\mathcal{A}_{0a-2},\mathcal{N}_{da-2},t_{a-2}))
\end{aligned}
$$

$$\times\left[\begin{array}{r}(k-a+1)^{\sigma}(k-a+3+2\sigma)\\-(k-a)^{\sigma}(k-a+3+3\sigma)\end{array}\right]+\frac{\sigma(\Delta t)^{\sigma}}{2Q\Gamma(\sigma+3)}$$

$$\times\sum_{a=2}^{k}[f_9(\mathcal{A}_{0a},\mathcal{N}_{da},t_a)-2f_9(\mathcal{A}_{0a-1},\mathcal{N}_{da-1},t_{a-1})+f_9(\mathcal{A}_{0a-2},\mathcal{N}_{da-2},t_{a-2})]$$

$$\times\left[\begin{array}{c}(k-a+1)^{\sigma}\left[\begin{array}{r}2(k-a)^2+(3\sigma+10)(k-a)\\+2\sigma^2+9\sigma+12\end{array}\right]\\-(k-a)^{\sigma}\left[\begin{array}{r}2(k-a)^2+(5\sigma+10)(k-a)\\6\sigma^2+18\sigma+12\end{array}\right]\end{array}\right]$$

$$+\frac{1-\sigma}{Q}G_9(\mathcal{A}_{0k},\mathcal{N}_{dk},t_k)\left\{B_9'(t_{k+1})-B_9'(t_k)\right\}$$

$$+\frac{\sigma(\Delta t)^{\sigma-1}}{Q\Gamma(\sigma+1)}\sum_{a=2}^{k}G_9(\mathcal{A}_{0a-2},\mathcal{N}_{da-2},t_{a-2})\left\{B_9'(t_{a-1})-B_9'(t_{a-2})\right\}[(k-a+1)^{\sigma}-(k-a)^{\sigma}]$$

$$+\frac{\sigma(\Delta t)^{\sigma-1}}{Q\Gamma(\sigma+2)}$$

$$\times\sum_{a=2}^{k}\left(G_9(\mathcal{A}_{0a-1},\mathcal{N}_{da-1},t_{a-1})\left\{B_9'(t_a)-B_9'(t_{a-1})\right\}-G_9(Z_{a-2},\mathcal{N}_{da-2},t_{a-2})\right)$$

$$\times\left\{B_9'(t_{a-1})-B_9'(t_{a-2})\right\}$$

$$\times\left[\begin{array}{r}(k-a+1)^{\sigma}(k-a+3+2\sigma)\\-(k-a)^{\sigma}(k-a+3+3\sigma)\end{array}\right]+\frac{\sigma(\Delta t)^{\sigma-1}}{2Q\Gamma(\sigma+3)}$$

$$\times\sum_{a=2}^{k}\Big[G_9(Z_a,\|_{da},t_a)\left\{B_9'(t_{a+1})-B_9'(t_a)\right\}-2G_9(Z_{a-1},\mathcal{N}_{da-1},t_{a-1})\left\{B_9'(t_a)-B_9'(t_{a-1})\right\}$$

$$+G_9(Z_{a-2},\mathcal{N}_{da-2},t_{a-2})\left\{B_9'(t_{a-1})-B_9'(t_{a-2})\right\}\Big]$$

$$\times\left[\begin{array}{c}(k-a+1)^{\sigma}\left[\begin{array}{r}2(k-a)^2+(3\sigma+10)(k-a)\\+2\sigma^2+9\sigma+12\end{array}\right]\\-(k-a)^{\sigma}\left[\begin{array}{r}2(k-a)^2+(5\sigma+10)(k-a)\\6\sigma^2+18\sigma+12\end{array}\right]\end{array}\right];\tag{7.69}$$

$$H(t_{k+1})=H(0)+\frac{1-\sigma}{Q}f_{10}(H(t_k),\mathcal{A}_0(t_k),t_k)$$

$$+\frac{\sigma(\Delta t)^{\sigma}}{Q\Gamma(\sigma+1)}\sum_{a=2}^{k}f_{10}(H_{a-2},\mathcal{A}_{0a-2},t_{a-2})[(k-a+1)^{\sigma}-(k-a)^{\sigma}]$$

$$+\frac{\sigma(\Delta t)^{\sigma}}{Q\Gamma(\sigma+2)}$$

$$\times\sum_{a=2}^{k}(f_{10}(H_{a-1},\mathcal{A}_{0a-1},t_{a-1})-f_{10}(H_{a-2},\mathcal{A}_{0a-2},t_{a-2}))$$

$$\times\begin{bmatrix}(k-a+1)^{\sigma}(k-a+3+2\sigma)\\-(k-a)^{\sigma}(k-a+3+3\sigma)\end{bmatrix}+\frac{\sigma(\Delta t)^{\sigma}}{2Q\Gamma(\sigma+3)}$$

$$\times\sum_{a=2}^{k}[f_{10}(H_a,\mathcal{A}_{0a},t_a)-2f_{10}(H_{a-1},\mathcal{A}_{0a-1},t_{a-1})+f_{10}(H_{a-2},\mathcal{A}_{0a-2},t_{a-2})]$$

$$\times\begin{bmatrix}(k-a+1)^{\sigma}\begin{bmatrix}2(k-a)^2+(3\sigma+10)(k-a)\\+2\sigma^2+9\sigma+12\end{bmatrix}\\-(k-a)^{\sigma}\begin{bmatrix}2(k-a)^2+(5\sigma+10)(k-a)\\6\sigma^2+18\sigma+12\end{bmatrix}\end{bmatrix}$$

$$+\frac{1-\sigma}{Q}G_{10}(H_k,\mathcal{A}_{0k},t_k)\left\{B'_{10}(t_{k+1})-B'_{10}(t_k)\right\}$$

$$+\frac{\sigma(\Delta t)^{\sigma-1}}{Q\Gamma(\sigma+1)}\sum_{a=2}^{k}G_{10}(H_{a-2},\mathcal{A}_{0a-2},t_{a-2})\left\{B'_{10}(t_{a-1})-B'_{10}(t_{a-2})\right\}[(k-a+1)^{\sigma}-(k-a)^{\sigma}]$$

$$+\frac{\sigma(\Delta t)^{\sigma-1}}{Q\Gamma(\sigma+2)}$$

$$\times\sum_{a=2}^{k}\left(G_{10}(H_{a-1},\mathcal{A}_{0a-1},t_{a-1})\left\{B'_{10}(t_a)-B'_{10}(t_{a-1})\right\}-G_{10}(Z_{a-2},\mathcal{A}_{0a-2},t_{a-2})\right)$$

$$\times\left\{B'_{10}(t_{a-1})-B'_{10}(t_{a-2})\right\}$$

$$\times\begin{bmatrix}(k-a+1)^{\sigma}(k-a+3+2\sigma)\\-(k-a)^{\sigma}(k-a+3+3\sigma)\end{bmatrix}+\frac{\sigma(\Delta t)^{\sigma-1}}{2Q\Gamma(\sigma+3)}$$

$$\times\sum_{a=2}^{k}\left[G_{10}(Z_a,\mathcal{A}_{0a},t_a)\left\{B'_{10}(t_{a+1})-B'_{10}(t_a)\right\}-2G_{10}(Z_{a-1},\mathcal{A}_{0a-1},t_{a-1})\left\{B'_{10}(t_a)-B'_{10}(t_{a-1})\right\}\right.$$

$$\left.+G_{10}(Z_{a-2},\mathcal{A}_{0a-2},t_{a-2})\left\{B'_{10}(t_{a-1})-B'_{10}(t_{a-2})\right\}\right]$$

$$\times\begin{bmatrix}(k-a+1)^{\sigma}\begin{bmatrix}2(k-a)^2+(3\sigma+10)(k-a)\\+2\sigma^2+9\sigma+12\end{bmatrix}\\-(k-a)^{\sigma}\begin{bmatrix}2(k-a)^2+(5\sigma+10)(k-a)\\6\sigma^2+18\sigma+12\end{bmatrix}\end{bmatrix};\tag{7.70}$$

$$\mathcal{M}_1(t_{k+1})=\mathcal{M}_1(0)+\frac{1-\sigma}{Q}f_{11}(\mathcal{M}_1(t_k),H(t_k),t_k)$$

$$+\frac{\sigma(\Delta t)^{\sigma}}{Q\Gamma(\sigma+1)}\sum_{a=2}^{k}f_{11}(\mathcal{M}_{1a-2},H_{a-2},t_{a-2})[(k-a+1)^{\sigma}-(k-a)^{\sigma}]$$

$$+\frac{\sigma(\Delta t)^{\sigma}}{Q\Gamma(\sigma+2)}$$

$$\times\sum_{a=2}^{k}(f_{11}(\mathcal{M}_{1a-1},H_{a-1},t_{a-1})-f_{11}(\mathcal{M}_{1a-2},H_{a-2},t_{a-2}))$$

$$\times\begin{bmatrix} (k-a+1)^{\sigma}(k-a+3+2\sigma) \\ -(k-a)^{\sigma}(k-a+3+3\sigma) \end{bmatrix} + \frac{\sigma(\Delta t)^{\sigma}}{2Q\Gamma(\sigma+3)}$$

$$\times\sum_{a=2}^{k}[f_{11}(\mathcal{M}_{1a},H_a,t_a)-2f_{11}(\mathcal{M}_{1a-1},H_{a-1},t_{a-1})+f_{11}(\mathcal{M}_{1a-2},H_{a-2},t_{a-2})]$$

$$\times\begin{bmatrix} (k-a+1)^{\sigma}\begin{bmatrix} 2(k-a)^2+(3\sigma+10)(k-a) \\ +2\sigma^2+9\sigma+12 \end{bmatrix} \\ -(k-a)^{\sigma}\begin{bmatrix} 2(k-a)^2+(5\sigma+10)(k-a) \\ 6\sigma^2+18\sigma+12 \end{bmatrix} \end{bmatrix}$$

$$+\frac{1-\sigma}{Q}G_{11}(\mathcal{M}_{1k},H_k,t_k)\left\{B'_{11}(t_{k+1})-B'_{11}(t_k)\right\}$$

$$+\frac{\sigma(\Delta t)^{\sigma-1}}{Q\Gamma(\sigma+1)}\sum_{a=2}^{k}G_{11}(\mathcal{M}_{1a-2},H_{a-2},t_{a-2})\left\{B'_{11}(t_{a-1})-B'_{11}(t_{a-2})\right\}[(k-a+1)^{\sigma}-(k-a)^{\sigma}]$$

$$+\frac{\sigma(\Delta t)^{\sigma-1}}{Q\Gamma(\sigma+2)}$$

$$\times\sum_{a=2}^{k}\Big(G_{11}(\mathcal{M}_{1a-1},H_{a-1},t_{a-1})\left\{B'_{11}(t_a)-B'_{11}(t_{a-1})\right\}-G_{11}(Z_{a-2},H_{a-2},t_{a-2})\Big)$$

$$\times\left\{B'_{11}(t_{a-1})-B'_{11}(t_{a-2})\right\}$$

$$\times\begin{bmatrix} (k-a+1)^{\sigma}(k-a+3+2\sigma) \\ -(k-a)^{\sigma}(k-a+3+3\sigma) \end{bmatrix} + \frac{\sigma(\Delta t)^{\sigma-1}}{2Q\Gamma(\sigma+3)}$$

$$\times\sum_{a=2}^{k}\Big[G_{11}(Z_a,H_a,t_a)\left\{B'_{11}(t_{a+1})-B'_{11}(t_a)\right\}-2G_{11}(Z_{a-1},H_{a-1},t_{a-1})\left\{B'_{11}(t_a)-B'_{11}(t_{a-1})\right\}$$

$$+G_{11}(Z_{a-2},H_{a-2},t_{a-2})\left\{B'_{11}(t_{a-1})-B'_{11}(t_{a-2})\right\}\Big]$$

$$\times\begin{bmatrix} (k-a+1)^{\sigma}\begin{bmatrix} 2(k-a)^2+(3\sigma+10)(k-a) \\ +2\sigma^2+9\sigma+12 \end{bmatrix} \\ -(k-a)^{\sigma}\begin{bmatrix} 2(k-a)^2+(5\sigma+10)(k-a) \\ 6\sigma^2+18\sigma+12 \end{bmatrix} \end{bmatrix}; \qquad (7.71)$$

$$\mathcal{M}_2(t_{k+1})=\mathcal{M}_2(0)+\frac{1-\sigma}{Q}f_{12}(\mathcal{M}_2(t_k),\mathcal{M}_1(t_k),t_k)$$

$$+\frac{\sigma(\Delta t)^{\sigma}}{Q\Gamma(\sigma+1)}\sum_{a=2}^{k}f_{12}(\mathcal{M}_{2a-2},\mathcal{M}_{1a-2},t_{a-2})[(k-a+1)^{\sigma}-(k-a)^{\sigma}]$$

$$+\frac{\sigma(\Delta t)^{\sigma}}{Q\Gamma(\sigma+2)}$$

$$\times\sum_{a=2}^{k}(f_{12}(\mathcal{M}_{2a-1},\mathcal{M}_{1a-1},t_{a-1})-f_{12}(\mathcal{M}_{2a-2},\mathcal{M}_{1a-2},t_{a-2}))$$

$$\times\begin{bmatrix}(k-a+1)^{\sigma}(k-a+3+2\sigma)\\-(k-a)^{\sigma}(k-a+3+3\sigma)\end{bmatrix}+\frac{\sigma(\Delta t)^{\sigma}}{2Q\Gamma(\sigma+3)}$$

$$\times\sum_{a=2}^{k}[f_{12}(\mathcal{M}_{2a},\mathcal{M}_{1a},t_a)-2f_{12}(\mathcal{M}_{2a-1},\mathcal{M}_{1a-1},t_{a-1})+f_{12}(\mathcal{M}_{2a-2},\mathcal{M}_{1a-2},t_{a-2})]$$

$$\times\begin{bmatrix}(k-a+1)^{\sigma}\begin{bmatrix}2(k-a)^2+(3\sigma+10)(k-a)\\+2\sigma^2+9\sigma+12\end{bmatrix}\\-(k-a)^{\sigma}\begin{bmatrix}2(k-a)^2+(5\sigma+10)(k-a)\\6\sigma^2+18\sigma+12\end{bmatrix}\end{bmatrix}$$

$$+\frac{1-\sigma}{Q}G_{12}(\mathcal{M}_{2k},\mathcal{M}_{1k},t_k)\left\{B'_{12}(t_{k+1})-B'_{12}(t_k)\right\}$$

$$+\frac{\sigma(\Delta t)^{\sigma-1}}{Q\Gamma(\sigma+1)}\sum_{a=2}^{k}G_{12}(\mathcal{M}_{2a-2},\mathcal{M}_{1a-2},t_{a-2})\left\{B'_{12}(t_{a-1})-B'_{12}(t_{a-2})\right\}[(k-a+1)^{\sigma}-(k-a)^{\sigma}]$$

$$+\frac{\sigma(\Delta t)^{\sigma-1}}{Q\Gamma(\sigma+2)}$$

$$\times\sum_{a=2}^{k}\Big(G_{12}(\mathcal{M}_{2a-1},\mathcal{M}_{1a-1},t_{a-1})\left\{B'_{12}(t_a)-B'_{12}(t_{a-1})\right\}-G_{12}(Z_{a-2},\mathcal{M}_{1a-2},t_{a-2})\Big)$$

$$\times\left\{B'_{12}(t_{a-1})-B'_{12}(t_{a-2})\right\}$$

$$\times\begin{bmatrix}(k-a+1)^{\sigma}(k-a+3+2\sigma)\\-(k-a)^{\sigma}(k-a+3+3\sigma)\end{bmatrix}+\frac{\sigma(\Delta t)^{\sigma-1}}{2Q\Gamma(\sigma+3)}$$

$$\times\sum_{a=2}^{k}\Big[G_{12}(Z_a,\mathcal{M}_{1a},t_a)\left\{B'_{12}(t_{a+1})-B'_{12}(t_a)\right\}-2G_{12}(Z_{a-1},\mathcal{M}_{1a-1},t_{a-1})\left\{B'_{12}(t_a)-B'_{12}(t_{a-1})\right\}$$

$$+G_{12}(Z_{a-2},\mathcal{M}_{1a-2},t_{a-2})\left\{B'_{12}(t_{a-1})-B'_{12}(t_{a-2})\right\}\Big]$$

$$\times\begin{bmatrix}(k-a+1)^{\sigma}\begin{bmatrix}2(k-a)^2+(3\sigma+10)(k-a)\\+2\sigma^2+9\sigma+12\end{bmatrix}\\-(k-a)^{\sigma}\begin{bmatrix}2(k-a)^2+(5\sigma+10)(k-a)\\6\sigma^2+18\sigma+12\end{bmatrix}\end{bmatrix}; \tag{7.72}$$

$$\hat{\mathcal{M}}_1(t_{k+1})=\hat{\mathcal{M}}_1(0)+\frac{1-\sigma}{Q}f_{13}\left(\hat{\mathcal{M}}_1(t_k),\mathcal{M}_2(t_k),t_k\right)$$

$$+\frac{\sigma(\Delta t)^{\sigma}}{Q\Gamma(\sigma+1)}\sum_{a=2}^{k}f_{13}\left(\hat{\mathcal{M}}_{1a-2},\mathcal{M}_{2a-2},t_{a-2}\right)[(k-a+1)^{\sigma}-(k-a)^{\sigma}]$$

$$+\frac{\sigma(\Delta t)^{\sigma}}{Q\Gamma(\sigma+2)}$$

$$\times\sum_{a=2}^{k}\Big(f_{13}(\hat{\mathcal{M}}_{1a-1},\mathcal{M}_{2a-1},t_{a-1})-f_{13}(\hat{\mathcal{M}}_{1a-2},\mathcal{M}_{2a-2},t_{a-2})\Big)$$

$$\times\left[\begin{array}{r}(k-a+1)^{\sigma}(k-a+3+2\sigma)\\-(k-a)^{\sigma}(k-a+3+3\sigma)\end{array}\right]+\frac{\sigma(\Delta t)^{\sigma}}{2Q\Gamma(\sigma+3)}$$

$$\times\sum_{a=2}^{k}\left[f_{13}(\hat{\mathcal{M}}_{1a},\mathcal{M}_{2a},t_a)-2f_{13}(\hat{\mathcal{M}}_{1a-1},\mathcal{M}_{2a-1},t_{a-1})+f_{13}\left(\hat{\mathcal{M}}_{1a-2},\mathcal{M}_{2a-2},t_{a-2}\right)\right]$$

$$\times\left[\begin{array}{r}(k-a+1)^{\sigma}\left[\begin{array}{r}2(k-a)^2+(3\sigma+10)(k-a)\\+2\sigma^2+9\sigma+12\end{array}\right]\\-(k-a)^{\sigma}\left[\begin{array}{r}2(k-a)^2+(5\sigma+10)(k-a)\\6\sigma^2+18\sigma+12\end{array}\right]\end{array}\right]$$

$$+\frac{1-\sigma}{Q}G_{13}\left(\hat{\mathcal{M}}_{1k},\mathcal{M}_{2k},t_k\right)\left\{B'_{13}(t_{k+1})-B'_{13}(t_k)\right\}$$

$$+\frac{\sigma(\Delta t)^{\sigma-1}}{Q\Gamma(\sigma+1)}\sum_{a=2}^{k}G_{13}\left(\hat{\mathcal{M}}_{1a-2},\mathcal{M}_{2a-2},t_{a-2}\right)\left\{B'_{13}(t_{a-1})-B'_{13}(t_{a-2})\right\}[(k-a+1)^{\sigma}-(k-a)^{\sigma}]$$

$$+\frac{\sigma(\Delta t)^{\sigma-1}}{Q\Gamma(\sigma+2)}$$

$$\times\sum_{a=2}^{k}\left(G_{13}(\hat{\mathcal{M}}_{1a-1},\mathcal{M}_{2a-1},t_{a-1})\left\{B'_{13}(t_a)-B'_{13}(t_{a-1})\right\}-G_{13}(Z_{a-2},\mathcal{M}_{2a-2},t_{a-2})\right)$$

$$\times\left\{B'_{13}(t_{a-1})-B'_{13}(t_{a-2})\right\}$$

$$\times\left[\begin{array}{r}(k-a+1)^{\sigma}(k-a+3+2\sigma)\\-(k-a)^{\sigma}(k-a+3+3\sigma)\end{array}\right]+\frac{\sigma(\Delta t)^{\sigma-1}}{2Q\Gamma(\sigma+3)}$$

$$\times\sum_{a=2}^{k}\left[G_{13}(Z_a,\mathcal{M}_{2a},t_a)\left\{B'_{13}(t_{a+1})-B'_{13}(t_a)\right\}-2G_{13}(Z_{a-1},\mathcal{M}_{2a-1},t_{a-1})\left\{B'_{13}(t_a)-B'_{13}(t_{a-1})\right\}\right.$$

$$\left.+G_{13}\left(Z_{a-2},\mathcal{M}_{2a-2},t_{a-2}\right)\left\{B'_{13}(t_{a-1})-B'_{13}(t_{a-2})\right\}\right]$$

$$\times\left[\begin{array}{r}(k-a+1)^{\sigma}\left[\begin{array}{r}2(k-a)^2+(3\sigma+10)(k-a)\\+2\sigma^2+9\sigma+12\end{array}\right]\\-(k-a)^{\sigma}\left[\begin{array}{r}2(k-a)^2+(5\sigma+10)(k-a)\\6\sigma^2+18\sigma+12\end{array}\right]\end{array}\right];\tag{7.73}$$

$$\hat{\mathcal{M}}_2(t_{k+1})=\hat{\mathcal{M}}_2(0)+\frac{1-\sigma}{Q}f_{14}\left(\hat{\mathcal{M}}_2(t_k),\hat{\mathcal{M}}_1(t_k),t_k\right)$$

$$+\frac{\sigma(\Delta t)^{\sigma}}{Q\Gamma(\sigma+1)}\sum_{a=2}^{k}f_{14}\left(\hat{\mathcal{M}}_{2a-2},\hat{\mathcal{M}}_{1a-2},t_{a-2}\right)[(k-a+1)^{\sigma}-(k-a)^{\sigma}]$$

$$+\frac{\sigma(\Delta t)^{\sigma}}{Q\Gamma(\sigma+2)}$$

$$\times\sum_{a=2}^{k}\left(f_{14}(\hat{\mathcal{M}}_{2a-1},\hat{\mathcal{M}}_{1a-1},t_{a-1})-f_{14}(\hat{\mathcal{M}}_{2a-2},\hat{\mathcal{M}}_{1a-2},t_{a-2})\right)$$

$$\times\left[\begin{array}{r}(k-a+1)^{\sigma}(k-a+3+2\sigma)\\-(k-a)^{\sigma}(k-a+3+3\sigma)\end{array}\right]+\frac{\sigma(\Delta t)^{\sigma}}{2Q\Gamma(\sigma+3)}$$

$$\times\sum_{a=2}^{k}\left[f_{14}(\hat{\mathcal{M}}_{2a},\hat{\mathcal{M}}_{1a},t_a)-2f_{14}(\hat{\mathcal{M}}_{2a-1},\hat{\mathcal{M}}_{1a-1},t_{a-1})+f_{14}\left(\hat{\mathcal{M}}_{2a-2},\hat{\mathcal{M}}_{1a-2},t_{a-2}\right)\right]$$

$$\times\left[\begin{array}{r}(k-a+1)^{\sigma}\left[\begin{array}{r}2(k-a)^2+(3\sigma+10)(k-a)\\+2\sigma^2+9\sigma+12\end{array}\right]\\-(k-a)^{\sigma}\left[\begin{array}{r}2(k-a)^2+(5\sigma+10)(k-a)\\6\sigma^2+18\sigma+12\end{array}\right]\end{array}\right]$$

$$+\frac{1-\sigma}{Q}G_{14}\left(\hat{\mathcal{M}}_{2k},\hat{\mathcal{M}}_{1k},t_k\right)\left\{B'_{14}(t_{k+1})-B'_{14}(t_k)\right\}$$

$$+\frac{\sigma(\Delta t)^{\sigma-1}}{Q\Gamma(\sigma+1)}\sum_{a=2}^{k}G_{14}\left(\hat{\mathcal{M}}_{2a-2},\hat{\mathcal{M}}_{1a-2},t_{a-2}\right)\left\{B'_{14}(t_{a-1})-B'_{14}(t_{a-2})\right\}[(k-a+1)^{\sigma}-(k-a)^{\sigma}]$$

$$+\frac{\sigma(\Delta t)^{\sigma-1}}{Q\Gamma(\sigma+2)}$$

$$\times\sum_{a=2}^{k}\left(G_{14}(\hat{\mathcal{M}}_{2a-1},\hat{\mathcal{M}}_{1a-1},t_{a-1})\left\{B'_{14}(t_a)-B'_{14}(t_{a-1})\right\}-G_{14}(Z_{a-2},\hat{\mathcal{M}}_{1a-2},t_{a-2})\right)$$

$$\times\left\{B'_{14}(t_{a-1})-B'_{14}(t_{a-2})\right\}$$

$$\times\left[\begin{array}{r}(k-a+1)^{\sigma}(k-a+3+2\sigma)\\-(k-a)^{\sigma}(k-a+3+3\sigma)\end{array}\right]+\frac{\sigma(\Delta t)^{\sigma-1}}{2Q\Gamma(\sigma+3)}$$

$$\times\sum_{a=2}^{k}\Big[G_{14}(Z_a,\hat{\mathcal{M}}_{1a},t_a)\left\{B'_{14}(t_{a+1})-B'_{14}(t_a)\right\}-2G_{14}(Z_{a-1},\hat{\mathcal{M}}_{1a-1},t_{a-1})\left\{B'_{14}(t_a)-B'_{14}(t_{a-1})\right\}$$

$$+G_{14}\left(Z_{a-2},\hat{\mathcal{M}}_{1a-2},t_{a-2}\right)\left\{B'_{14}(t_{a-1})-B'_{14}(t_{a-2})\right\}\Big]$$

$$\times\left[\begin{array}{r}(k-a+1)^{\sigma}\left[\begin{array}{r}2(k-a)^2+(3\sigma+10)(k-a)\\+2\sigma^2+9\sigma+12\end{array}\right]\\-(k-a)^{\sigma}\left[\begin{array}{r}2(k-a)^2+(5\sigma+10)(k-a)\\6\sigma^2+18\sigma+12\end{array}\right]\end{array}\right]; \tag{7.74}$$

$$\mathcal{T}_{\beta}(t_{k+1})=\mathcal{T}_{\beta}(0)+\frac{1-\sigma}{Q}f_{15}\left(\mathcal{T}_{\beta}(t_k),\hat{\mathcal{M}}_1(t_k),t_k\right)$$

$$+\frac{\sigma(\Delta t)^{\sigma}}{Q\Gamma(\sigma+1)}\sum_{a=2}^{k}f_{15}\left(\mathcal{T}_{\beta a-2},\hat{\mathcal{M}}_{1a-2},t_{a-2}\right)[(k-a+1)^{\sigma}-(k-a)^{\sigma}]$$

$$+\frac{\sigma(\Delta t)^{\sigma}}{Q\Gamma(\sigma+2)}$$

$$\times\sum_{a=2}^{k}\left(f_{15}(\mathcal{T}_{\beta a-1},\hat{\mathcal{M}}_{1a-1},t_{a-1})-f_{15}(\mathcal{T}_{\beta a-2},\hat{\mathcal{M}}_{1a-2},t_{a-2})\right)$$

$$\times \begin{bmatrix} (k-a+1)^{\sigma}(k-a+3+2\sigma) \\ -(k-a)^{\sigma}(k-a+3+3\sigma) \end{bmatrix} + \frac{\sigma(\Delta t)^{\sigma}}{2Q\Gamma(\sigma+3)}$$

$$\times \sum_{a=2}^{k} \left[ f_{15}(\mathcal{T}_{\beta a}, \hat{\mathcal{M}}_{1a}, t_a) - 2 f_{15}(\mathcal{T}_{\beta a-1}, \hat{\mathcal{M}}_{1a-1}, t_{a-1}) + f_{15}\left(\mathcal{T}_{\beta a-2}, \hat{\mathcal{M}}_{1a-2}, t_{a-2}\right)\right]$$

$$\times \begin{bmatrix} (k-a+1)^{\sigma} \begin{bmatrix} 2(k-a)^2 + (3\sigma+10)(k-a) \\ +2\sigma^2 + 9\sigma + 12 \end{bmatrix} \\ -(k-a)^{\sigma} \begin{bmatrix} 2(k-a)^2 + (5\sigma+10)(k-a) \\ 6\sigma^2 + 18\sigma + 12 \end{bmatrix} \end{bmatrix}$$

$$+ \frac{1-\sigma}{Q} G_{15}\left(\mathcal{T}_{\beta k}, \hat{\mathcal{M}}_{1k}, t_k\right)\left\{B_{15}'(t_{k+1}) - B_{15}'(t_k)\right\}$$

$$+ \frac{\sigma(\Delta t)^{\sigma-1}}{Q\Gamma(\sigma+1)} \sum_{a=2}^{k} G_{15}\left(\mathcal{T}_{\beta a-2}, \hat{\mathcal{M}}_{1a-2}, t_{a-2}\right)\left\{B_{15}'(t_{a-1}) - B_{15}'(t_{a-2})\right\}\left[(k-a+1)^{\sigma} - (k-a)^{\sigma}\right]$$

$$+ \frac{\sigma(\Delta t)^{\sigma-1}}{Q\Gamma(\sigma+2)}$$

$$\times \sum_{a=2}^{k} \left( G_{15}(\mathcal{T}_{\beta a-1}, \hat{\mathcal{M}}_{1a-1}, t_{a-1})\left\{B_{15}'(t_a) - B_{15}'(t_{a-1})\right\} - G_{15}(Z_{a-2}, \hat{\mathcal{M}}_{1a-2}, t_{a-2})\right)$$

$$\times \left\{B_{15}'(t_{a-1}) - B_{15}'(t_{a-2})\right\}$$

$$\times \begin{bmatrix} (k-a+1)^{\sigma}(k-a+3+2\sigma) \\ -(k-a)^{\sigma}(k-a+3+3\sigma) \end{bmatrix} + \frac{\sigma(\Delta t)^{\sigma-1}}{2Q\Gamma(\sigma+3)}$$

$$\times \sum_{a=2}^{k} \Big[ G_{15}(Z_a, \hat{\mathcal{M}}_{1a}, t_a)\left\{B_{15}'(t_{a+1}) - B_{15}'(t_a)\right\} - 2G_{15}(Z_{a-1}, \hat{\mathcal{M}}_{1a-1}, t_{a-1})\left\{B_{15}'(t_a) - B_{15}'(t_{a-1})\right\}$$

$$+ G_{15}\left(Z_{a-2}, \hat{\mathcal{M}}_{1a-2}, t_{a-2}\right)\left\{B_{15}'(t_{a-1}) - B_{15}'(t_{a-2})\right\}\Big]$$

$$\times \begin{bmatrix} (k-a+1)^{\sigma} \begin{bmatrix} 2(k-a)^2 + (3\sigma+10)(k-a) \\ +2\sigma^2 + 9\sigma + 12 \end{bmatrix} \\ -(k-a)^{\sigma} \begin{bmatrix} 2(k-a)^2 + (5\sigma+10)(k-a) \\ 6\sigma^2 + 18\sigma + 12 \end{bmatrix} \end{bmatrix}; \tag{7.75}$$

$$\mathcal{I}_{10}(t_{k+1}) = \mathcal{I}_{10}(0) + \frac{1-\sigma}{Q} f_{16}\left(\mathcal{I}_{10}(t_k), \mathcal{T}_{\beta}(t_k), t_k\right)$$

$$+ \frac{\sigma(\Delta t)^{\sigma}}{Q\Gamma(\sigma+1)} \sum_{a=2}^{k} f_{16}\left(\mathcal{I}_{10a-2}, \mathcal{T}_{\beta a-2}, t_{a-2}\right)\left[(k-a+1)^{\sigma} - (k-a)^{\sigma}\right]$$

$$+ \frac{\sigma(\Delta t)^{\sigma}}{Q\Gamma(\sigma+2)}$$

$$\times \sum_{a=2}^{k} \left( f_{16}(\mathcal{I}_{10a-1}, \mathcal{T}_{\beta a-1}, t_{a-1}) - f_{16}(\mathcal{I}_{10a-2}, \mathcal{T}_{\beta a-2}, t_{a-2})\right)$$

$$\times\left[\begin{array}{c}(k-a+1)^{\sigma}(k-a+3+2\sigma)\\-(k-a)^{\sigma}(k-a+3+3\sigma)\end{array}\right]+\frac{\sigma(\Delta t)^{\sigma}}{2Q\Gamma(\sigma+3)}$$
$$\times\sum_{a=2}^{k}\left[f_{16}(\mathcal{I}_{10a},\mathcal{T}_{\beta a},t_a)-2f_{16}(\mathcal{I}_{10a-1},\mathcal{T}_{\beta a-1},t_{a-1})+f_{16}\left(\mathcal{I}_{10a-2},\mathcal{T}_{\beta a-2},t_{a-2}\right)\right]$$
$$\times\left[\begin{array}{c}(k-a+1)^{\sigma}\left[\begin{array}{r}2(k-a)^2+(3\sigma+10)(k-a)\\+2\sigma^2+9\sigma+12\end{array}\right]\\-(k-a)^{\sigma}\left[\begin{array}{r}2(k-a)^2+(5\sigma+10)(k-a)\\6\sigma^2+18\sigma+12\end{array}\right]\end{array}\right]$$
$$+\frac{1-\sigma}{Q}G_{16}\left(\mathcal{I}_{10k},\mathcal{T}_{\beta k},t_k\right)\left\{B_{16}'(t_{k+1})-B_{16}'(t_k)\right\}$$
$$+\frac{\sigma(\Delta t)^{\sigma-1}}{Q\Gamma(\sigma+1)}\sum_{a=2}^{k}G_{16}\left(\mathcal{I}_{10a-2},\mathcal{T}_{\beta a-2},t_{a-2}\right)\left\{B_{16}'(t_{a-1})-B_{16}'(t_{a-2})\right\}\left[(k-a+1)^{\sigma}-(k-a)^{\sigma}\right]$$
$$+\frac{\sigma(\Delta t)^{\sigma-1}}{Q\Gamma(\sigma+2)}$$
$$\times\sum_{a=2}^{k}\left(G_{16}(\mathcal{I}_{10a-1},\mathcal{T}_{\beta a-1},t_{a-1})\left\{B_{16}'(t_a)-B_{16}'(t_{a-1})\right\}-G_{16}(Z_{a-2},\mathcal{T}_{\beta a-2},t_{a-2})\right)$$
$$\times\left\{B_{16}'(t_{a-1})-B_{16}'(t_{a-2})\right\}$$
$$\times\left[\begin{array}{c}(k-a+1)^{\sigma}(k-a+3+2\sigma)\\-(k-a)^{\sigma}(k-a+3+3\sigma)\end{array}\right]+\frac{\sigma(\Delta t)^{\sigma-1}}{2Q\Gamma(\sigma+3)}$$
$$\times\sum_{a=2}^{k}\left[G_{16}(Z_a,\mathcal{T}_{\beta a},t_a)\left\{B_{16}'(t_{a+1})-B_{16}'(t_a)\right\}-2G_{16}(Z_{a-1},\mathcal{T}_{\beta a-1},t_{a-1})\left\{B_{16}'(t_a)-B_{16}'(t_{a-1})\right\}\right.$$
$$\left.+G_{16}\left(Z_{a-2},\mathcal{T}_{\beta a-2},t_{a-2}\right)\left\{B_{16}'(t_{a-1})-B_{16}'(t_{a-2})\right\}\right]$$
$$\times\left[\begin{array}{c}(k-a+1)^{\sigma}\left[\begin{array}{r}2(k-a)^2+(3\sigma+10)(k-a)\\+2\sigma^2+9\sigma+12\end{array}\right]\\-(k-a)^{\sigma}\left[\begin{array}{r}2(k-a)^2+(5\sigma+10)(k-a)\\6\sigma^2+18\sigma+12\end{array}\right]\end{array}\right];\tag{7.76}$$

$$\mathcal{T}_{\alpha}(t_{k+1})=\mathcal{T}_{\alpha}(0)+\frac{1-\sigma}{Q}f_{17}\left(\mathcal{T}_{\alpha}(t_k),\mathcal{I}_{10}(t_k),t_k\right)$$
$$+\frac{\sigma(\Delta t)^{\sigma}}{Q\Gamma(\sigma+1)}\sum_{a=2}^{k}f_{17}\left(\mathcal{T}_{\alpha a-2},\mathcal{I}_{10a-2},t_{a-2}\right)\left[(k-a+1)^{\sigma}-(k-a)^{\sigma}\right]$$
$$+\frac{\sigma(\Delta t)^{\sigma}}{Q\Gamma(\sigma+2)}$$
$$\times\sum_{a=2}^{k}\left(f_{17}(\mathcal{T}_{\alpha a-1},\mathcal{I}_{10a-1},t_{a-1})-f_{17}(\mathcal{T}_{\alpha a-2},\mathcal{I}_{10a-2},t_{a-2})\right)$$

$$\times\left[\begin{array}{c}(k-a+1)^{\sigma}(k-a+3+2\sigma)\\-(k-a)^{\sigma}(k-a+3+3\sigma)\end{array}\right]+\frac{\sigma(\Delta t)^{\sigma}}{2Q\Gamma(\sigma+3)}$$

$$\times\sum_{a=2}^{k}[f_{17}(\mathcal{T}_{\alpha a},\mathcal{I}_{10a},t_a)-2f_{17}(\mathcal{T}_{\alpha a-1},\mathcal{I}_{10a-1},t_{a-1})+f_{17}(\mathcal{T}_{\alpha a-2},\mathcal{I}_{10a-2},t_{a-2})]$$

$$\times\left[\begin{array}{c}(k-a+1)^{\sigma}\left[\begin{array}{r}2(k-a)^2+(3\sigma+10)(k-a)\\+2\sigma^2+9\sigma+12\end{array}\right]\\-(k-a)^{\sigma}\left[\begin{array}{r}2(k-a)^2+(5\sigma+10)(k-a)\\6\sigma^2+18\sigma+12\end{array}\right]\end{array}\right]$$

$$+\frac{1-\sigma}{Q}G_{17}(\mathcal{T}_{\alpha k},\mathcal{I}_{10k},t_k)\left\{B'_{17}(t_{k+1})-B'_{17}(t_k)\right\}$$

$$+\frac{\sigma(\Delta t)^{\sigma-1}}{Q\Gamma(\sigma+1)}\sum_{a=2}^{k}G_{17}(\mathcal{T}_{\alpha a-2},\mathcal{I}_{10a-2},t_{a-2})\left\{B'_{17}(t_{a-1})-B'_{17}(t_{a-2})\right\}[(k-a+1)^{\sigma}-(k-a)^{\sigma}]$$

$$+\frac{\sigma(\Delta t)^{\sigma-1}}{Q\Gamma(\sigma+2)}$$

$$\times\sum_{a=2}^{k}\left(G_{17}(\mathcal{T}_{\alpha a-1},\mathcal{I}_{10a-1},t_{a-1})\left\{B'_{17}(t_a)-B'_{17}(t_{a-1})\right\}-G_{17}(Z_{a-2},\mathcal{I}_{10a-2},t_{a-2})\right)$$

$$\times\left\{B'_{17}(t_{a-1})-B'_{17}(t_{a-2})\right\}$$

$$\times\left[\begin{array}{c}(k-a+1)^{\sigma}(k-a+3+2\sigma)\\-(k-a)^{\sigma}(k-a+3+3\sigma)\end{array}\right]+\frac{\sigma(\Delta t)^{\sigma-1}}{2Q\Gamma(\sigma+3)}$$

$$\times\sum_{a=2}^{k}\left[G_{17}(Z_a,\mathcal{I}_{10a},t_a)\left\{B'_{17}(t_{a+1})-B'_{17}(t_a)\right\}-2G_{17}(Z_{a-1},\mathcal{I}_{10a-1},t_{a-1})\left\{B'_{17}(t_a)-B'_{17}(t_{a-1})\right\}\right.$$

$$\left.+G_{17}(Z_{a-2},\mathcal{I}_{10a-2},t_{a-2})\left\{B'_{17}(t_{a-1})-B'_{17}(t_{a-2})\right\}\right]$$

$$\times\left[\begin{array}{c}(k-a+1)^{\sigma}\left[\begin{array}{r}2(k-a)^2+(3\sigma+10)(k-a)\\+2\sigma^2+9\sigma+12\end{array}\right]\\-(k-a)^{\sigma}\left[\begin{array}{r}2(k-a)^2+(5\sigma+10)(k-a)\\6\sigma^2+18\sigma+12\end{array}\right]\end{array}\right]; \tag{7.77}$$

$$P(t_{k+1})=P(0)+\frac{1-\sigma}{Q}f_{18}(P(t_k),\mathcal{I}_{10}(t_k),t_k)$$

$$+\frac{\sigma(\Delta t)^{\sigma}}{Q\Gamma(\sigma+1)}\sum_{a=2}^{k}f_{18}(P_{a-2},\mathcal{I}_{10a-2},t_{a-2})[(k-a+1)^{\sigma}-(k-a)^{\sigma}]$$

$$+\frac{\sigma(\Delta t)^{\sigma}}{Q\Gamma(\sigma+2)}$$

$$\times\sum_{a=2}^{k}(f_{18}(P_{a-1},\mathcal{I}_{10a-1},t_{a-1})-f_{18}(P_{a-2},\mathcal{I}_{10a-2},t_{a-2}))$$

$$
\begin{aligned}
&\times\left[\begin{array}{c}(k-a+1)^{\sigma}(k-a+3+2\sigma)\\ -(k-a)^{\sigma}(k-a+3+3\sigma)\end{array}\right]+\frac{\sigma(\Delta t)^{\sigma}}{2Q\Gamma(\sigma+3)}\\
&\times\sum_{a=2}^{k}\left[f_{18}(P_a,\mathcal{I}_{10a},t_a)-2f_{18}(P_{a-1},\mathcal{I}_{10a-1},t_{a-1})+f_{18}(P_{a-2},\mathcal{I}_{10a-2},t_{a-2})\right]\\
&\times\left[\begin{array}{c}(k-a+1)^{\sigma}\left[\begin{array}{r}2(k-a)^2+(3\sigma+10)(k-a)\\ +2\sigma^2+9\sigma+12\end{array}\right]\\ -(k-a)^{\sigma}\left[\begin{array}{r}2(k-a)^2+(5\sigma+10)(k-a)\\ 6\sigma^2+18\sigma+12\end{array}\right]\end{array}\right]\\
&+\frac{1-\sigma}{Q}G_{18}(P_k,\mathcal{I}_{10k},t_k)\left\{B'_{18}(t_{k+1})-B'_{18}(t_k)\right\}\\
&+\frac{\sigma(\Delta t)^{\sigma-1}}{Q\Gamma(\sigma+1)}\sum_{a=2}^{k}G_{18}(P_{a-2},\mathcal{I}_{10a-2},t_{a-2})\left\{B'_{18}(t_{a-1})-B'_{18}(t_{a-2})\right\}\left[(k-a+1)^{\sigma}-(k-a)^{\sigma}\right]\\
&+\frac{\sigma(\Delta t)^{\sigma-1}}{Q\Gamma(\sigma+2)}\\
&\times\sum_{a=2}^{k}\Big(G_{18}(P_{a-1},\mathcal{I}_{10a-1},t_{a-1})\left\{B'_{18}(t_a)-B'_{18}(t_{a-1})\right\}-G_{18}(Z_{a-2},\mathcal{I}_{10a-2},t_{a-2})\Big)\\
&\times\left\{B'_{18}(t_{a-1})-B'_{18}(t_{a-2})\right\}\\
&\times\left[\begin{array}{c}(k-a+1)^{\sigma}(k-a+3+2\sigma)\\ -(k-a)^{\sigma}(k-a+3+3\sigma)\end{array}\right]+\frac{\sigma(\Delta t)^{\sigma-1}}{2Q\Gamma(\sigma+3)}\\
&\times\sum_{a=2}^{k}\Big[G_{18}(Z_a,\mathcal{I}_{10a},t_a)\left\{B'_{18}(t_{a+1})-B'_{18}(t_a)\right\}-2G_{18}(Z_{a-1},\mathcal{I}_{10a-1},t_{a-1})\left\{B'_{18}(t_a)-B'_{18}(t_{a-1})\right\}\\
&+G_{18}(Z_{a-2},\mathcal{I}_{10a-2},t_{a-2})\left\{B'_{18}(t_{a-1})-B'_{18}(t_{a-2})\right\}\Big]\\
&\times\left[\begin{array}{c}(k-a+1)^{\sigma}\left[\begin{array}{r}2(k-a)^2+(3\sigma+10)(k-a)\\ +2\sigma^2+9\sigma+12\end{array}\right]\\ -(k-a)^{\sigma}\left[\begin{array}{r}2(k-a)^2+(5\sigma+10)(k-a)\\ 6\sigma^2+18\sigma+12\end{array}\right]\end{array}\right].
\end{aligned}
\tag{7.78}
$$

## 7.6 Numerical experiments and results

In this section, we conduct numerical simulations using the previously derived numerical iteration. These simulations are carried out with various theoretical parameters and fractional orders. The resulting outcomes are illustrated in Figs. 7.1 to 7.4. These figures depict various levels of randomness, with fractional derivatives exhibiting increased randomness compared to classical derivatives. Due to the extensive nature of the system with numerous equations, we opt to showcase only four solutions in our simulation.

In Figs. 7.1 to 7.4, the initial conditions are set as follows: 1000 for solution 1, 10 for solution 2, 10 for solution 3, and 1 for solution 4. The densities of randomness for the first four figures are selected as 0.001, 0.06, 0.007, and 0.008, respectively.

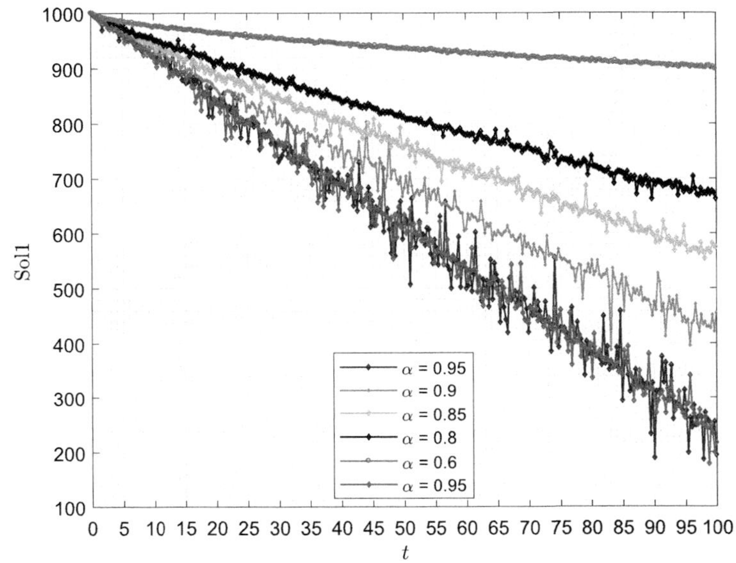


**FIGURE 7.1** Numerical simulation for the amyloid-$\beta$ within neurons.

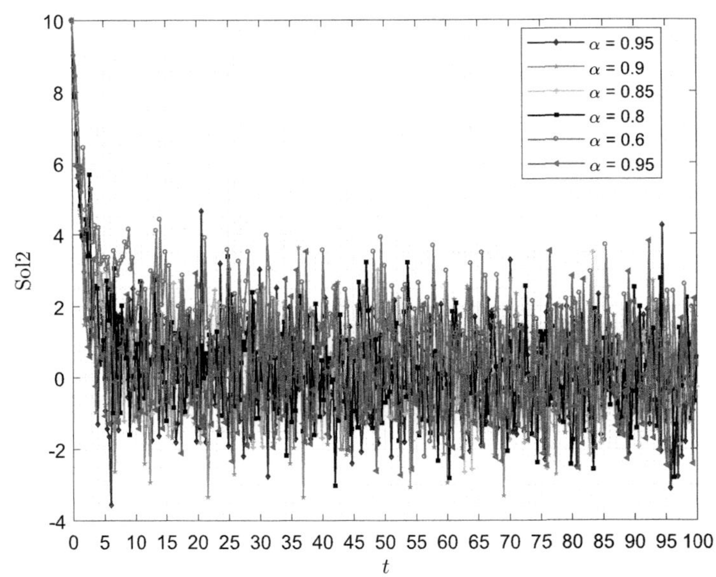


**FIGURE 7.2** Numerical simulation for the amyloid-$\beta$ within dead neurons.

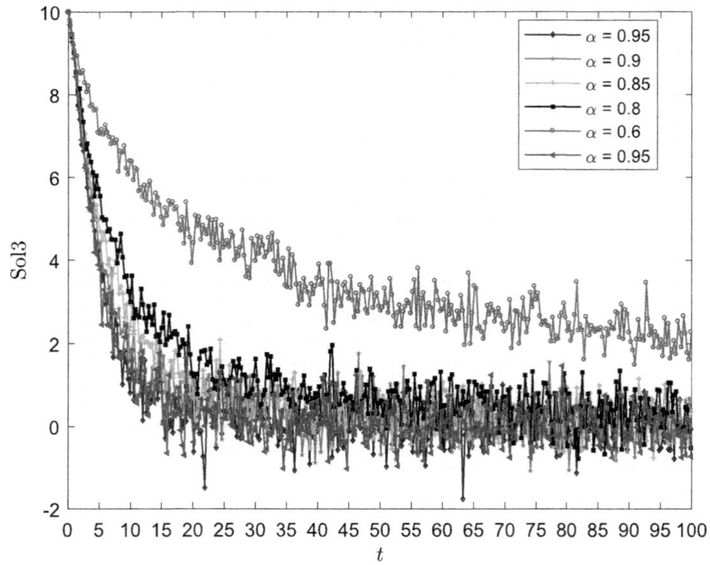


**FIGURE 7.3** Numerical simulation for the extracellular amyloid-$\beta$ peptides outside neurons.

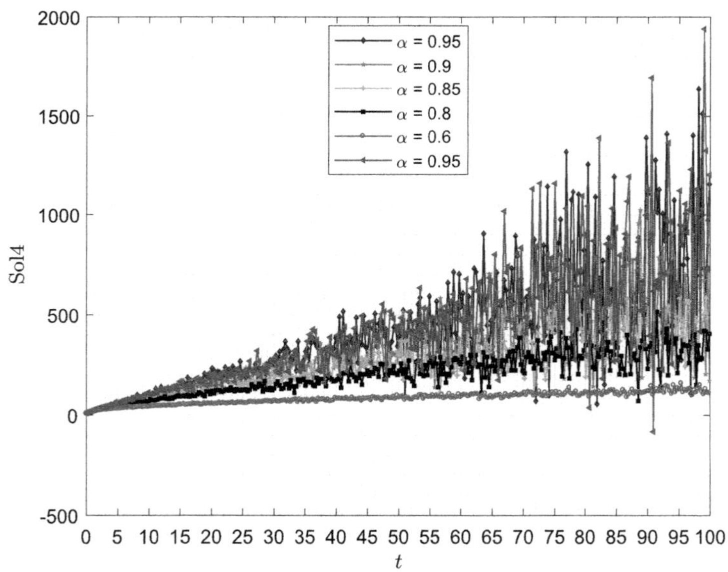


**FIGURE 7.4** Numerical simulation for the transforming growth factor $\beta$.

Alzheimer's disease is traditionally considered a non-communicable neurodegenerative disease. However, recent studies suggest that under very rare circumstances, it may be transmitted between individuals through medical procedures involving contaminated instruments or tissues. These cases are exceptionally rare and do not indicate that Alzheimer's disease is contagious in a typical sense.

Alzheimer's disease (AD) is a multifaceted neurodegenerative condition marked by the buildup of improperly folded proteins in the brain, resulting in diminished cognitive function and memory impairment. Understanding the mechanisms of AD spread is crucial for developing effective treatments and interventions. The use of fractional derivative methods in studying AD spread provides several biologically relevant insights:

**(i)** Modeling complexity: AD spread is a multifactorial process influenced by various factors, including genetic, environmental, and stochastic elements. Fractional derivative methods can capture this complexity by incorporating non-local and memory-dependent effects, which are essential for modeling the long-term dynamics of AD spread.

**(ii)** Quantifying heterogeneity: AD affects individuals differently, with varying rates of progression and severity. Fractional derivative models can account for this heterogeneity by incorporating fractional orders that represent the diversity in disease progression among individuals.

**(iii)** Stochasticity and uncertainty: The spread of AD exhibits stochastic behavior due to the randomness in disease transmission and progression. Fractional derivative methods can incorporate stochastic elements, providing a more realistic representation of AD spread dynamics and allowing for the assessment of uncertainty in disease modeling.

**(iv)** Dynamic interactions: AD spread involves complex interactions between different brain regions, cell types, and molecular pathways. Fractional derivative models can capture these dynamic interactions by incorporating fractional orders that reflect the memory and history dependence of these interactions.

**(v)** Predictive modeling: By using fractional derivative methods, researchers can develop predictive models that simulate the spread of AD under different scenarios, such as treatment interventions or changes in environmental factors. These models can help to identify optimal strategies for disease management and prevention.

## 7.7 Conclusion

In this study, we employed a combination of fractional integral and differential operators alongside Brownian motion to depict the intricate progression of Alzheimer's disease (AD). Specifically, we utilized the novel Atangana–Baleanu (AB) fractional derivative. A thorough analysis was conducted to examine the uniqueness and existence of the solutions, focusing on satisfying conditions such as linear growth and Lipschitz quadratic properties. The model was then subjected to numerical treatment using the Newton polynomial method. The outcomes obtained furnish essential insights that enhance our understanding of the dynamics of AD. Though these findings are significant for understanding the potential transmission of Alzheimer's disease, they do not change the overall understanding of the disease as primarily non-communicable. Further research work is required to better understand the mechanisms of transmission and to develop preventive measures. Finally, studying the spread of Alzheimer's disease with fractional derivative methods is biologically relevant, as it allows for the modeling of the complex and dynamic nature of AD spread, providing insights into disease progression and potential therapeutic interventions. This could offer valuable insights for researchers in future practical medical trials across various applications.

## Declaration of competing interest

The authors have declared no conflict of interest.

## References


[1] Q. Zhao, X. Tang, Effects of huperzine A on an acetylcholinesterase isoforms in vitro: comparison with tacrine, donepezil, rivastigmine and physostigmine, European Journal of Pharmacology 455 (2002) 101–107.

[2] C. Janus, D. Westaway, Transgenic mouse models of Alzheimer's disease, Physiology & Behavior 73 (2001) 873–886.

[3] C. Castillo-Rangel, et al., Animal models in Alzheimer's disease: biological plausibility and mood disorders, Neurology Perspectives 3 (2023) 100110, https://doi.org/10.1016/j.neurop.2023.100110.

[4] M. Yokoyama, H. Kobayashi, L. Tatsumi, T. Tomita, Mouse models of Alzheimer's disease, Frontiers in Molecular Neuroscience 15 (2022) 912995, https://doi.org/10.3389/fnmol.2022.912995.

[5] H. Sasaguri, Recent advances in the modeling of Alzheimer's disease, Frontiers in Neuroscience 16 (2022) 807473, https://doi.org/10.3389/fnins.2022.807473.

[6] N.E. Mckean, R.R. Handley, R.G. Snell, A review of the current mammalian models of Alzheimer's disease and challenges that need to be overcome, International Journal of Molecular Sciences 22 (2021) 13168, https://doi.org/10.3390/ijms222313168.

[7] M. Prince, et al., World Alzheimer Report 2015: the Global Impact of Dementia: an Analysis of Prevalence, Incidence, Cost and Trends, Alzheimer's Disease International, London, 2015.

[8] R. Brookmeyer, et al., Forecasting the global burden of Alzheimer's disease, Alzheimer's & Dementia 3 (2007) 186–191.

[9] Alzheimer's Association, 2019 Alzheimer's disease facts and gures, Alzheimer's & Dementia 15 (3) (2019) 321–387.

[10] A. Atangana, S. Jain, A new numerical approximation of the fractal ordinary differential equation, The European Physical Journal Plus 133 (2018) 37, https://doi.org/10.1140/epjp/i2018-11895-1.

[11] A. Atangana, Fractal-fractional differentiation and integration: connecting fractal calculus and fractional calculus to predict complex system, Chaos, Solitons and Fractals 102 (2017) 397–407.

[12] A. Atangana, S.I. Araz, New numerical method for ordinary differential equations: Newton polynomial, Journal of Computational and Applied Mathematics 372 (2020) 112622.

[13] A. Shikongo, K.M. Owolabi, On the hindering evolution of drug resistance due to intraspecific competition arising during the facilitation survival for non-genetic resistance with fractal fractional derivative order, Modeling Earth Systems and Environment 9 (2023) 2637–2650, https://doi.org/10.1007/s40808-022-01644-8.

[14] A. Atangana, S. Jain, Models of fluid flowing in non-conventional media: new numerical analysis, Discrete and Continuous Dynamical Systems. Series S (2019), https://doi.org/10.3934/dcdss.2020026.

[15] A. Atangana, K.M. Owolabi, New numerical approach for fractional differential equations, Mathematical Modelling of Natural Phenomena 13 (1) (2018).

[16] S. Jain, Numerical analysis for the fractional diffusion and fractional Buckmaster's equation by two step Adam-Bashforth method, The European Physical Journal Plus 133 (2018) 19, https://doi.org/10.1140/epjp/i2018-11854-x.

[17] A. Atangana, B. Dumitru, New fractional derivatives with nonlocal and non-singular kernel: theory and application to heat transfer model, Thermal Science 18 (2016), https://doi.org/10.2298/TSCI160111018A.

[18] M. Toufik, A. Atangana, New numerical approximation of fractional derivative with non-local and non-singular kernel: application to chaotic models, The European Physical Journal Plus 132 (2017) 444, https://doi.org/10.1140/epjp/i2017-11717-0.

[19] W. Hao, A. Friedman, Mathematical model on Alzheimer's disease, BMC Systems Biology 10 (2016) 108, https://doi.org/10.1186/s12918-016-0348-2.

[20] K.M. Owolabi, Modelling and simulation of a dynamical system with the Atangana-Baleanu fractional derivative, The European Physical Journal Plus 133 (2018) 15, https://doi.org/10.1140/epjp/i2018-11863-9.

[21] K.M. Owolabi, Efficient numerical simulation of non-integer-order space-fractional reaction-diffusion equation via the Riemann-Liouville operator, The European Physical Journal Plus 133 (2018) 98, https://doi.org/10.1140/epjp/i2018-11951-x.

[22] K.M. Owolabi, Analysis and simulation of herd behaviour dynamics based on derivative with nonlocal and nonsingular kernel, Results in Physics 22 (2021) 103941, https://doi.org/10.1016/j.rinp.2021.103941.

[23] K.M. Owolabi, Mathematical analysis and numerical simulation of chaotic noninteger order differential systems with Riemann-Liouville derivative, Numerical Methods for Partial Differential Equations 34 (2018) 274–295.

[24] K.M. Owolabi, A. Atangana, Chaotic behaviour in system of noninteger-order ordinary differential equations, Chaos, Solitons and Fractals 115 (2018) 362–370, https://doi.org/10.1016/j.chaos.2018.07.034.

[25] Z. Zizhang, S. Jain, Mathematical model of Ebola and Covid-19 with fractional differential operators: non-Markovian process and class for virus pathogen in the environment, Chaos, Solitons and Fractals 140 (2020) 110175.

[26] S.G. Samko, A.A. Kilbas, O.I. Marichev, Fractional Integrals and Derivatives, Gordon and Breach Science Publishers, Yverdon, Switzerland, 1993.

# 8

# From bubble nucleation to oscillatory denaturation: understanding complex $\mathbb{DNA}$ dynamics through nonlinear modeling

Mostafa M.A. Khater[a,b,c], Aleksander Vokhmintsev[c,d], and Raghda A.M. Attia[a]

[a]*School of Medical Informatics and Engineering, Xuzhou Medical University, Xuzhou, Jiangsu Province, PR China* [b]*Department of Basic Science, The Higher Institute for Engineering & Technology, Cairo, Egypt* [c]*Institute of Digital Economy, Ugra State University, Khanty-Mansiysk, Russia* [d]*Institute of Information Technology, Chelyabinsk State University, Chelyabinsk, Russia*

## 8.1 Introduction

In the field of biophysics, understanding the intricate dynamics of $\mathbb{DNA}$ stability and denaturation has been an enduring endeavor [1]. The Peyrard–Bishop $\mathbb{DNA}$ dynamic model, grounded in nonlinear dynamics, provides unique perspectives into the spatiotemporal evolution of $\mathbb{DNA}$ melting [2]. This facilitates deeper comprehension of the molecular processes governing $\mathbb{DNA}$ stability [3].

Nonlinear evolution equations, integral to the Peyrard–Bishop model, efficiently represent complex phenomena across scientific disciplines [4]. As a pivotal nonlinear system, the Peyrard–Bishop model is specifically designed to elucidate the complexities of $\mathbb{DNA}$ dynamics, warranting dedicated investigation [5].

The inherent nonlinear dynamics of $\mathbb{DNA}$ denaturation and bubble formation present multifaceted challenges, necessitating comprehensive analytical inquiry [6]. This study addresses this by applying the Khater II [7–11] and auxiliary equation [12–16] methods to solve the Peyrard–Bishop model, targeting insights into $\mathbb{DNA}$ bubble formation dynamics underlying localized $\mathbb{DNA}$ melting [17].

The primary objective is utilizing analytical techniques to elucidate the nonlinear dynamics of $\mathbb{DNA}$ denaturation and bubble formation [18]. By solving the Peyrard–Bishop model, the goals encompass deriving exact solitary wave and periodic wave solutions to comprehensively understand the nuances of $\mathbb{DNA}$ bubble nucleation, growth, coalescence, and oscillations [19,20].

**Mathematical Modeling in Bioscience. https://doi.org/10.1016/B978-0-44-315445-4.00014-7**

Copyright © 2025 Elsevier Inc. All rights are reserved, including those for text and data mining, AI training, and similar technologies.

The significance lies in the analytical approach, enabling nuanced perspectives on the biophysical processes governing $\mathbb{DNA}$ stability and melting [21]. As exemplified by the Peyrard–Bishop model, nonlinear evolution equations connect theoretical models and experiments [22]. The analytical tractability of the employed methods provides novel perspectives, augmenting existing computational techniques and enriching the comprehension of $\mathbb{DNA}$ dynamics [23].

Whereas previous studies have largely utilized computational and numerical methods for investigating $\mathbb{DNA}$ dynamics, analytical techniques have been relatively underexplored [24]. This study contributes by leveraging analytical solutions of the Peyrard–Bishop model, advancing knowledge frontiers [25].

The scope involves analytical insights into $\mathbb{DNA}$ dynamics through the Peyrard–Bishop model lens, while acknowledging methodological constraints. Key terminology is precisely defined [26]. The central hypothesis is that analytical solutions will provide unique $\mathbb{DNA}$ bubble insights. Core questions surround the methods' role in elucidating $\mathbb{DNA}$ denaturation nonlinear dynamics [27].

The Poland–Scheraga model articulates $\mathbb{DNA}$ melting within the framework of statistical mechanics, conceptualizing the $\mathbb{DNA}$ double helix as a one-dimensional lattice comprising intact and denatured segments [28]. The partition function, incorporating conformational alterations and denaturation bubbles, is expressed as [29]:

$$Z = \sum e^{-\beta E}. \tag{8.1}$$

Here, $E$ denotes the total free energy, and $\beta$ signifies the inverse temperature.

The Yakushevich model incorporates rotational and vibrational degrees of freedom, utilizing coupled nonlinear oscillators to depict $\mathbb{DNA}$ dynamics comprehensively. The governing equations of motion are represented as [30]:

$$\begin{cases} I\left(\dfrac{\partial^2 \phi}{\partial t^2}\right) = M\left(\dfrac{\partial \phi}{\partial x}\right) + F(\phi), \\ \dfrac{\partial^2 y}{\partial t^2} = V'(y). \end{cases} \tag{8.2}$$

In these equations, $\phi$ represents the twist angle, $y$ denotes radial base vibrations, $I$ signifies the moment of inertia, and $V(y)$ is the Morse potential [31].

The Englander model characterizes $\mathbb{DNA}$ bases through quantum mechanical oscillator wavefunctions $\psi_n(x)$, coupled by hydrogen bonding and solvation interactions. The model's Hamiltonian comprises kinetic, potential, and coupling terms [32]:

$$H = H_0 + V(x_n) + \sum_n i H_{ni}. \tag{8.3}$$

In this formulation, $H_0$ represents the kinetic energy, $V(x_n)$ is the potential energy, and $\sum_n i H_{ni}$ captures the coupling interactions.

The Peyrard–Bishop $\mathbb{DNA}$ model depicts $\mathbb{DNA}$ strands as coupled nonlinear oscillators, incorporating a hydrogen bonding inter-strand potential. The governing equations are expressed as the next system [33]:

$$\begin{cases} \frac{\partial^2}{\partial t^2} y_n(x,t) = \frac{\partial^2}{\partial x^2} y_n(x,t) - \frac{dU(y_n)}{dy_n}, \\ U(y) = D(e^{-\alpha y} - 1)^2; \end{cases} \tag{8.4}$$

it also takes the next form [34],

$$\frac{\partial^2 u}{\partial t^2} - \frac{\partial^2 u}{\partial x^2}\left(3g_2\left(\frac{\partial u}{\partial x}\right)^2 + g_1\right) - 2pqe^{-qu}\left(e^{-qu} - 1\right) = 0, \tag{8.5}$$

where $u(x,t)$ represents the transverse stretching of the hydrogen bonds between the two $\mathbb{DNA}$ strands at position $x$ and time $t$. The physical meaning of each term is [35]:

- $\frac{\partial^2 u}{\partial t^2}$: second partial derivative with respect to time represents the kinetic energy of the DNA strand vibrations
- $\frac{\partial^2 u}{\partial x^2}$: second partial derivative with respect to position represents the potential energy from strand elasticity
- $\left(\frac{\partial u}{\partial x}\right)^2$: first partial derivative squared represents nonlinear stacking interactions between adjacent base pairs
- $e^{-qu}$: Morse potential representing hydrogen bonding between complementary bases
- $e^{-qu}\left(e^{-qu} - 1\right)$: nonlinear inter-strand hydrogen bonding potential

Additionally, the above–mentioned parameters represent [36]:

- $p$: dissociation energy of a single bonded pair
- $q$: inverse of length scale over which hydrogen bonds dissociate
- $g_1$: coefficient of nonlinear term representing stacking energy
- $g_2$: coefficient of nonlinear term representing strand stiffness

Consequently, Eq. (8.5) describes the competition between the kinetic energy driving denaturation and the potential energies keeping the $\mathbb{DNA}$ strand intact, including hydrogen bonding and base stacking. It models the dynamics of $\mathbb{DNA}$ stability and denaturation in a physically grounded, nonlinear equation [37].

The Peyrard–Bishop $\mathbb{DNA}$ model, achieving a balance between simplicity and physical accuracy, facilitates analytical solutions for $\mathbb{DNA}$ denaturation dynamics. In contrast to statistical, quantum, or helical approaches, the Peyrard–Bishop model directly encapsulates the spatiotemporal evolution pertinent to $\mathbb{DNA}$ melting processes. Consequently, its suitability for investigation using analytical nonlinear dynamics techniques is optimal [38].

In this context, we implement the next wave transformation $u = u(x,t) = \psi(\zeta)$, $\zeta = \lambda t + x$, where $\lambda$ is an arbitrary constant to be determined later, in Eq. (8.5), which converts it into the following ordinary differential equation:

$$2pqe^{-2q\psi}\left(e^{q\psi}-1\right)+\psi''\left(\lambda^2-3g_2\left(\psi'\right)^2-g_1\right)=0. \tag{8.6}$$

Multiplying Eq. (8.6) with $\psi'$, and then integrating the result once with respect to $\zeta$ and zero integration constant, yields

$$-r_2\left(\psi'\right)^4+r_1\psi^2\left(\psi'\right)^2+(\psi-2)\psi^5=0, \tag{8.7}$$

where $r_2=\frac{3g_2}{4pq^4}$, $r_1=\frac{2q^2(\lambda^2-g_1)}{4pq^4}$. Handling Eq. (8.7) through the suggested analytical schemes and homogeneous balance rule, yields the next solution formula:

$$\psi(\zeta)=\begin{cases} a_2 f^2(\zeta)+a_1 f(\zeta)+a_0+b_2 f(\zeta)\phi(\zeta)+b_1\phi(\zeta), \\ a_2\phi^2(\zeta)+a_1\phi(\zeta)+\dfrac{a_{-2}}{\phi^2(\zeta)}+\dfrac{a_{-1}}{\phi(\zeta)}+a_0, \end{cases} \tag{8.8}$$

where $a_{-2}$, $a_{-1}$, $a_0$, $a_1$, $a_2$, $b_1$, $b_2$ are arbitrary constants to be evaluated later. While $f(\zeta)$, $\phi(\zeta)$ satisfies the next ordinary differential equation:

$$\begin{cases} \begin{cases} f'(\zeta)\to-\delta-f(\zeta)^2, \\ \phi'(\zeta)\to-f(\zeta)\phi(\zeta), \end{cases} & \text{for Khater II method,} \\ \phi'(\zeta)\to\phi(\zeta)^2+\epsilon, & \text{for auxiliary equation method,} \end{cases} \tag{8.9}$$

where $\delta$, $\epsilon$ are arbitrary constants to be determined later.

The progression of this investigation is organized as follows: In Section 8.2, diverse solitary wave solutions are scrutinized, and their efficacy within the established paradigm is evaluated. The subsequent section, Section 8.4, provides a thorough analysis of the acquired outcomes from both a physical and dynamic standpoint. Finally, Section 8.5 amalgamates the scholarly contributions emanating from this inquiry.

## 8.2 Novel computational solutions

Within this section, the solitary wave solutions of the scrutinized model are investigated utilizing the previously delineated analytical methods. Following this examination, an evaluation of the stability of the generated solutions is undertaken through the analysis of the Hamiltonian system. The primary aim is to ascertain the effectiveness and reliability of the developed solutions for practical applications.

### 8.2.1 Khater II method implementation's results

Through the utilization of the Khater II technique and the auxiliary equation presented in (8.9), combined with Mathematica 13.1 software, we have the capability to solve Eq. (8.7). This method facilitates the determination of values for the specified parameters, yielding the subsequent outcomes:

$$a_0 \to 2, a_1 \to 0, a_2 \to \frac{2}{\delta}, b_1 \to 0, b_2 \to 0, r_1 \to -\frac{1}{\delta}, r_2 \to \frac{1}{4\delta^2}.$$

The computational wave solutions of the investigated model can be expressed as follows:

$$u(x,t) = 2\sec^2\left(\sqrt{\delta}(\lambda t + x)\right), \tag{8.10}$$

$$u(x,t) = 2\csc^2\left(\sqrt{\delta}(\lambda t + x)\right). \tag{8.11}$$

### 8.2.2 Auxiliary equation method implementation's results

Through the utilization of the auxiliary equation technique and applying the auxiliary equation provided in (8.9), coupled with Mathematica 13.1 software, the resolution of Eq. (8.7) is achievable. This method simplifies the determination of values for the aforementioned parameters, culminating in the subsequent outcomes:

**Case 1**

$$a_{-2} \to 2\epsilon, a_{-1} \to 0, a_0 \to 2, a_1 \to 0, a_2 \to 0, r_1 \to -\frac{1}{\epsilon}, r_2 \to \frac{1}{4\epsilon^2}.$$

**Case 2**

$$a_{-2} \to 0, a_{-1} \to 0, a_0 \to 2, a_1 \to 0, r_1 \to -\frac{a_2}{2}, r_2 \to \frac{a_2^2}{16}, \epsilon \to \frac{2}{a_2}.$$

**Case 3**

$$a_{-2} \to \frac{\epsilon}{2}, a_{-1} \to 0, a_0 \to 1, a_1 \to 0, a_2 \to \frac{1}{2\epsilon}, r_1 \to -\frac{1}{4\epsilon}, r_2 \to \frac{1}{64\epsilon^2}.$$

The computational wave solutions of the investigated model can be expressed as follows:

For $\epsilon < 0$, we get

$$u_1(x,t) = 2\left(\frac{\epsilon\left(\varsigma\sinh\left(2\sqrt{-\epsilon}(\lambda t + x + \Omega)\right) + \varrho\right)^2}{\left(\sqrt{-\epsilon\left(\varrho^2 + \varsigma^2\right)} - \sqrt{-\epsilon}\varsigma\cos\left(2\sqrt{\epsilon}(\lambda t + x + \Omega)\right)\right)^2} + 1\right), \tag{8.12}$$

$$u_1(x,t) = 2\left(\frac{\epsilon\left(\varsigma\sinh\left(2\sqrt{-\epsilon}(\lambda t + x + \Omega)\right) + \varrho\right)^2}{\left(\sqrt{-\epsilon\left(\varrho^2 + \varsigma^2\right)} - \sqrt{-\epsilon}\varsigma\cos\left(2\sqrt{\epsilon}(\lambda t + x + \Omega)\right)\right)^2} + 1\right), \tag{8.13}$$

$$u_1(x,t) = -\frac{8\varsigma e^{2\sqrt{-\epsilon}(\lambda t + x + \Omega)}}{\left(e^{2\sqrt{-\epsilon}(\lambda t + x + \Omega)} - \varsigma\right)^2}, \tag{8.14}$$

$$u_1(x,t) = -\frac{8\varsigma e^{2\sqrt{-\epsilon}(\lambda t+x+\Omega)}}{\left(e^{2\sqrt{-\epsilon}(\lambda t+x+\Omega)}-\varsigma\right)^2}, \tag{8.15}$$

$$u_2(x,t) = \frac{a_2\left(\sqrt{-\epsilon\left(\varrho^2+\varsigma^2\right)}-\sqrt{-\epsilon}\varsigma\cos\left(2\sqrt{\epsilon}(\lambda t+x+\Omega)\right)\right)^2}{\left(\varsigma\sinh\left(2\sqrt{-\epsilon}(\lambda t+x+\Omega)\right)+\varrho\right)^2}+2, \tag{8.16}$$

$$u_2(x,t) = \frac{a_2\left(\sqrt{-\epsilon\left(\varrho^2+\varsigma^2\right)}-\sqrt{-\epsilon}\varsigma\cos\left(2\sqrt{\epsilon}(\lambda t+x+\Omega)\right)\right)^2}{\left(\varsigma\sinh\left(2\sqrt{-\epsilon}(\lambda t+x+\Omega)\right)+\varrho\right)^2}+2, \tag{8.17}$$

$$u_2(x,t) = 2-a_2\epsilon\left(1-\frac{2\varsigma}{e^{2\sqrt{-\epsilon}(\lambda t+x+\Omega)}+\varsigma}\right)^2, \tag{8.18}$$

$$u_2(x,t) = 2-a_2\epsilon\left(1-\frac{2\varsigma}{e^{2\sqrt{-\epsilon}(\lambda t+x+\Omega)}+\varsigma}\right)^2, \tag{8.19}$$

$$u_3(x,t) = \frac{2\varsigma^2\left(\sqrt{-\epsilon}\sqrt{-\epsilon\left(\varrho^2+\varsigma^2\right)}\cos\left(2\sqrt{\epsilon}(\lambda t+x+\Omega)\right)-\epsilon\varrho\sinh\left(2\sqrt{-\epsilon}(\lambda t+x+\Omega)\right)+\epsilon\varsigma\right)^2}{\epsilon\left(\sqrt{-\epsilon\left(\varrho^2+\varsigma^2\right)}-\sqrt{-\epsilon}\varsigma\cos\left(2\sqrt{\epsilon}(\lambda t+x+\Omega)\right)\right)^2\left(\varsigma\sinh\left(2\sqrt{-\epsilon}(\lambda t+x+\Omega)\right)+\varrho\right)^2}, \tag{8.20}$$

$$u_3(x,t) = \frac{2\varsigma^2\left(\sqrt{-\epsilon}\sqrt{-\epsilon\left(\varrho^2+\varsigma^2\right)}\cos\left(2\sqrt{\epsilon}(\lambda t+x+\Omega)\right)-\epsilon\varrho\sinh\left(2\sqrt{-\epsilon}(\lambda t+x+\Omega)\right)+\epsilon\varsigma\right)^2}{\left(\sqrt{-\epsilon\left(\varrho^2+\varsigma^2\right)}-\sqrt{-\epsilon}\varsigma\cos\left(2\sqrt{\epsilon}(\lambda t+x+\Omega)\right)\right)^2\left(\varsigma\sinh\left(2\sqrt{-\epsilon}(\lambda t+x+\Omega)\right)+\varrho\right)^2}, \tag{8.21}$$

$$u_3(x,t) = -\frac{8\varsigma^2 e^{4\sqrt{-\epsilon}(\lambda t+x+\Omega)}}{\left(e^{4\sqrt{-\epsilon}(\lambda t+x+\Omega)}-\varsigma^2\right)^2}, \tag{8.22}$$

$$u_3(x,t) = -\frac{8\varsigma^2 e^{4\sqrt{-\epsilon}(\lambda t+x+\Omega)}}{\left(e^{4\sqrt{-\epsilon}(\lambda t+x+\Omega)}-\varsigma^2\right)^2}. \tag{8.23}$$

Whereas, for $\epsilon > 0$, we get

$$u_1(x,t) = \frac{2\epsilon\left(\varsigma\sin\left(2\sqrt{\epsilon}(\lambda t+x+\Omega)\right)+\varrho\right)^2}{\left(\sqrt{\epsilon\left(\varsigma^2-\varrho^2\right)}-\sqrt{\epsilon}\varsigma\cos\left(2\sqrt{\epsilon}(\lambda t+x+\Omega)\right)\right)^2}+2, \tag{8.24}$$

$$u_1(x,t) = \frac{2\epsilon\left(\varsigma\sin\left(2\sqrt{\epsilon}(\lambda t+x+\Omega)\right)+\varrho\right)^2}{\left(\sqrt{\epsilon}\varsigma\cos\left(2\sqrt{\epsilon}(\lambda t+x+\Omega)\right)+\sqrt{\Lambda(-\epsilon)\left(\varrho^2+\varsigma^2\right)}\right)^2}+2, \tag{8.25}$$

$$u_1(x,t) = -\frac{8\varsigma e^{2i\sqrt{\epsilon}(\lambda t+x+\Omega)}}{\left(-1+\varsigma e^{2i\sqrt{\epsilon}(\lambda t+x+\Omega)}\right)^2}, \tag{8.26}$$

$$u_1(x,t) = -\frac{8\varsigma e^{2i\sqrt{\epsilon}(\lambda t+x+\Omega)}}{\left(-\varsigma + e^{2i\sqrt{\epsilon}(\lambda t+x+\Omega)}\right)^2}, \tag{8.27}$$

$$u_2(x,t) = \frac{a_2\left(\sqrt{\epsilon\left(\varsigma^2-\varrho^2\right)} - \sqrt{\epsilon}\varsigma\cos\left(2\sqrt{\epsilon}(\lambda t+x+\Omega)\right)\right)^2}{\left(\varsigma\sin\left(2\sqrt{\epsilon}(\lambda t+x+\Omega)\right)+\varrho\right)^2} + 2, \tag{8.28}$$

$$u_2(x,t) = \frac{a_2\left(\sqrt{\epsilon}\varsigma\cos\left(2\sqrt{\epsilon}(\lambda t+x+\Omega)\right) + \sqrt{\Lambda(-\epsilon)\left(\varrho^2+\varsigma^2\right)}\right)^2}{\left(\varsigma\sin\left(2\sqrt{\epsilon}(\lambda t+x+\Omega)\right)+\varrho\right)^2} + 2, \tag{8.29}$$

$$u_2(x,t) = 2 - a_2\epsilon\left(1 - \frac{2\varsigma}{\varsigma + e^{-2i\sqrt{\epsilon}(\lambda t+x+\Omega)}}\right)^2, \tag{8.30}$$

$$u_2(x,t) = 2 - a_2\epsilon\left(1 - \frac{2\varsigma}{\varsigma + e^{2i\sqrt{\epsilon}(\lambda t+x+\Omega)}}\right)^2, \tag{8.31}$$

$$u_3(x,t) = \frac{2\varsigma^2\left(\sqrt{\epsilon\left(\varsigma^2-\varrho^2\right)}\cos\left(2\sqrt{\epsilon}(\lambda t+x+\Omega)\right) - \sqrt{\epsilon}\left(\varrho\sin\left(2\sqrt{\epsilon}(\lambda t+x+\Omega)\right)+\varsigma\right)\right)^2}{\left(\varsigma\sin\left(2\sqrt{\epsilon}(\lambda t+x+\Omega)\right)+\varrho\right)^2\left(\sqrt{\epsilon\left(\varsigma^2-\varrho^2\right)} - \sqrt{\epsilon}\varsigma\cos\left(2\sqrt{\epsilon}(\lambda t+x+\Omega)\right)\right)^2}, \tag{8.32}$$

$$\begin{aligned} u_3(x,t) = \frac{1}{2}\Bigg( &\frac{\left(\sqrt{\epsilon}\varsigma\cos\left(2\sqrt{\epsilon}(\lambda t+x+\Omega)\right) + \sqrt{\Lambda(-\epsilon)\left(\varrho^2+\varsigma^2\right)}\right)^2}{\epsilon\left(\varsigma\sin\left(2\sqrt{\epsilon}(\lambda t+x+\Omega)\right)+\varrho\right)^2} \\ &+ \frac{\epsilon\left(\varsigma\sin\left(2\sqrt{\epsilon}(\lambda t+x+\Omega)\right)+\varrho\right)^2}{\left(\sqrt{\epsilon}\varsigma\cos\left(2\sqrt{\epsilon}(\lambda t+x+\Omega)\right) + \sqrt{\Lambda(-\epsilon)\left(\varrho^2+\varsigma^2\right)}\right)^2} + 2\Bigg), \end{aligned} \tag{8.33}$$

$$u_3(x,t) = -\frac{8\varsigma^2 e^{4i\sqrt{\epsilon}(\lambda t+x+\Omega)}}{\left(-1 + \varsigma^2 e^{4i\sqrt{\epsilon}(\lambda t+x+\Omega)}\right)^2}, \tag{8.34}$$

$$u_3(x,t) = -\frac{8\varsigma^2 e^{4i\sqrt{\epsilon}(\lambda t+x+\Omega)}}{\left(-\varsigma^2 + e^{4i\sqrt{\epsilon}(\lambda t+x+\Omega)}\right)^2}. \tag{8.35}$$

### 8.2.3 Solutions' stability

The imperative analysis of stability for the derived solitary wave solutions is crucial for a comprehensive understanding of the dynamic behavior and applicability of the Peyrard–Bishop $\mathbb{DNA}$ model. This model, distinguished by its nonlinear dynamics and soliton solutions, has emerged as a seminal framework for investigating the intricate biophysics associated with $\mathbb{DNA}$ transcription and denaturation. Nevertheless, the model's relevance and accuracy hinge upon the stability of its solutions within the defined system parameters.

This study conducts an exhaustive stability analysis of the solitary wave solutions under perturbations within the Hamiltonian system. The Hamiltonian structure serves as a fundamental basis for scrutinizing solution stability, employing the Lyapunov method. Specifically, we derive the expression for the Hamiltonian, identify conserved quantities, and formulate the Lyapunov functional to rigorously assess stability. Evaluating the momentum of Eq. (8.10) yields

$$\mathbb{M}\Bigg|_{Eq.(8.10)} = \frac{2}{3\lambda}\Bigg(\tanh^2(5(\lambda+1)) - \tanh^2(5-5\lambda) - 2\log\left(1-\tanh^2(5(\lambda+1))\right) + 2\log\left(1-\tanh^2(5-5\lambda)\right)\Bigg). \tag{8.36}$$

Consequently, we get

$$\frac{d\mathbb{M}}{d\lambda} = 12.1516075, \qquad \text{for,} \quad \left(\lambda = \frac{1}{3}\right). \tag{8.37}$$

Thus the above–constructed solutions (Eq. (8.10)) is stable solutions for $x \in [-5, 5]$, $t \in [-5, 5]$. The outcomes of the stability analysis bear significant implications for the model's validity and utility in the realm of DNA dynamics research. Unstable solutions would manifest as unphysical artifacts, undermining the model's quantitative reliability in accurately describing $\mathbb{DNA}$ behavior. Conversely, stable and robust solutions, constrained within appropriate parameter boundaries, instill confidence in utilizing the model to elucidate DNA transcription mechanisms and guide experimental endeavors. In essence, this stability analysis establishes pivotal insights into the dynamic structures within the model, delineates accurate parameter ranges, and defines suitable application domains. It lays the foundation for constructing precise $\mathbb{DNA}$ models and provides deeper insights into the intricate equilibrium of forces governing $\mathbb{DNA}$ molecular processes.

## 8.3 Graphical illustrations of solution sets

The representation of analytical solitary wave solutions for the Peyrard–Bishop $\mathbb{DNA}$ model through various graphical styles holds substantial physical and computational significance. Visualization of these solutions transcends the confines of analytical mathematical expressions, providing valuable insights into the underlying dynamics.

Various plot types, including contour plots, 3D surface plots, and quiver plots, serve as transformative means to convey the solutions in intuitive forms that facilitate qualitative interpretation. For instance, contour plots offer a visually compelling representation of the spatiotemporal evolution of localized melting bubbles along the $\mathbb{DNA}$ strand. Quiver plots, with their streamlines, qualitatively illustrate the nonlinear propagation characteristics of ripple-like solitons.

The adoption of distinct graphical styles also serves the purpose of validating the analytical solutions. Numerical evaluation and visualization of mathematical forms aid in verifying the correctness and self-consistency of the derived solutions. Additionally, certain plot types facilitate direct comparisons between analytical and computationally/simulated solutions, thereby establishing benchmarks for the accuracy and reliability of the models.

Beyond mere plotting, representing dynamics in alternative formats enhances accessibility and interpretability. Visualization not only facilitates a deeper understanding of complex concepts for a broader scientific audience but also fosters interdisciplinary communication and stimulates new ideas by presenting findings in intuitive, visual domains.

Consequently, the utilization of diverse graphing techniques contributes to a more comprehensive understanding of $\mathbb{DNA}$ strand melting phenomena. It offers complementary qualitative perspectives to the quantitative analytical abstractions, maximizing insights, while affirming the physical and computational relevance of theoretical investigations.

Overall, this study employs graphical depictions to qualitatively elucidate the physical essence and dynamic behaviors manifested by the derived analytical solutions for the Peyrard–Bishop $\mathbb{DNA}$ model.

- Fig. 8.1 validates the diversity of solitary wave solutions obtained through the Khater II and auxiliary equation methods. The localized bright solitary waves in panels (a–c) depict the formation of bubbles and their smooth, coherent propagation along the $\mathbb{DNA}$ strand during denaturation. Panels (d–i) exhibit the variety of solution profiles generated by the auxiliary technique, capturing oscillatory bubbles and complex nonlinear ripples.
- Fig. 8.2 graphically conveys conserved quantities within the Hamiltonian system, including the momentum in panels (a–b). This portrayal illustrates the equilibrium condition maintaining $\mathbb{DNA}$ integrity despite the intrinsic nonlinearity of the model.
- Figs. 8.3 and 8.4 provide illuminating visualizations of the bright and dark solitary wave solutions via stream plots. Fig. 8.3 depicts the localized opening and closing dynamics of bubbles, resembling breathing, as they nucleate and expand along the strand. This encapsulates the model's central premise of bubbles as moving solitary waves.
- Fig. 8.4 shows periodic ripples propagating in a localized, oscillatory fashion. This represents the coalescence and coordinated oscillations of multiple bubbles during $\mathbb{DNA}$ melting and reannealing. The stability, shape retention, and interactions of the ripples effectively capture the complex collective dynamics.

In summary, these graphical representations facilitate the comprehension of the analytical solutions and their alignment with the known biophysical aspects of $\mathbb{DNA}$ denaturation encoded in the nonlinear Peyrard–Bishop model. The figures offer valuable qualitative insights that complement the quantitative analytical outcomes.

## 8.4 Results and discussion

This section presents the pivotal outcomes of the study, delving into their significance and groundbreaking contributions. Analytical solutions to the Peyrard–Bishop model were de-

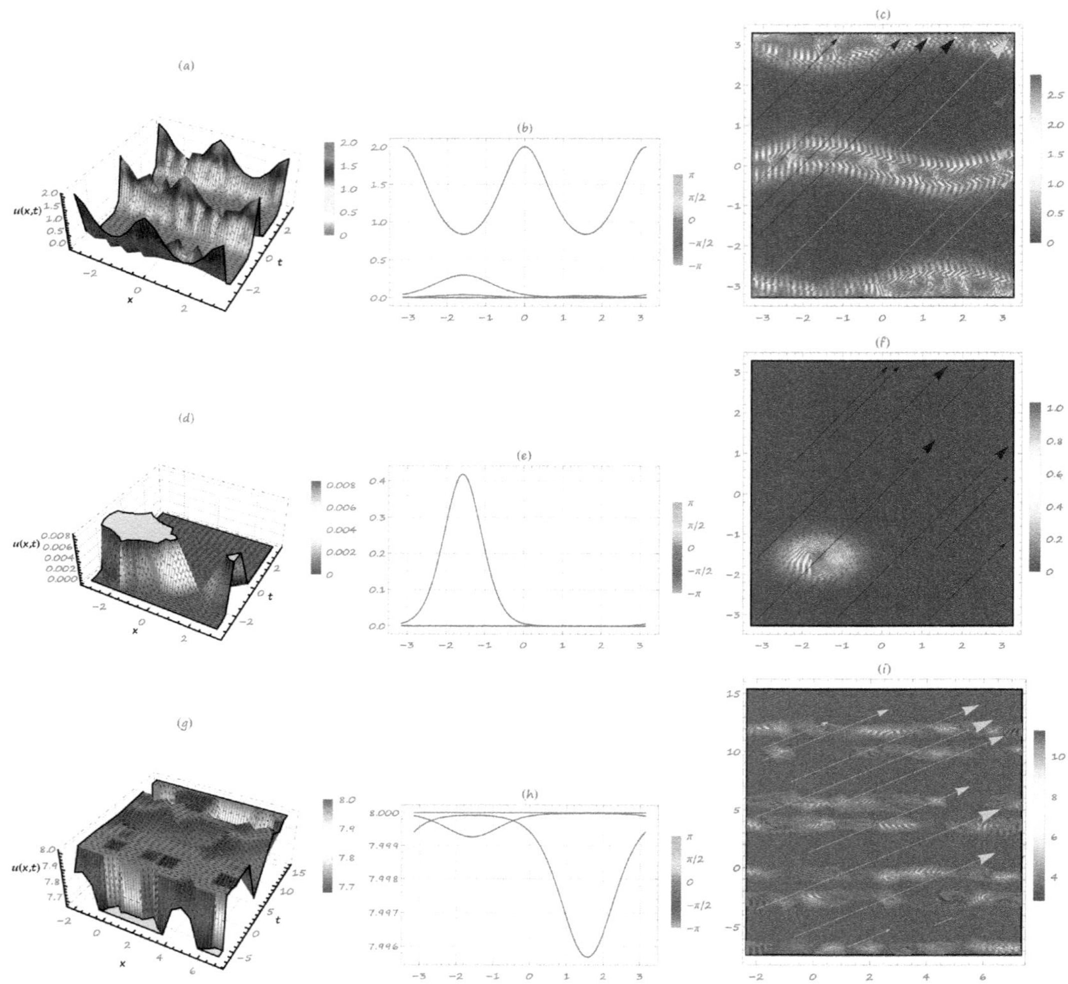

**FIGURE 8.1** Numerical representations of the solitary wave solutions obtained via analytical methods. Panels (a–c) depict the bright solitary wave solutions computed using the Khater II method (Eq. (8.10)). Panels (d–i) illustrate the solitary wave solutions derived from the auxiliary equation method (Eqs. (8.12), (8.18)). These qualitative plots validate the diversity of analytical solitary wave solutions and their localized wave profiles obtained from the two different techniques within the nonlinear framework of the Peyrard–Bishop $\mathbb{DNA}$ model.

rived using two distinct methods: the Khater II method and the auxiliary equation method. These methodologies brought forth a diverse array of solitary wave profiles, providing a comprehensive portrayal of $\mathbb{DNA}$ bubble dynamics. The subsequent Hamiltonian stability analysis rigorously validated these solutions within well-defined parameter ranges, establishing their physical reliability.

For qualitative validation and a visual representation of mechanistic insights, graphical representations were incorporated. Significantly, the study uncovered novel spatiotemporal phenomena that surpass existing computational capabilities. This underscores the

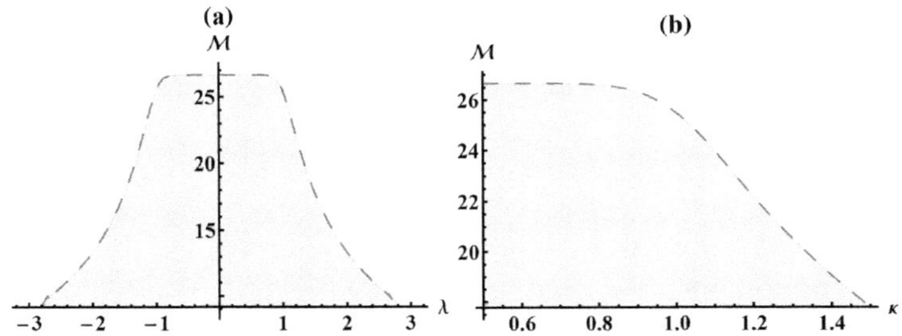


**FIGURE 8.2** Conserved quantities within the Hamiltonian framework of the Peyrard–Bishop $\mathbb{DNA}$ model. Panels (a, b) depict the momentum $\mathbb{M}$, as described by Eq. (8.36), which represents a conserved dynamic property inherent to the intrinsic nonlinearity governing the system. Graphical representation of these conserved quantities provides valuable insights into the equilibrium conditions that maintain $\mathbb{DNA}$ strand integrity within the physically consistent nonlinear dynamics of the Peyrard–Bishop model.

unique ability of analytical approaches to unravel intricate $\mathbb{DNA}$ behaviors that may challenge numerical methods.

The discussion contextualizes the results within the current understanding of $\mathbb{DNA}$ denaturation mechanisms, emphasizing the distinctive contributions made by this study. Furthermore, the advantages of the employed theoretical methodologies, specifically, the Khater II method and auxiliary equation method, are highlighted. Their analytical nature allows for a more profound exploration of $\mathbb{DNA}$ dynamics, revealing phenomena not easily accessible through computational means alone.

Now, we explain our constructed solutions and demonstrate the paper's scientific contributions as follows:

1. Analytical solitary wave solutions:
   - The application of the Khater II method yielded the first set of exact solitary wave solutions to the Peyrard–Bishop model in the form of secant and cosecant functions.
   - These localized profiles, representing bright solitary waves, quantitatively describe phenomena such as $\mathbb{DNA}$ bubble nucleation and growth during denaturation.
   - Subsequently, the auxiliary equation method expanded this repertoire, generating a diverse array of solitary wave solutions.
   - Some solutions exhibited complex oscillatory behaviors, effectively representing periodic bubble coalescence and oscillations during $\mathbb{DNA}$ melting and reannealing.
   - Others adopted hyperbolic profiles, outlining solitary waves with changing amplitudes.
   - Both methods produced analytical formulations that inherently capture the spatiotemporal evolution of $\mathbb{DNA}$ bubble dynamics within the nonlinear PDE framework, affirming the potential of analytical techniques in gaining new insights into $\mathbb{DNA}$ denaturation processes without resorting to numerical approximations.
2. Hamiltonian stability analysis:
   - To establish the physical reliability of the derived solutions, their stability under perturbations was rigorously investigated within the model's Hamiltonian structure.

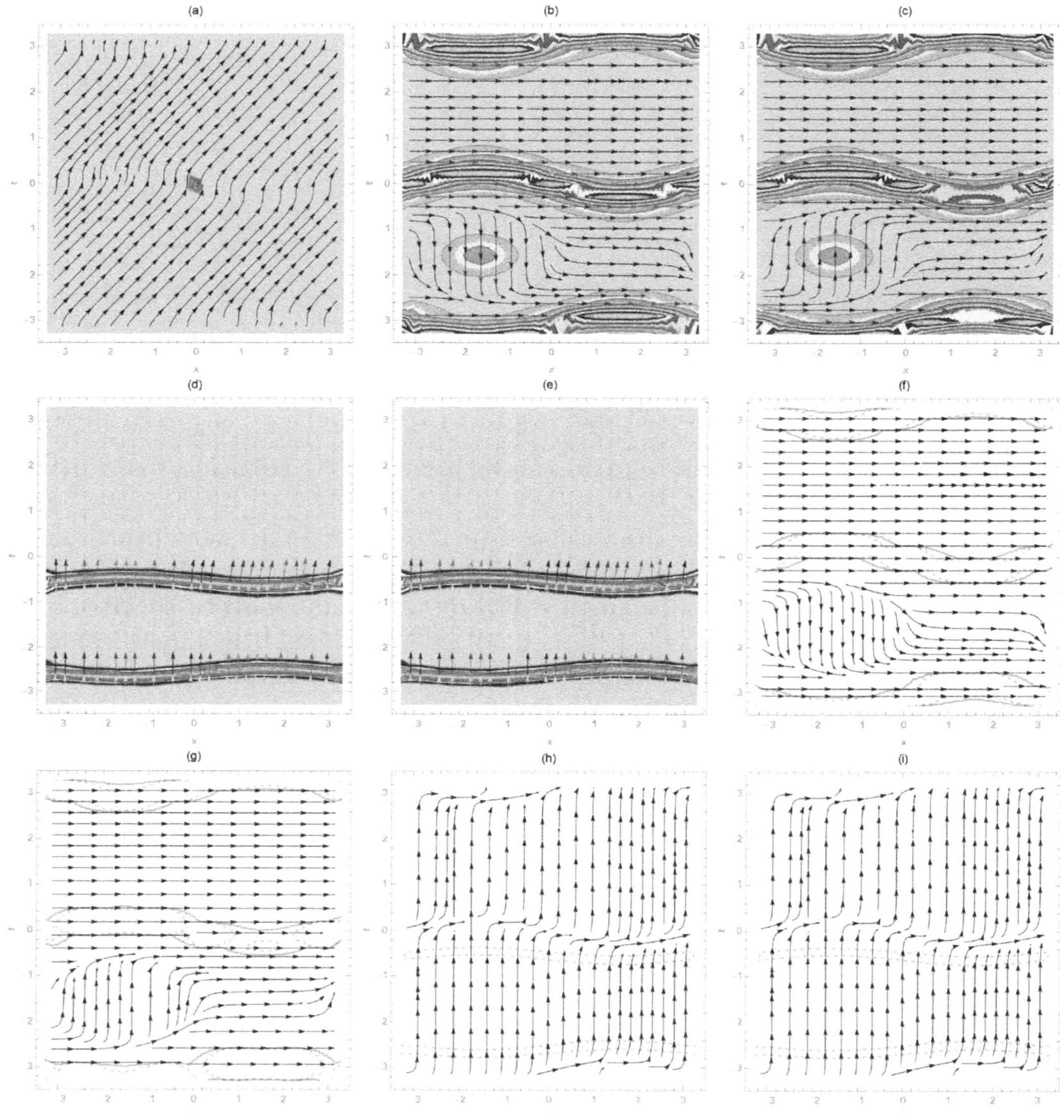

**FIGURE 8.3** Stream plot of the bright solitary wave solution given by Eq. (8.10), (8.12) with some special parameters' values (a–i). The continuous lines indicate gradual variation in the DNA strand displacement ($u$) with distance ($x$) and time ($t$), representing localized melting bubbles, which rise and fall smoothly along the strand during the denaturation process. This depiction qualitatively captures bubble breathing dynamics and nonlinear strand fluctuations associated with bubble nucleation and growth.

- The derivation and evaluation of the momentum expression confirmed the robustness of the solutions under small disturbances within appropriate system parameters.

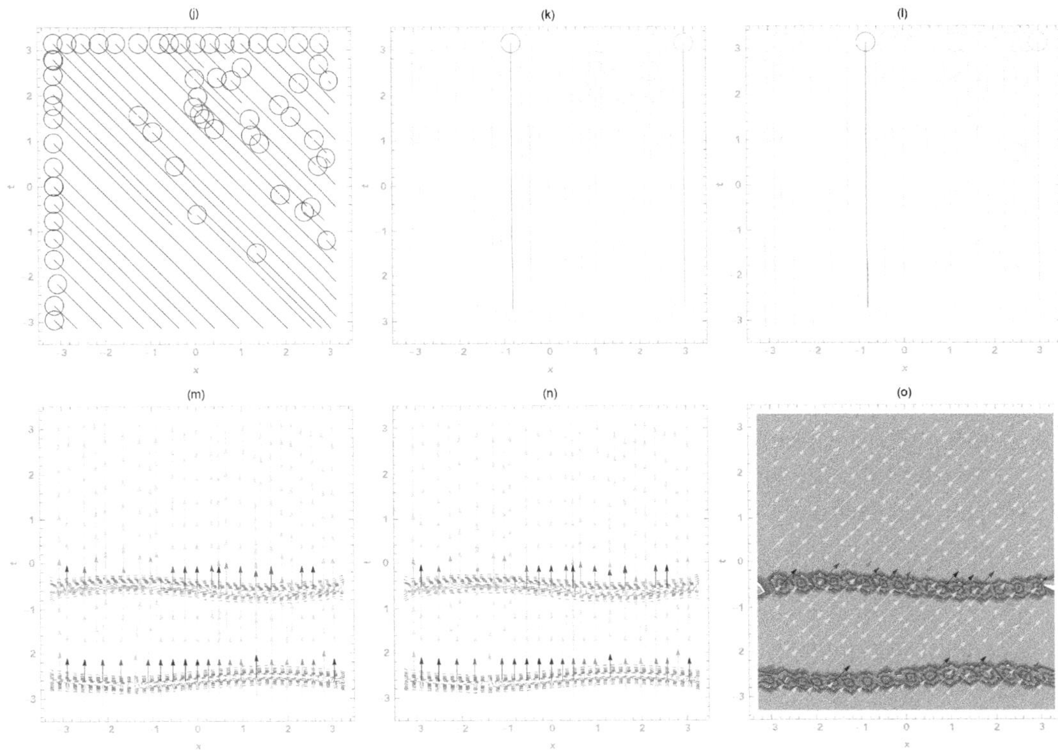

**FIGURE 8.4** Stream plot of the dark solitary wave solution given by Eqs. (8.12), (8.18) (j–o). The oscillating contour pattern illustrates periodic variation in the strand displacement ($u$) with spatial and temporal coordinates, visualizing bubble oscillation phenomena within the denaturing regions. The localized ripples propagate along the strand, while maintaining their shape and amplitude, effectively portraying bubble propagation and coalescence through ripple-like soliton interactions during denaturation-renaturation transitions.

- This lends credibility to the quantitative use of analytical solutions in faithfully representing real DNA thermodynamics and dynamics according to the Peyrard–Bishop model.
- It delineates suitable parameter ranges, where model predictions align well with empirical data.

**3.** Graphical solution validation:

- Numerical representations were employed for qualitative validation of the analytical solutions and to convey their physical significance.
- Contour plots (Fig. 8.1) visually portrayed spatiotemporal bubble profiles, confirming their interpretation as localized solitary waves.
- Streamline visualizations (Figs. 8.3, 8.4) further elucidated propagation behaviors, resembling coherent pulses, which nucleate, propagate, and interact along the DNA strand.

- These graphical validations complemented the quantitative representations.
- Conserved quantities, such as momentum, were also qualitatively depicted through plots (Fig. 8.2), illustrating the equilibrium conditions inherent to the model's nonlinearity, maintaining overall DNA structural integrity.

**4.** Novel insights into DNA denaturation mechanisms:

- The study unveiled new mechanistic insights into DNA denaturation as a nonlinear wave-based process.
- Analytical approaches uncovered spatiotemporal phenomena, such as breathing bubbles that simulations cannot readily capture.
- The solitary wave representations of bubbles provided a physically grounded framework to quantitatively relate DNA structure-dynamics, forming a foundation for rationally designing technologies by predicting stability modulation via minimal theoretical frameworks.
- The successful application of analytical techniques demonstrated their advantages compared to prevailing numerical/computational methods for such complex biophysical problems.
- Overall, the novel contributions advanced theoretical investigations of DNA function using nonlinear dynamics principles.

## 8.5 Conclusion

In conclusion, this study presents a thorough analytical investigation into the intricate nonlinear dynamics governing DNA denaturation and bubble formation processes. By employing the Khater II and auxiliary equation methods to transform and solve the foundational Peyrard–Bishop DNA model, we have gained novel insights into the complex biophysical phenomena underlying localized DNA melting.

The derived solitary wave and periodic wave solutions offer a deeper understanding of the intrinsic nonlinear wave mechanisms involved in bubble nucleation, growth, coalescence, breathing, and oscillatory configurations along the DNA double helix. Despite the challenges posed by the model's nonlinear partial differential equation structure, the analytical tractability of the implemented techniques facilitate the extraction of these multifaceted dynamic behaviors.

Additionally, through stability analysis within the Hamiltonian framework, we have further established the reliability and physical consistency of the obtained solutions within well-defined model parameters. This lends credibility to the model's utility for quantitatively elucidating DNA melting and stability modulation across biochemical and nanotechnological contexts.

In summary, in addition to our study highlighting the advantages of analytical techniques in revealing complex DNA behaviors, it also acknowledges some limitations. Future research could explore extensions of analytical approaches to incorporate greater biochemical realism, presenting an enticing avenue for further investigation.

# References

[1] É.L. Vogt, J.F.A. Model, M.V. Lima, S.K. de Souza, D.S. Rocha, R.B. Fabres, M. de Amaral, L.A.R. Simões, A.S. Vinagre, The impact of chasing stress on the metabolism of the Atlantic Ghost Crab Ocypode quadrata (Fabricius, 1787), Journal of Experimental Zoology - A 339 (9) (2023) 887–897.

[2] C.L. Paloschi, M.H.F. Tavares, E.A. Berte, K. Model, K.M. Rosa, F.G.d. Conceição, F.R. Domanski, E. de Souza Vismara, P.F. Montanher, R.M.A. Maciel, L. d, S. Ribeiro, N. Ramos Mertz, S.C. Sampaio, F.M. Costa, E.R. Lozano, M. Potrich, Imidacloprid: impact on Africanized Apis mellifera L. (Hymenoptera: Apidae) workers and honey contamination, Chemosphere 338 (2023) 139591.

[3] H. Ngoubi, R.y. Mbakob, I. Sali, G.H. Ben-Bolie, T.C. Kofané, Complex Ginzburg-Landau equation in the modified Peyrard-Bishop-Dauxois model, The European Physical Journal Plus 138 (9) (2023) 842.

[4] M. Vivas-Cortez, S. Arshed, M. Sadaf, Z. Perveen, G. Akram, Numerical simulations of the soliton dynamics for a nonlinear biological model: modulation instability analysis, PLoS ONE 18 (2) (2023) e0281318.

[5] A. Rani, M. Ashraf, M. Shakeel, Q. Mahmood-Ul-Hassan, J. Ahmad, Analysis of some new wave solutions of DNA-Peyrard-Bishop equation via mathematical method, Modern Physics Letters B 36 (21) (2022) 2250047.

[6] G. Tremblay, R. Kraft, P. Nulsen, D. Patnaude, T. Gauron, A. Kenter, K. Gage, B. Bissell, J. Model, G. Austin, T. Aldcroft, S. Hurley, HRC update, Chandra News 31 (2022) 4.

[7] M.M. Khater, Multi-vector with nonlocal and non-singular kernel ultrashort optical solitons pulses waves in birefringent fibers, Chaos, Solitons and Fractals 167 (2023) 113098.

[8] M.M. Khater, In solid physics equations, accurate and novel soliton wave structures for heating a single crystal of sodium fluoride, International Journal of Modern Physics B 37 (07) (2023) 2350068.

[9] M.M. Khater, S.H. Alfalqi, J.F. Alzaidi, R.A. Attia, Analytically and numerically, dispersive, weakly nonlinear wave packets are presented in a quasi-monochromatic medium, Results in Physics 46 (2023) 106312.

[10] M.M. Khater, Y. Xia, X. Zhang, R.A. Attia, Waves propagation of optical waves through nonlinear media; modified Kawahara equation, Results in Physics 52 (2023) 106796.

[11] M.M. Khater, Hybrid accurate simulations for constructing some novel analytical and numerical solutions of three-order gnls equation, International Journal of Geometric Methods in Modern Physics 20 (09) (2023) 2350159.

[12] M.M. Khater, Nonparaxial pulse propagation in a planar waveguide with Kerr–like and quintic nonlinearities; computational simulations, Chaos, Solitons and Fractals 157 (2022) 111970.

[13] M.M. Khater, Physics of crystal lattices and plasma; analytical and numerical simulations of the Gilson–pickering equation, Results in Physics 44 (2023) 106193.

[14] M.M. Khater, Advancements in computational techniques for precise solitary wave solutions in the (1+1)-dimensional Mikhailov-Novikov-Wang equation, International Journal of Theoretical Physics 62 (7) (2023) 152.

[15] D. Zhao, D. Lu, M.M. Khater, Ultra-short pulses generation's precise influence on the light transmission in optical fibers, Results in Physics 37 (2022) 105411.

[16] M.M. Khater, De Broglie waves and nuclear element interaction; abundant waves structures of the nonlinear fractional phi-four equation, Chaos, Solitons and Fractals 163 (2022) 112549.

[17] L. Ouahid, M.A. Abdou, S. Kumar, S. Owyed, S.S. Ray, A plentiful supply of soliton solutions for DNA Peyrard-Bishop equation by means of a new auxiliary equation strategy, International Journal of Modern Physics B 35 (26) (2021) 2150265.

[18] T.E.P. Sutantyo, A. Ripai, Z. Abdullah, W. Hidayat, Nonlinear dynamics of modified Peyrard-Bishop DNA model in nosé-hoover thermostat, Journal of Physics. Conference Series 1876 (2021) 012021.

[19] L. Ouahid, Plenty of soliton solutions to the DNA Peyrard-Bishop equation via two distinctive strategies, Physica Scripta 96 (3) (2021) 035224.

[20] M. Hillebrand, Chaotic dynamics of polyatomic systems with an emphasis on DNA models, arXiv e-prints, arXiv:2102.11020, 2021.

[21] L.B. Baker, J.B. Model, K.A. Barnes, M.L. Anderson, S.P. Lee, K.A. Lee, S.D. Brown, A.J. Reimel, T.J. Roberts, R.P. Nuccio, J.L. Bonsignore, C.T. Ungaro, J.M. Carter, W. Li, M.S. Seib, J.T. Reeder, A.J. Aranyosi, J.A. Rogers, R. Ghaffari, Skin-interfaced microfluidic system with personalized sweating rate and sweat chloride analytics for sports science applications, Science Advances 6 (50) (2020) eabe3929.

[22] D. Chevizovich, D. Michieletto, A. Mvogo, F. Zakiryanov, S. Zdravković, A review on nonlinear DNA physics, Royal Society Open Science 7 (11) (2020) 200774.

[23] S. Kim, B. Lee, J.T. Reeder, S.H. Seo, S.-U. Lee, A. Hourlier-Fargette, J. Shin, Y. Sekine, H. Jeong, Y.S. Oh, A.J. Aranyosi, S.P. Lee, J.B. Model, G. Lee, M.-H. Seo, S.S. Kwak, S. Jo, G. Park, S. Han, I. Park, H.-I. Jung, R. Ghaffari, J. Koo, P.V. Braun, J.A. Rogers, Soft, skin-interfaced microfluidic systems with integrated immunoassays, fluorometric sensors, and impedance measurement capabilities, Proceedings of the National Academy of Science 117 (45) (2020) 27906–27915.

[24] L. Lenzini, F.D. Patti, S. Lepri, R. Livi, S. Luccioli, Thermodynamics of DNA denaturation in a model of bacterial intergenic sequences, Chaos, Solitons and Fractals 130 (2020) 109446.

[25] J.B. Okaly, I. Ndzana Fabien, R.L. Woulaché, T.C. Kofané, Solitary wavelike solutions in nonlinear dynamics of damped DNA systems, The European Physical Journal Plus 134 (12) (2019) 598.

[26] I.V. Likhachev, V.D. Lakhno, Investigation of DNA denaturation in Peyrard-Bishop-Dauxois model by molecular dynamics method, The European Physical Journal B 92 (11) (2019) 253.

[27] J.B. Okaly, I. Ndzana Fabien, R.L. Woulaché, C.B. Tabi, T.C. Kofané, Base pairs opening and bubble transport in damped DNA dynamics with transport memory effects, Chaos 29 (9) (2019) 093103.

[28] T.E.P. Sutantyo, D. Dwiputra, W. Hidayat, F.P. Zen, Chaotic behaviour of modified Hamiltonian Peyrard-Bishop-Dauxois model on DNA system, Journal of Physics. Conference Series 1245 (2019) 012070.

[29] S. Zdravković, D. Chevizovich, A.N. Bugay, A. Maluckov, Stationary solitary and kink solutions in the helicoidal Peyrard-Bishop model of DNA molecule, Chaos 29 (5) (2019) 053118.

[30] S.F. Rohmah, I.A. Ramadhan, S. Latifah, D. Dwiputra, W. Hidayat, F.P. Zen, Time-dependent damping effect for the dynamics of DNA transcription, Journal of Physics. Conference Series 1204 (2019) 012012.

[31] M. Model, The significance of cell water content in cell biology, in: G. Popescu, Y. Park (Eds.), Quantitative Phase Imaging V, in: Society of Photo-Optical Instrumentation Engineers (SPIE) Conference Series, vol. 10887, 2019, p. 108871H.

[32] M. Hillebrand, G. Kalosakas, A. Schwellnus, C. Skokos, Heterogeneity and chaos in the Peyrard-Bishop-Dauxois DNA model, Physical Review E 99 (2) (2019) 022213.

[33] A.J. Bandodkar, P. Gutruf, J. Choi, K. Lee, Y. Sekine, J.T. Reeder, W.J. Jeang, A.J. Aranyosi, S.P. Lee, J.B. Model, R. Ghaffari, C.-J. Su, J.P. Leshock, T. Ray, A. Verrillo, K. Thomas, V. Krishnamurthi, S. Han, J. Kim, S. Krishnan, T. Hang, J.A. Rogers, Battery-free, skin-interfaced microfluidic/electronic systems for simultaneous electrochemical, colorimetric, and volumetric analysis of sweat, Science Advances 5 (1) (2019) eaav3294.

[34] Y. Ma, J. Choi, A. Hourlier-Fargette, Y. Xue, H.U. Chung, J.Y. Lee, X. Wang, Z. Xie, D. Kang, H. Wang, S. Han, S.-K. Kang, Y. Kang, X. Yu, M.J. Slepian, M.S. Raj, J.B. Model, X. Feng, R. Ghaffari, J.A. Rogers, Y. Huang, Relation between blood pressure and pulse wave velocity for human arteries, Proceedings of the National Academy of Science 115 (44) (2018) 11144–11149.

[35] J.B. Okaly, A. Mvogo, R.L. Woulaché, T.C. Kofané, Nonlinear dynamics of damped DNA systems with long-range interactions, Communications in Nonlinear Science and Numerical Simulations 55 (2018) 183–193.

[36] L. Ouahid, M.A. Abdou, S. Owyed, S. Kumar, New optical soliton solutions via two distinctive schemes for the DNA Peyrard-Bishop equation in fractal order, Modern Physics Letters B 35 (26) (2021) 2150444.

[37] J.C.M. Model, H.M. Veit, Development of a more sustainable hybrid process for lithium and cobalt recovery from lithium-ion batteries, Minerals 13 (6) (2023) 798.

[38] A. Djine, G.R. Deffo, S.B. Yamgoué, Bifurcation of backward and forward solitary waves in helicoidal Peyrard-Bishop-Dauxois model of DNA, Chaos, Solitons and Fractals 170 (2023) 113334.

# 9

# Inertial projective Mann algorithm for split equilibrium problems in Parkinson's screening

Watcharaporn Yajai[a], Arunee Promsri[b], Shilpa Das[c], and Watcharaporn Cholamjiak[a]

[a]*Department of Mathematics, School of Science, University of Phayao, Phayao, Thailand* [b]*Department of Physical Therapy, School of Allied Health Sciences, University of Phayao, Phayao, Thailand* [c]*Department of Mathematics, R.G. Baruah College, Fatasil Ambari, Guwahati, India*

## 9.1 Introduction

In this work, we assume that $\mathbb{H}_1$ and $\mathbb{H}_2$ are real Hilbert spaces. Let $\mathbb{C}$ and $\mathbb{Q}$ be a nonempty subset of $\mathbb{H}_1$ and $\mathbb{H}_2$, respectively. The equilibrium problems (EP) [11] is to find a point $\vartheta^* \in \mathbb{C}$ such that

$$f(\vartheta^*, \delta) \geqslant 0$$

for all $\delta \in \mathbb{C}$. Given $f : \mathbb{C} \times \mathbb{C} \to \mathbb{R}$ and $g : \mathbb{Q} \times \mathbb{Q} \to \mathbb{R}$ be two bifunctions. Suppose that $\mathbb{A} : \mathbb{H}_1 \to \mathbb{H}_2$ is a bounded linear operator. In 2011, the split equilibrium problems (SEP) introduced by Moudafi [17] is to find $\vartheta^* \in \mathbb{C}$ such that

$$f(\vartheta^*, \delta) \geqslant 0, \text{ for all } \delta \in \mathbb{C} \tag{9.1}$$

and such that

$$\mu^* = \mathbb{A}\vartheta^* \in \mathbb{Q} \text{ solves } g(\mu^*, \psi) \geqslant 0 \text{ for all } \psi \in \mathbb{Q}. \tag{9.2}$$

We denote problem (9.1) is the classical equilibrium problem by solution set $EP(f)$, and denote that $EP(g)$ is solution set of (9.2). The solution set of the split equilibrium problem (9.1) and (9.2) is denoted by $\Omega = \{\zeta \in EP(f) : \mathbb{A}\zeta \in EP(g)\}$. The solution approximation of the SEP has been studied by many authors; see [13,26]. The famous algorithm, called Mann's iteration [16], is in the form of the SEP as follows:

$$\vartheta^{k+1} = d^k \vartheta^k + (1 - d^k) T^f_{r^k}(I + \gamma \mathbb{A}^T (T^g_{r^k} - I)\mathbb{A})\vartheta^k$$

for each $k \in \mathbb{N}$, where $\{d^k\}$ is a real sequence on [0,1], $T^f_{r^k}$, $T^g_{r^k}$, defined in Lemma 9.2.4, and $\mathbb{A}^T$ is the adjoint of matrix $\mathbb{A}$. In 2016, Suantai [20] used Mann's algorithm [16] combined

Mathematical Modeling in Bioscience. https://doi.org/10.1016/B978-0-44-315445-4.00015-9

Copyright © 2025 Elsevier Inc. All rights are reserved, including those for text and data mining, AI training, and similar technologies.

with a non-spreading multi-valued mapping that the algorithm can solve, a common solution of fixed point problem of a multi-valued mapping $W$ and the SEP. The algorithm was generated as follows:

$$\begin{cases} \vartheta^1 \in \mathbb{C}, \ \forall k \geq 1, \\ \psi^k = T^f_{r^k}(I + \gamma \mathbb{A}^T(T^g_{r^k} - I)\mathbb{A})\vartheta^k, \\ \vartheta^{k+1} = o^k \vartheta^k + (1 - o^k) W \psi^k, \end{cases} \tag{9.3}$$

where $k \in (0, \frac{1}{2}])$, $\gamma \in (0, \frac{1}{L})$ such that $L$ is the spectral radius of $\mathbb{A}^T\mathbb{A}$, $\{r^k\} \subset (0, \infty)$, and $\{o^k\} \subset (0, 1)$. Algorithm modification to speed up convergence is another interesting method for many authors. One technique is called the inertial technique, part of the heavy ball method, which Polyak introduced [18] in 1964.

Inspired by the above research, we modify the inertial technique with Mann's algorithm [16] to find a solution for the SEP. The algorithm of optimization problems currently used in the machine learning application for Parkinson's disease is presented; see [7]. This work motivates us to apply the projection metric to focus the solution on the constrained set, enhancing the model for Parkinson's disease screening applications. This work is organized as follows: Section 9.2 presents the necessary lemmas and properties for proving weak convergence in the theorem; Section 9.3 introduces the modified projective Mann algorithm for solving split equilibrium problems in machine learning; Section 9.4 discusses the application of data classification using Parkinson's disease.

## 9.2 Preliminaries

To investigate the convergence of our algorithm, we start by providing basic results and some definitions. Let $\mathbb{C}$ be a nonempty, closed, and convex subset of Hilbert space $\mathbb{H}_1$. For every point $\vartheta \in \mathbb{H}_1$, there exists a unique nearest point of $\mathbb{C}$, defined by $P_{\mathbb{C}}\vartheta$ such that $\|\vartheta - P_{\mathbb{C}}\vartheta\| \leq \|\vartheta - \delta\|$ for all $\delta \in \mathbb{C}$. Such a $P_{\mathbb{C}}$ is called the metric projection from $\mathbb{H}_1$ onto $\mathbb{C}$. We know that $P_{\mathbb{C}}$ is a firmly nonexpansive mapping from $\mathbb{H}_1$ onto $\mathbb{C}$, i.e.,

$$\|P_{\mathbb{C}}\vartheta - P_{\mathbb{C}}\delta\|^2 \leq \langle P_{\mathbb{C}}\vartheta - P_{\mathbb{C}}\delta, \vartheta - \delta\rangle, \ \forall \vartheta, \delta \in \mathbb{H}_1.$$

Furthermore, for any $\vartheta \in \mathbb{H}_1$ and $\zeta \in \mathbb{C}$, $\zeta = P_{\mathbb{C}}\vartheta$ if and only if

$$\langle \vartheta - \zeta, \zeta - \delta\rangle \geqslant 0, \ \forall \delta \in \mathbb{C}.$$

A mapping $\mathbb{A} : \mathbb{C} \to \mathbb{H}_1$ is called $\Upsilon$-inverse strongly monotone if there exists $\Upsilon > 0$ such that

$$\langle \vartheta - \delta, \mathbb{A}\vartheta - \mathbb{A}\delta\rangle \geq \Upsilon\|\mathbb{A}\vartheta - \mathbb{A}\delta\|^2, \ \forall \vartheta, \delta \in \mathbb{C}.$$

**Proposition 9.2.1.** [27] *Let $\mathbb{A} : \mathbb{H} \to \mathbb{H}$ be a $d$-inverse strongly monotone mapping, then*

**(i)** *If $\beta$ is any constant in $(0, 2d]$, then the mapping $I - \beta\mathbb{A}$ is nonexpansive, where $I$ is the identity mapping on $\mathbb{H}$.*

**(ii)** *$\mathbb{A}$ is an $\frac{1}{d}$-Lipschitz continuous and monotone mapping.*

**Lemma 9.2.2.** [21] *Let $\mathbb{X}$ be a Banach space that satisfies Opial's condition and $\{\vartheta^k\}$ a sequence in $\mathbb{X}$. Let $\theta, \psi \in \mathbb{X}$ be such that $\lim_{k\to\infty} \|\vartheta^k - \theta\|$ and $\lim_{k\to\infty} \|\vartheta^k - \psi\|$ exist. If $\{\vartheta^{k^n}\}$ and $\{\vartheta^{l^n}\}$ are subsequences of $\{\vartheta^k\}$, which converge weakly to $\theta$ and $\psi$, respectively, then $\theta = \psi$.*

**Assumption 9.2.3.** [4] Let $f : \mathbb{C} \times \mathbb{C} \to \mathbb{R}$ be a bifunction satisfying the following assumptions:

**(1)** $f$ is monotone, i.e., $f(\vartheta, \delta) + f(\delta, \vartheta) \le 0$ for all $\vartheta \in \mathbb{C}$;
**(2)** $f(\vartheta, \vartheta) = 0$ for all $\vartheta \in \mathbb{C}$;
**(3)** for each $\vartheta \in \mathbb{C}$, $\delta \to f(\vartheta, \delta)$ is convex and lower semi-continuous;
**(4)** for each $\vartheta, \delta, \zeta \in \mathbb{C}$, $\limsup_{\beta\to 0^+} f(\beta\zeta + (1-\beta)\vartheta, \delta) \le f(\vartheta, \delta)$.

**Lemma 9.2.4.** [8] *Let $f : \mathbb{C} \times \mathbb{C} \to \mathbb{R}$ satisfy Assumption 9.2.3. Denote a mapping $T_r^f : \mathbb{H}_1 \to \mathbb{C}$ for each $r > 0$ and $e \in \mathbb{H}_1$ as follows:*

$$T_r^f(e) = \{\alpha \in \mathbb{C} : f(\alpha, \nu) + \frac{1}{r}\langle \nu - \alpha, \alpha - e\rangle \geqslant 0, \ \forall \nu \in \mathbb{C}\}.$$

*Then the following hold:*

**(1)** *$T_r^f$ is nonempty and single-valued;*
**(2)** *$T_r^f$ is firmly nonexpansive, i.e., $\exists\, \nu, e \in \mathbb{H}_1$,*

$$\|T_r^f e - T_r^f \nu\|^2 \le \langle T_r^f e - T_r^f \nu, e - \nu\rangle;$$

**(3)** *$EP(f, \mathbb{C}) = F(T_r^f)$;*
**(4)** *$EP(f, \mathbb{C})$ is convex and closed.*

*Furthermore, let $g : \mathbb{Q} \times \mathbb{Q} \to \mathbb{R}$ satisfy Assumption 9.2.3. Define a mapping $T_s^g : \mathbb{H}_2 \to \mathbb{Q}$ for each $s > 0$ and $b \in \mathbb{H}_2$ as follows:*

$$T_s^g(b) = \{h \in \mathbb{Q} : g(h, a) + \frac{1}{s}\langle a - h, h - b\rangle \geqslant 0, \ \forall a \in \mathbb{Q}\}.$$

*Then we easily observe that the following hold:*

**(1)** *$T_s^g$ is nonempty and single-valued;*
**(2)** *$T_s^g$ is firmly nonexpansive;*
**(3)** *$F(T_s^g) = EP(g, \mathbb{Q})$;*
**(4)** *$EP(g, \mathbb{Q})$ is closed and convex.*

**Lemma 9.2.5.** [23] *Let $\{a^k\}$, $\{b^k\}$ be sequences of nonnegative real numbers such that*

$$a^{k+1} \le a^k + b^k, \ k \in \mathbb{N}.$$

*If $\sum_{k=1}^{\infty} b^k < \infty$, then $\lim_{k\to\infty} a^k$ exists.*

## 9.3 Main results

In this part, let $\mathbb{C}$, $E \subset \mathbb{H}_1$, $\mathbb{Q} \subset \mathbb{H}_2$ be nonempty, closed, and convex subsets of real Hilbert spaces $\mathbb{H}_1$ and $\mathbb{H}_2$, respectively. Assume that $\mathbb{A} : \mathbb{H}_1 \to \mathbb{H}_2$ is a bounded linear operator. Let $f : \mathbb{C} \times \mathbb{C} \to \mathbb{R}$, $g : \mathbb{Q} \times \mathbb{Q} \to \mathbb{R}$ be a bifunction satisfying Assumption 9.2.3 and $g$ is upper semi-continuous in the first argument. Let $L$ be the spectral radius of $\mathbb{A}^T\mathbb{A}$, and $\mathbb{A}^T$ the adjoint of $\mathbb{A}$. $\Omega = \{\vartheta^* \in E : \vartheta^* \in EP(f) \text{ and } \mathbb{A}\vartheta^* \in EP(g)\}$.

---

**Algorithm 9.3.1** An inertial projective Mann algorithm.

---

**Initialization.** Select $\vartheta^1 \in \mathbb{C}$, $\{d^k\} \subset (0,1)$, $r^k \subset (0,\infty)$, and $\beta^k \in (0, \frac{1}{L})$
**Step 1.** Compute

$$\zeta^k = T^f_{r^k}(I - \beta^k \mathbb{A}^T(I - T^g_{r^k})\mathbb{A})\delta^k,$$

where $\delta^k = \vartheta^k + p^k(\vartheta^k - \vartheta^{k-1})$.
**Step 2.** Compute

$$\vartheta^{k+1} = P_E(d^k\delta^k + (1 - d^k)\zeta^k).$$

Replace $k = k+1$ and return to **Step 1.**

---

**Theorem 9.3.2.** *Let the sequence $\{\vartheta^k\}$ be defined by Algorithm 9.3.1 and assume that the following conditions hold:*

**(C1)** $\beta^k \subset (0, \frac{1}{L})$;
**(C2)** $\sum_{k=1}^{\infty} p^k \|\vartheta^k - \vartheta^{k-1}\| < +\infty$;
**(C3)** $0 < \liminf_{k\to\infty} d^k \le \limsup_{k\to\infty} d^k < 1$;
**(C4)** $0 < \liminf_{k\to\infty} \beta^k \le \limsup_{k\to\infty} \beta^k < \frac{1}{L}$.

*Then the sequence $\{\vartheta^k\}$ converges weakly to $\vartheta^* \in \Omega$.*

*Proof.* Since $T^g_{r^k}$ is firmly nonexpansive and $I - T^g_{r^k}$ is 1-inverse strongly monotone, $\mathbb{A}^T(I - T^g_{r^k})\mathbb{A}$ is an $\frac{1}{L}$-inverse strongly and monotone mapping. Indeed,

$$\begin{aligned}
\|\mathbb{A}^T(I - T^g_{r^k})\mathbb{A}\vartheta - \mathbb{A}^T(I - T^g_{r^k})\mathbb{A}\delta\|^2 &= \langle \mathbb{A}^T(I - T^g_{r^k})(\mathbb{A}\vartheta - \mathbb{A}\delta), \mathbb{A}^T(I - T^g_{r^k})(\mathbb{A}\vartheta - \mathbb{A}\delta)\rangle \\
&= \langle (I - T^g_{r^k})(\mathbb{A}\vartheta - \mathbb{A}\delta), \mathbb{A}\mathbb{A}^T(I - T^g_{r^k})(\mathbb{A}\vartheta - \mathbb{A}\delta)\rangle \\
&\le L\langle (I - T^g_{r^k})(\mathbb{A}\vartheta - \mathbb{A}\delta), (I - T^g_{r^k})(\mathbb{A}\vartheta - \mathbb{A}\delta)\rangle \\
&= L\|(I - T^g_{r^k})(\mathbb{A}\vartheta - \mathbb{A}\delta)\|^2 \\
&\le L\langle \mathbb{A}\vartheta - \mathbb{A}\delta, (I - T^g_{r^k})(\mathbb{A}\vartheta - \mathbb{A}\delta)\rangle \\
&= L\langle \vartheta - \delta, \mathbb{A}^T(I - T^g_{r^k})\mathbb{A}\vartheta - \mathbb{A}^T(I - T^g_{r^k})\mathbb{A}\delta\rangle
\end{aligned}$$

for all $\vartheta, \delta \in \mathbb{H}_1$.

This implies that $\mathbb{A}^T(I - T^g_{r^k})\mathbb{A}$ is an $\frac{1}{L}$-inverse strongly and monotone mapping. As $\beta^k \in (0, \frac{1}{L})$, $I - \beta^k\mathbb{A}^T(I - T^g_{r^k})\mathbb{A}$ is nonexpansive by Proposition 9.2.1.

**Step 1**. Show that $\{\vartheta^k\}$ is bounded.

Let $\vartheta^* \in \Omega$, then $\vartheta^* = T^f_{r^k}\vartheta^*$ and $(I - \beta^k\mathbb{A}^T(I - T^g_{r^k})\mathbb{A})\vartheta^* = \vartheta^*$. Since $T^f_{r^k}$ and $P_E$ are nonexpansive in a Hilbert space, we have

$$\begin{aligned}
\|\vartheta^{k+1} - \vartheta^*\| &= \|P_E(d^k\delta^k + (1 - d^k)\zeta^k) - \vartheta^*\| \\
&\le d^k\|\delta^k - \vartheta^*\| + (1 - d^k)\|\zeta^k - \vartheta^*\| \\
&\le \|\delta^k - \vartheta^*\| \\
&\le \|\vartheta^k - \vartheta^*\| + p^k\|\vartheta^k - \vartheta^{k-1}\|.
\end{aligned}$$

From Lemma 9.2.5 and condition (C2), we then have that $\lim_{k\to\infty}\|\vartheta^k - \vartheta^*\|$ exists. This implies that $\{\vartheta^k\}$ bounded, by the definition of $\{\delta^k\}$, $\{\delta^k\}$ is also bound.

**Step 2**. Show that $\lim_{k\to\infty}\|\zeta^k - \delta^k\| = 0$. Since $\mathbb{A}^T(I - T^g_{r^k})\mathbb{A}$ is $\frac{1}{L}$-Lipschitz continuous monotone, $I - \beta^k\mathbb{A}^T(I - T^g_{r^k})\mathbb{A}$ is nonexpansive, and $T^f_{r^k}$ is firmly nonexpansive, we have that

$$\begin{aligned}
\|\vartheta^{k+1} - \vartheta^*\|^2 &= \|P_E(d^k\delta^k + (1 - d^k)\zeta^k) - \vartheta^*\|^2 \\
&\le \|d^k(\delta^k - \vartheta^*) + (1 - d^k)(\zeta^k - \vartheta^*)\|^2 \\
&\le d^k\|\delta^k - \vartheta^*\|^2 + (1 - d^k)\|\zeta^k - \vartheta^*\|^2 - d^k(1 - d^k)\|\delta^k - \zeta^k\|^2 \\
&\le \|\delta^k - \vartheta^*\|^2 - d^k(1 - d^k)\|\delta^k - \zeta^k\|^2 \\
&\le \|\vartheta^k + p^k(\vartheta^k - \vartheta^{k-1}) - \vartheta^*\|^2 - d^k(1 - d^k)\|\delta^k - \zeta^k\|^2 \\
&\le \|\vartheta^k - \vartheta^*\|^2 + 2\langle p^k(\vartheta^k - \vartheta^{k-1}), \delta^k - \vartheta^*\rangle - d^k(1 - d^k)\|\delta^k - \zeta^k\|^2 \\
&\le \|\vartheta^k - \vartheta^*\|^2 + 2p^k\|\vartheta^k - \vartheta^{k-1}\|\|\delta^k - \vartheta^*\| - d^k(1 - d^k)\|\delta^k - \zeta^k\|^2.
\end{aligned}$$

This implies that

$$d^k(1 - d^k)\|\delta^k - \zeta^k\|^2 \le \|\vartheta^k - \vartheta^*\|^2 - \|\vartheta^{k+1} - \vartheta^*\|^2 + 2p^k\|\vartheta^k - \vartheta^{k-1}\|\|\delta^k - \vartheta^*\| \tag{9.4}$$

by condition (C2)–(C3), and $\lim_{k\to\infty}\|\vartheta^k - \vartheta^*\|$ exist, therefore we have

$$\lim_{k\to\infty}\|\delta^k - \zeta^k\| = 0. \tag{9.5}$$

**Step 3**. Show that $\lim_{k\to\infty}\|\mathbb{A}\delta^k - T^g_{r^k}\mathbb{A}\delta^k\| = 0$. Again by $T^f_{r^k}$ being firmly nonexpansive, we have

$$\begin{aligned}
\|\zeta^k - \vartheta^*\|^2 &= \|T^f_{r^k}(I - \beta^k\mathbb{A}^T(I - T^g_{r^k})\mathbb{A})\delta^k - \vartheta^*\|^2 \\
&= \|T^f_{r^k}(I - \beta^k\mathbb{A}^T(I - T^g_{r^k})\mathbb{A})\delta^k - T^f_{r^k}\vartheta^*\|^2 \\
&\le \|\delta^k - \beta^k\mathbb{A}^T(I - T^g_{r^k})\mathbb{A}\delta^k - \vartheta^*\|^2
\end{aligned}$$

$$
\begin{aligned}
&= \|\delta^k - \vartheta^*\|^2 + \beta^{k^2} \|\mathbb{A}^T (I - T^g_{r^k})\mathbb{A}\delta^k\|^2 + 2\beta^k \langle \vartheta^* - \delta^k, \mathbb{A}^T (I - T^g_{r^k})\mathbb{A}\delta^k \rangle \\
&\leq \|\delta^k - \vartheta^*\|^2 + \beta^{k^2} \langle \mathbb{A}\delta^k - T^g_{r^k}\mathbb{A}\delta^k, \mathbb{A}\mathbb{A}^T (I - T^g_{r^k})\mathbb{A}\delta^k \rangle \\
&\quad + 2\beta^k \langle \vartheta^* - \delta^k, \mathbb{A}^T (I - T^g_{r^k})\mathbb{A}\delta^k \rangle. \qquad (9.6)
\end{aligned}
$$

On the other hand, we have

$$
\begin{aligned}
\beta^{k^2} \langle \mathbb{A}\delta^k - T^g_{r^k}\mathbb{A}\delta^k, \mathbb{A}\mathbb{A}^T (I - T^g_{r^k})\mathbb{A}\delta^k \rangle &\leq L\beta^{k^2} \langle \mathbb{A}\delta^k - T^g_{r^k}\mathbb{A}\delta^k, \mathbb{A}\delta^k - T^g_{r^k}\mathbb{A}\delta^k \rangle \\
&= L\beta^{k^2} \|\mathbb{A}\delta^k - T^g_{r^k}\mathbb{A}\delta^k\|^2, \qquad (9.7)
\end{aligned}
$$

and

$$
\begin{aligned}
&2\beta^k \langle \vartheta^* - \delta^k, \mathbb{A}^T (I - T^g_{r^k})\mathbb{A}\delta^k \rangle \\
&= 2\beta^k \langle \mathbb{A}(\vartheta^* - \delta^k), \mathbb{A}\delta^k - T^g_{r^k}\mathbb{A}\delta^k \rangle \\
&= 2\beta^k \langle \mathbb{A}(\vartheta^* - \delta^k) + (\mathbb{A}\delta^k - T^g_{r^k}\mathbb{A}\delta^k) - (\mathbb{A}\delta^k - T^g_{r^k}\mathbb{A}\delta^k), \mathbb{A}\delta^k - T^g_{r^k}\mathbb{A}\delta^k \rangle \\
&= 2\beta^k \{\langle \mathbb{A}\vartheta^* - T^g_{r^k}\mathbb{A}\delta^k, \mathbb{A}\delta^k - T^g_{r^k}\mathbb{A}\delta^k \rangle - \|\mathbb{A}\delta^k - T^g_{r^k}\mathbb{A}\delta^k\|^2\} \\
&\leq 2\beta^k \{\frac{1}{2}\|\mathbb{A}\delta^k - T^g_{r^k}\mathbb{A}\delta^k\|^2 - \|\mathbb{A}\delta^k - T^g_{r^k}\mathbb{A}\delta^k\|^2\} \\
&< -\beta^k \|\mathbb{A}\delta^k - T^g_{r^k}\mathbb{A}\delta^k\|^2. \qquad (9.8)
\end{aligned}
$$

Using (9.6), (9.7), and (9.8), we have

$$
\begin{aligned}
\|\zeta^k - \vartheta^*\|^2 &\leq \|\delta^k - \vartheta^*\|^2 + L\beta^{k^2} \|\mathbb{A}\delta^k - T^g_{r^k}\mathbb{A}\delta^k\|^2 - \beta^k \|\mathbb{A}\delta^k - T^g_{r^k}\mathbb{A}\delta^k\|^2 \\
&= \|\delta^k - \vartheta^*\|^2 + \beta^k (L\beta^k - 1)\|\mathbb{A}\delta^k - T^g_{r^k}\mathbb{A}\delta^k\|^2. \qquad (9.9)
\end{aligned}
$$

It follows that

$$
\begin{aligned}
\|\vartheta^{k+1} - \vartheta^*\|^2 &= \|P_E(d^k\delta^k + (1 - d^k)\zeta^k) - \vartheta^*\|^2 \\
&\leq d^k \|\delta^k - \vartheta^*\|^2 + (1 - d^k)\|\zeta^k - \vartheta^*\| \\
&= d^k \|\delta^k - \vartheta^*\|^2 + (1 - d^k)(\|\delta^k - \vartheta^*\|^2 + \beta^k (L\beta^k - 1)\|\mathbb{A}\delta^k - T^g_{r^k}\mathbb{A}\delta^k\|^2) \\
&\leq \|\delta^k - \vartheta^*\|^2 + (1 - d^k)\beta^k (L\beta^k - 1)\|\mathbb{A}\delta^k - T^g_{r^k}\mathbb{A}\delta^k\|^2 \\
&\leq \|\vartheta^k - \vartheta^*\|^2 + 2\langle p^k(\vartheta^k - \vartheta^{k-1}), \delta^k - \vartheta^* \rangle \\
&\quad + (1 - d^k)\beta^k (L\beta^k - 1)\|\mathbb{A}\delta^k - T^g_{r^k}\mathbb{A}\delta^k\|^2.
\end{aligned}
$$

Therefore, we have

$$
\begin{aligned}
&(1 - d^k)\beta^k (1 - L\beta^k)\|\mathbb{A}\delta^k - T^g_{r^k}\mathbb{A}\delta^k\|^2 \\
&\qquad \leq \|\vartheta^k - \vartheta^*\|^2 - \|\vartheta^{k+1} - \vartheta^*\|^2 + 2\langle p^k(\vartheta^k - \vartheta^{k-1}), \delta^k - \vartheta^* \rangle.
\end{aligned}
$$

Since condition (C3)–(C4) and $\lim_{k\to\infty} \|\vartheta^k - \vartheta^*\|$ exists, we obtain

$$
\lim_{k\to\infty} \|\mathbb{A}\delta^k - T^g_{r^k}\mathbb{A}\delta^k\| = 0. \qquad (9.10)
$$

**Step 4**. Show that $\|\zeta^k - \vartheta^k\| \to 0$ as $k \to \infty$.

We have

$$\begin{aligned}\|\delta^k - \vartheta^k\| &= \|\vartheta^k + p^k(\vartheta^k - \vartheta^{k-1}) - \vartheta^k\| \\ &\leq p^k\|\vartheta^k - \vartheta^{k-1}\| \to 0\end{aligned} \tag{9.11}$$

as $k \to \infty$.

And by combining (9.5) and (9.11), we then have

$$\|\zeta^k - \vartheta^k\| \leq \|\zeta^k - \delta^k\| + \|\delta^k - \vartheta^k\| \to 0 \tag{9.12}$$

as $k \to \infty$.

**Step 5**. Let $\omega_W(\vartheta^k) = \{\vartheta \in \mathbb{H} : \vartheta^{k^i} \rightharpoonup \vartheta, \{\vartheta^{k^i}\} \subset \{\vartheta^k\}\}$. From the reflexiveness of $\mathbb{H}$, we have $\omega_W(\vartheta^k) \neq \emptyset$. Let $l \in \omega_W(\vartheta^k)$; there exists a subsequence $\{\vartheta^{k^i}\} \subset \{\vartheta^k\}$ converging weakly to $l$. Since $\{\vartheta^k\}$ is a sequence in $E$ for all $k \geqslant 2$ and $E$ is closed convex set, $l \in E$. From (9.12), it follows that $\zeta^{k^i} \rightharpoonup l$ as $i \to \infty$.

Next show that $l \in EP(f)$. From $\zeta^k = T^f_{r^k}(I - \beta^k\mathbb{A}^T(I - T^g_{r^k})\mathbb{A})\delta^k$, we have

$$f(\zeta^k, \delta) + \frac{1}{r^k}\langle \delta - \zeta^k, \zeta^k - \delta^k - \beta^k\mathbb{A}^T(I - T^g_{r^k})\mathbb{A}\delta^k\rangle \geqslant 0$$

for all $\delta \in \mathbb{C}$, which implies that

$$f(\zeta^k, \delta) + \frac{1}{r^k}\langle \delta - \zeta^k, \zeta^k - \delta^k\rangle - \frac{1}{r^k}\langle \delta - \zeta^k, \beta^k\mathbb{A}^T(I - T^g_{r^k})\mathbb{A}\delta^k\rangle \geqslant 0$$

for all $\delta \in \mathbb{C}$. By Assumption 9.2.3(1), we have

$$\frac{1}{r^{k^i}}\langle \delta - \zeta^{k^i}, \zeta^{k^i} - \delta^{k^i}\rangle - \frac{1}{r^{k^i}}\langle \delta - \zeta^{k^i}, \beta^{k^i}\mathbb{A}^T(I - T^g_{r^{k^i}})\mathbb{A}\delta^{k^i}\rangle \geqslant f(\delta, \zeta^{k^i})$$

for all $\delta \in \mathbb{C}$. From $\liminf_{k\to\infty} r^k > 0$, from Assumption 9.2.3(3), (9.5), and (9.10), we gain

$$f(\delta, l) \leq 0$$

for all $\delta \in \mathbb{C}$. For each $0 < c \leq 1$ and $\delta \in \mathbb{C}$, define $\delta^c = c\delta + (1-c)l$. As $\delta \in \mathbb{C}$ and $l \in \mathbb{C}$, $\delta^c \in \mathbb{C}$; therefore $f(\delta^c, l) \leq 0$. By Assumption 9.2.3(2), we have

$$0 = f(\delta^c, \delta^c) \leq cf(\delta^c, \delta) + (1-c)f(\delta^c, l) \leq cf(\delta^c, \delta),$$

and hence $f(\delta^c, \delta) \geqslant 0$. By Assumption 9.2.3(4), $f(l, \delta) \geqslant 0$ for all $\delta \in \mathbb{C}$; therefore $l \in EP(f)$. As $\mathbb{A}$ is bounded linear operator, $\mathbb{A}\vartheta^{k^i} \rightharpoonup \mathbb{A}l$. Then following from (9.10),

$$T^g_{r^{k^i}}\mathbb{A}\vartheta^{k^i} \rightharpoonup \mathbb{A}l \tag{9.13}$$

as $i \to \infty$. By the definition of $T^{g}_{r^{k^i}} \mathbb{A}\vartheta^{k^i}$, we have

$$g(T^{g}_{r^{k^i}} \mathbb{A}\vartheta^{k^i}, \delta) + \frac{1}{r^{k^i}} \langle \delta - T^{g}_{r^{k^i}} \mathbb{A}\vartheta^{k^i}, T^{g}_{r^{k^i}} \mathbb{A}\vartheta^{k^i} - \mathbb{A}\vartheta^{k^i} \rangle \geqslant 0$$

for all $\delta \in \mathbb{C}$. Given that the first argument of $g$ is upper semi-continuous, it follows from (9.13) that

$$g(\mathbb{A}l, \delta) \geqslant 0$$

for all $\delta \in \mathbb{C}$. This shows that $\mathbb{A}l \in EP(g)$. Thus $l \in \Omega$.

**Step 6**. We will show that $\{\vartheta^k\}$ and $\{\zeta^k\}$ converge weakly to an element of $\Omega$. It is sufficient to show that $\omega_W(\vartheta^k)$ is single point set. Let $\vartheta^*, l \in \omega_W(\vartheta^k)$ and $\{\vartheta^{k^n}\}, \{\vartheta^{k^m}\} \subset \{\vartheta^k\}$ be such that $\vartheta^{k^n} \rightharpoonup \vartheta^*$ and $\vartheta^{k^m} \rightharpoonup l$. From (9.12), we also have $\zeta^{k^n} \rightharpoonup \vartheta^*$ and $\zeta^{k^m} \rightharpoonup l$. Using Lemma 9.2.2, we gain $\vartheta^* = l$. This completes the proof. □

## 9.4 Application to data classification problem

Parkinson's disease is the most common age-related neurodegenerative disorder that primarily affects movement control by the loss of dopaminergic neurons in the substantia nigra pars compacta and by an accumulation of misfolded a-synuclein found in intracytoplasmic inclusions called Lewy bodies [3], [1]. Though motor-related symptoms, such as bradykinesia (slow movement), tremors, and rigidity are well-known, non-motor symptoms, e.g., vocal or speech impairments, can also significantly impact the quality of life for individuals with PD. Vocal problems in individuals with PD are often called "Parkinson's speech disorder" or "hypokinetic dysarthria," which occur due to the degeneration of neurons in the brain's basal ganglia and related motor control pathways [9]. This motor speech disorder is characterized by a reduced range of motion and muscle weakness in the muscles responsible for speech production, leading to various issues affecting speech and voice, e.g., reduced volume (hypophonia), monotone speech, reduced articulation, imprecise consonants, stuttering or stammering, and decreased expressiveness [22]. A supportive method to diagnose or predict PD is clinically helpful, for example, focusing on the early speech dysfunction observed in people with PD [9], as the earliest stages of PD can be difficult to recognize, as reflected by the long delay of an average of 10 years that typically separates the timing of the diagnosis from the person's first noticeable symptom [12].

We use database in UCI (https://archive.ics.uci.edu/dataset/301/parkinson+speech+dataset+with+multiple+types+of+sound+recordings), which belongs to 20 PWP (6 female, 14 male) and 20 healthy individuals (10 female, 10 male), who came to the Department of Neurology in Cerrahpasa Faculty of Medicine, Istanbul University. From all subjects, multiple types of sound recordings (26 voice samples, including sustained vowels, numbers, words, and short sentences) are taken as follows: $1^{st}$ to $3^{rd}$ samples represent the sustained vowels /a/, /o/, and /u/, respectively; $4^{th}$ to $13^{th}$ samples represent numbers from 1 to 10,

$14^{th}$ to $17^{th}$ samples represent short sentences; $18^{th}$ to $26^{th}$ samples represent individual words. See Table 9.1.

**Table 9.1** Data description for Parkinson's disease dataset in Microsoft Excel.

| Column 1 | Subject id | |
|---|---|---|
| Column 2–27 | Group: Frequency parameters | Features |
| | | 1. Jitter (local) |
| | | 2. Jitter (local, absolute) |
| | | 3. Jitter (rap) |
| | | 4. Jitter (ppq5) |
| | | 5. Jitter (ddp) |
| | Group: Amplitude parameters | Features |
| | | 6. Shimmer (local) |
| | | 7. Shimmer (local, dB) |
| | | 8. Shimmer (apq3) |
| | | 9. Shimmer (apq5) |
| | | 10. Shimmer (apq11) |
| | | 11. Shimmer (dda) |
| | Group: Harmonicity parameters | Features |
| | | 12. Autocorrelation |
| | | 13. Noise to Harmonic |
| | | 14. Harmonic to Noise |
| | Group: Pitch parameters | Features |
| | | 15. Median pitch |
| | | 16. Mean pitch |
| | | 17. Standard deviation |
| | | 18. Minimum pitch |
| | | 19. Maximum pitchh |
| | Group: Pulse Parameters | Features |
| | | 20. Number of pulses |
| | | 21. Number of periods |
| | | 22. Mean period |
| | | 23. Standard deviation of period |
| | Group: Voicing parameters | Features |
| | | 24. Fraction of locally unvoiced frames |
| | | 25. Number of voice breaks |
| | | 26. Degree of voice breaks |
| Column 28 | UPDRS | |
| Column 29 | class information | |

We have 1,040 observations with no missing data and 27 variables. Our experiment considered 70% of the training data and 30% of the testing data. Eight boxes were split to cover all datasets; see splitting in Fig. 9.1.

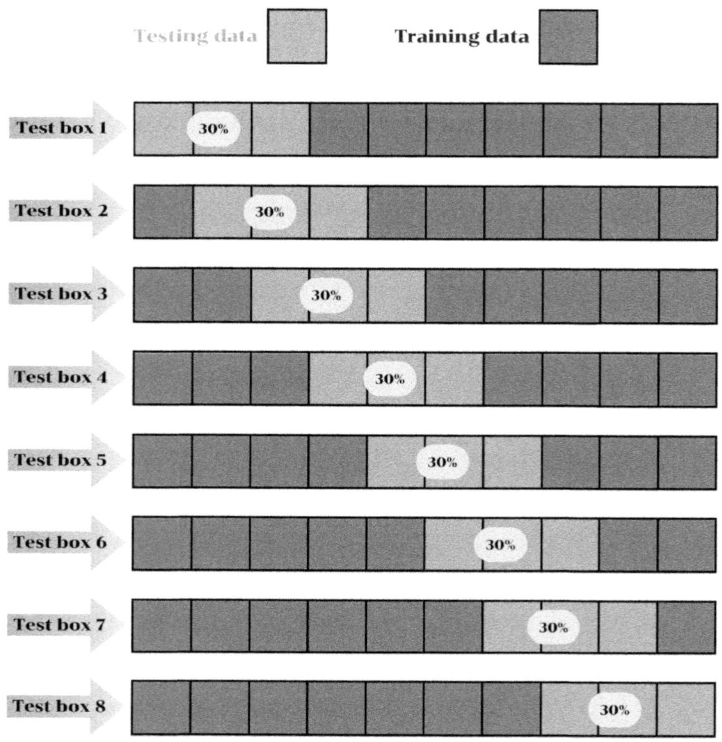


**FIGURE 9.1** Groups of dataset.

For our application in machine learning, following Huang et al. [14], we use *extreme learning machine* (ELM). This concept was explained by letting $E := \{(\mathbf{s}_n, \mathbf{r}_n) : \mathbf{s}_n \in \mathbb{R}^n, \mathbf{r}_n \in \mathbb{R}^q, n = 1, 2, 3, ..., \iota\}$ to be a training set, where $\mathbf{r}_n$ is a training target, and $\mathbf{s}_n$ is an input training data. For single-hidden layer feedforward neural networks (SLFNs) with $M$ hidden nodes, the output function is in the form

$$\mathbf{Q}_n = \sum_{k=1}^{M} \Xi_k V(c_k \mathbf{s}_n + d_k),$$

where $V$ is an activated function, $c_k$ is parameters of weight, and $d_k$ is parameters of the final bias, and $\Xi_k$ is the optimal output weight at the $k$-th hidden node. We can define the hidden layer as follows:

$$L = \begin{bmatrix} V(c_1\mathbf{s}_1 + d_1) & \dots & V(c_M\mathbf{s}_1 + d_M) \\ \vdots & \ddots & \vdots \\ V(c_1\mathbf{s}_\iota + d_1) & \dots & V(c_M\mathbf{s}_\iota + d_M) \end{bmatrix}.$$

To solve ELM is to find optimal output weight $\Xi = [\Xi_1^T, ..., \Xi_M^T]^T$ such that $L\Xi = T$, where $T = [\mathbf{r}_1^T, ..., \mathbf{r}_\iota^T]^T$ is the training data. In some cases, in finding $\Xi = L^\dagger T$, $L^\dagger$ is the

*Moore–Penrose generalized inverse* of $L$. When $L^\dagger$ is difficult to calculate, we can solve $L\Xi = T$ by the following least square problem:

$$\min_{\Xi \in \mathbb{R}^M} \{\|L\Xi - T\|_2^2\}. \tag{9.14}$$

The problem (9.14) is called *the least absolute shrinkage and selection operator* (LASSO) [25]. For applying our algorithm to solve (9.14), we set

$$f(\vartheta, \delta) = \begin{cases} 0, & \forall \vartheta, \delta \in \mathbb{C}, \\ -1, & \text{otherwise}, \end{cases} \tag{9.15}$$

$$g(\vartheta, \delta) = \begin{cases} 0, & \forall \vartheta, \delta \in \mathbb{Q}, \\ -1, & \text{otherwise}. \end{cases} \tag{9.16}$$

Then, our algorithm can be reduced to solve split feasibility problem (SFP), which was introduced by Censor and Elfving [5], as follows: finding a point $\vartheta^* \in \mathbb{C}$ such that

$$\mathbb{A}\vartheta^* \in \mathbb{Q}, \tag{9.17}$$

where $\mathbb{C}$, $\mathbb{Q}$ is nonempty closed and convex subsets of $\mathbb{R}^B$ and $\mathbb{R}^D$, respectively, nd $\mathbb{A}$ is an $B \times D$ real matrix.

We create 3 models for our Algorithm 9.3.1 of split feasibility problem (9.17) to obtain results in machine learning (9.14).

**1. Least square model:** ($L$)

$$\min_{\vartheta \in \mathbb{H}_1} \frac{1}{2} \|P_{\mathbb{Q}} L\Xi - L\Xi\|_2^2. \tag{9.18}$$

Setting $\mathbb{A} = L$, $\mathbb{C} = \mathbb{H}_1$ and $\mathbb{Q} = \{T\}$.

**2. Least square on constraint set $L_1$:** ($LL_1$)

$$\min_{\vartheta \in \mathbb{C}} \frac{1}{2} \|P_{\mathbb{Q}} L\Xi - L\Xi\|_2^2. \tag{9.19}$$

Setting $\mathbb{A} = L$, $\mathbb{Q} = \{T\}$ and $\mathbb{C} = \{\vartheta \in \mathbb{H}_1 : \|\vartheta\|_1 \leq \lambda\}$, where $\lambda > 0$.

**3. Least square on constraint set $L_2$:** ($LL_2$)

$$\min_{\vartheta \in \mathbb{C}} \frac{1}{2} \|P_{\mathbb{Q}} L\Xi - L\Xi\|_2^2. \tag{9.20}$$

Setting $\mathbb{A} = L$, $\mathbb{Q} = \{T\}$ and $\mathbb{C} = \{\vartheta \in \mathbb{H}_1 : \|\vartheta\|_2^2 \leq \lambda\}$, where $\lambda > 0$.

In this research, we utilize the binary cross-entropy loss function, where $\hat{z}_k$ is the $k$-th scalar value in the model output, $z_k$ is the corresponding target value, and $W$ is the number of scalar values in the model output, and the function as following:

$$Loss = -\frac{1}{W} \sum_{k=1}^{W} z_k \log \hat{z}_k + (1 - z_k) \log(1 - \hat{z}_k).$$

In this paper, precision defines the potential of positive prediction, recall defines the model's quality to detect pertinent data, and accuracy denotes the rate of correct predictions by the model. The formulation of 3 measures [24] are denoted as follows:

$$\text{Accuracy(Acc)} = \frac{TP+TN}{TP+FP+TN+FN} \times 100\%, \tag{9.21}$$

$$\text{Precision(Pre)} = \frac{TP}{TP+FP} \times 100\%, \tag{9.22}$$

$$\text{Recall(Rec)} = \frac{TP}{TP+FN} \times 100\%, \tag{9.23}$$

where a confusion matrix for original and predicted classes are displayed in terms of $TP$: true positive (healthy subjects who was correctly classified), $TN$ is true negative (patients with PD who was correctly classified), $FP$ is false positives (Patients with PD who was incorrectly classified) and $FN$ is false negatives (healthy subjects who were incorrectly classified). Likewise, $N$ and $P$ are the negative, and positive population of malignant and benign cases, respectively, and we define F as the harmonic mean value between recall and precision, as follows:

$$\text{F1-Score} = 2\left(\frac{\text{Precision} \times \text{Recall}}{\text{Precision} + \text{Recall}}\right) \times 100\%. \tag{9.24}$$

In our calculation, we begin by setting the activation function as sigmoid; we stop the number of iterations at 500 (iterations = 500); hidden nodes $M = 150$, regularization parameter $d^k = 0.5$, $p^k = 0.5$ for Algorithm $L$, Algorithm $LL_1$, and Algorithm $LL_2$. We gain the outcomes of the different parameters $P$ by $\beta^k = \frac{P}{max(eigenvalue(A^TA))}$ for Algorithm $L$, Algorithm $LL_1$ and Algorithm $LL_2$ as seen in Table 9.2.

We can see that $\beta^k = 0.85$ greatly improves the performance of Algorithm $L$, Algorithm $LL_1$, and Algorithm $LL_2$. As a result, we select it as the default inertial parameter for next calculation.

Next, we will select $\beta^k = \frac{0.85}{max(eigenvalue(A^TA))}$, $d^k = 0.5$, $p^k = 0.5$ for Algorithm $L$, Algorithm $LL_1$, and Algorithm $LL_2$; we stop number of iterations at 500 (iterations = 500). We gain the numeric outcomes as seen in Table 9.3.

We can see that $d^k = 0.5$ greatly improves the performance of Algorithm $L$, Algorithm $LL_1$, and Algorithm $LL_2$. As a result, we select it as the default inertial parameter for next calculation.

Next, we will select $\beta^k = \frac{0.85}{max(eigenvalue(A^TA))}$, $d^k = 0.5$, $p^k = 0.5$ for Algorithm $L$, Algorithm $LL_1$, and Algorithm $LL_2$; we stop the number of iterations at 500 (iterations = 500). We gain the numeric outcomes can see in Table 9.4.

We can see that $p^k = 0.5$ greatly improves the performance of Algorithm $L$, Algorithm $LL_1$, and Algorithm $LL_2$. Table 9.5 shows accuracy of another method.

For comparing, we set activation function as sigmoid function and hidden nodes M = 150 in Table 9.6.

**Table 9.2** Training time and numerical results of both of training and validation loss when $\beta^k$ is different.

| | $P$ | Training time | Loss | |
|---|---|---|---|---|
| | | | Training | Validation |
| Algorithm $L$ | 0.15 | 0.4416 | 0.2721 | 0.2725 |
| | 0.3 | 0.4411 | 0.2575 | 0.2536 |
| | 0.45 | 0.4398 | 0.2477 | 0.2405 |
| | 0.6 | 0.4409 | 0.2406 | 0.2312 |
| | 0.85 | 0.4390 | 0.2327 | 0.2211 |
| Algorithm $LL_1$ | 0.15 | 0.4394 | 0.2721 | 0.2725 |
| | 0.3 | 0.4393 | 0.2575 | 0.2536 |
| | 0.45 | 0.4337 | 0.2477 | 0.2405 |
| | 0.6 | 0.4295 | 0.2406 | 0.2312 |
| | 0.85 | 0.4261 | 0.2327 | 0.2211 |
| Algorithm $LL_2$ | 0.15 | 0.6433 | 0.2721 | 0.2725 |
| | 0.3 | 0.6320 | 0.2575 | 0.2536 |
| | 0.45 | 0.6287 | 0.2477 | 0.2405 |
| | 0.6 | 0.6345 | 0.2406 | 0.2312 |
| | 0.85 | 0.5213 | 0.2327 | 0.2211 |

**Table 9.3** Training time and numerical results of both of training and validation loss when $d^k$ is different.

| | $d^k$ | Training time | Loss | |
|---|---|---|---|---|
| | | | Training | Validation |
| Algorithm $L$ | 0.5 | 0.4273 | 0.2327 | 0.2211 |
| | 0.6 | 0.4378 | 0.2377 | 0.2274 |
| | 0.7 | 0.4403 | 0.2446 | 0.2364 |
| | 0.8 | 0.4338 | 0.2546 | 0.2496 |
| | 0.9 | 0.4739 | 0.2698 | 0.2695 |
| Algorithm $LL_1$ | 0.5 | 0.4420 | 0.2327 | 0.2211 |
| | 0.6 | 0.4841 | 0.2377 | 0.2274 |
| | 0.7 | 0.4521 | 0.2446 | 0.2364 |
| | 0.8 | 0.5063 | 0.2546 | 0.2496 |
| | 0.9 | 0.4437 | 0.2698 | 0.2695 |
| Algorithm $LL_2$ | 0.5 | 0.5122 | 0.2327 | 0.2211 |
| | 0.6 | 0.7036 | 0.2377 | 0.2274 |
| | 0.7 | 0.6256 | 0.2446 | 0.2364 |
| | 0.8 | 0.6271 | 0.2546 | 0.2496 |
| | 0.9 | 0.6177 | 0.2698 | 0.2695 |

**Table 9.4** Training time and numerical results of both of training and validation loss when $p^k$ is different.

| | $p^k$ | Training time | Loss | |
|---|---|---|---|---|
| | | | Training | Validation |
| Algorithm $L$ | 0.1 | 0.4282 | 0.2465 | 0.2389 |
| | 0.3 | 0.4355 | 0.2403 | 0.2309 |
| | 0.5 | 0.4244 | 0.2327 | 0.2211 |
| | $\frac{1}{k}$ | 0.4342 | 0.2487 | 0.2419 |
| | $\frac{1}{10k+2}$ | 0.4293 | 0.2490 | 0.2423 |
| Algorithm $LL_1$ | 0.1 | 0.4390 | 0.2465 | 0.2389 |
| | 0.3 | 0.4334 | 0.2403 | 0.2309 |
| | 0.5 | 0.4313 | 0.2327 | 0.2211 |
| | $\frac{1}{k}$ | 0.4714 | 0.2487 | 0.2419 |
| | $\frac{1}{10k+2}$ | 0.4352 | 0.2490 | 0.2423 |
| Algorithm $LL_2$ | 0.1 | 0.6094 | 0.2465 | 0.2389 |
| | 0.3 | 0.6133 | 0.2403 | 0.2309 |
| | 0.5 | 0.5149 | 0.2327 | 0.2211 |
| | $\frac{1}{k}$ | 0.6099 | 0.2487 | 0.2419 |
| | $\frac{1}{10k+2}$ | 0.5218 | 0.2490 | 0.2423 |

**Table 9.5** Accuracy of another methods and our algorithm.

| Authors | Methods | Accuracy (%) |
|---|---|---|
| Sakar et al. [19] | KNN+SVM | 55.00 |
| Eskidere et al. [10] | Random subspace classifier ensemble | 74.17 |
| Celik et al. [6] | LR | 76.03 |
| Celik et al. [6] | SVM | 75.49 |
| Benba et al. [2] | MFCC+SVM | 82.50 |
| Li et al. [15] | Ensemble learning algorithm | 86.50 |
| **Algorithm** $L$ | | **90.38** |
| **Algorithm** $LL_1$ | | **90.38** |
| **Algorithm** $LL_2$ | | **90.38** |

Table 9.6 shows that our Algorithm $L$, Algorithm $LL_1$, and Algorithm $LL_2$ in Test box 2 perform at a high level of accuracy, F1-score, recall, and precision at iteration number 782 and when Algorithm $L$ use low of training time. It is the highest possibility of exactly classifying comparison to algorithms investigations. Both of training-validation loss and accuracy plots show that our algorithm has good fitting model for Parkinson's disease dataset.

From Figs. 9.2–9.3, we saw that our Algorithm $L$ has good fit; this means that our Algorithm $L$ appropriately learns the training dataset and generalizes well in relation to classification the Parkinson's disease dataset.

**Table 9.6** Efficiency comparative of existing algorithms and our algorithms.

| | Iter | Time | Pre (%) | Rec (%) | F1 (%) | Acc (%) |
|---|---|---|---|---|---|---|
| **Test box 1** | | | | | | |
| Algorithm $L$ | 227 | 0.2046 | 79.57 | 94.87 | 86.55 | 85.26 |
| Algorithm $LL_1$ | 227 | 0.2062 | 79.57 | 94.87 | 86.55 | 85.26 |
| Algorithm $LL_2$ | 227 | 0.2419 | 79.57 | 94.87 | 86.55 | 85.26 |
| Algorithm of Suantai $L$ [20] | 227 | 0.2039 | 79.10 | 89.74 | 84.08 | 83.01 |
| Algorithm of Suantai $LL_1$ [20] | 227 | 0.2042 | 79.10 | 89.74 | 84.08 | 83.01 |
| Algorithm of Suantai $LL_2$ [20] | 227 | 0.2337 | 79.10 | 89.74 | 84.08 | 83.01 |
| **Test box 2** | | | | | | |
| Algorithm $L$ | 782 | 0.6618 | 86.21 | 96.15 | 90.91 | 90.38 |
| Algorithm $LL_1$ | 782 | 0.6879 | 86.21 | 96.15 | 90.91 | 90.38 |
| Algorithm $LL_2$ | 782 | 0.8087 | 86.21 | 96.15 | 90.91 | 90.38 |
| Algorithm of Suantai $L$ [20] | 782 | 0.6741 | 83.62 | 94.87 | 88.89 | 83.01 |
| Algorithm of Suantai $LL_1$ [20] | 782 | 0.6659 | 83.62 | 94.87 | 88.89 | 83.01 |
| Algorithm of Suantai $LL_2$ [20] | 782 | 0.7850 | 83.62 | 94.87 | 88.89 | 83.01 |
| **Test box 3** | | | | | | |
| Algorithm $L$ | 175 | 0.1691 | 85.80 | 96.79 | 90.96 | 90.38 |
| Algorithm $LL_1$ | 175 | 0.1637 | 85.80 | 96.79 | 90.96 | 90.38 |
| Algorithm $LL_2$ | 175 | 0.1925 | 85.80 | 96.79 | 90.96 | 90.38 |
| Algorithm of Suantai $L$ [20] | 175 | 0.1617 | 86.50 | 90.38 | 88.40 | 88.14 |
| Algorithm of Suantai $LL_1$ [20] | 175 | 0.1627 | 86.50 | 90.38 | 88.40 | 88.14 |
| Algorithm of Suantai $LL_2$ [20] | 175 | 0.1873 | 86.50 | 90.38 | 88.40 | 88.14 |
| **Test box 4** | | | | | | |
| Algorithm $L$ | 487 | 0.4272 | 83.70 | 98.72 | 90.59 | 89.74 |
| Algorithm $LL_1$ | 487 | 0.4348 | 83.70 | 98.72 | 90.59 | 89.74 |
| Algorithm $LL_2$ | 487 | 0.5091 | 83.70 | 98.72 | 90.59 | 89.74 |
| Algorithm of Suantai $L$ [20] | 487 | 0.4179 | 83.15 | 98.08 | 90.00 | 89.10 |
| Algorithm of Suantai $LL_1$ [20] | 487 | 0.4267 | 83.15 | 98.08 | 90.00 | 89.10 |
| Algorithm of Suantai $LL_2$ [20] | 487 | 0.5964 | 83.15 | 98.08 | 90.00 | 89.10 |
| **Test box 5** | | | | | | |
| Algorithm $L$ | 464 | 0.4136 | 72.56 | 100.00 | 84.10 | 81.09 |
| Algorithm $LL_1$ | 464 | 0.4108 | 72.56 | 100.00 | 84.10 | 81.09 |
| Algorithm $LL_2$ | 464 | 0.4876 | 72.56 | 100.00 | 84.10 | 81.09 |
| Algorithm of Suantai $L$ [20] | 464 | 0.4029 | 71.76 | 99.36 | 83.33 | 80.13 |
| Algorithm of Suantai $LL_1$ [20] | 464 | 0.4014 | 71.76 | 99.36 | 83.33 | 80.13 |
| Algorithm of Suantai $LL_2$ [20] | 464 | 0.4657 | 71.76 | 99.36 | 83.33 | 80.13 |

*continued on next page*

## 9.5 Conclusion

In this work, we introduce an inertial projective Mann algorithm to solve split equilibrium problems, and then we prove the weak convergence theorem of the proposed algorithm

**Table 9.6** (continued)

| | Iter | Time | Pre (%) | Rec (%) | F1 (%) | Acc (%) |
|---|---|---|---|---|---|---|
| **Test box 6** | | | | | | |
| Algorithm $L$ | 571 | 0.5063 | 75.61 | 99.36 | 85.87 | 83.65 |
| Algorithm $LL_1$ | 571 | 0.5009 | 75.61 | 99.36 | 85.87 | 83.65 |
| Algorithm $LL_2$ | 571 | 0.5829 | 75.61 | 99.36 | 85.87 | 83.65 |
| Algorithm of Suantai $L$ [20] | 571 | 0.4912 | 73.56 | 98.08 | 84.07 | 81.41 |
| Algorithm of Suantai $LL_1$ [20] | 571 | 0.4957 | 73.56 | 98.08 | 84.07 | 81.41 |
| Algorithm of Suantai $LL_2$ [20] | 571 | 0.5688 | 73.56 | 98.08 | 84.07 | 81.41 |
| **Test box 7** | | | | | | |
| Algorithm $L$ | 350 | 0.3092 | 71.50 | 98.08 | 82.70 | 79.49 |
| Algorithm $LL_1$ | 350 | 0.3143 | 71.50 | 98.08 | 82.70 | 79.49 |
| Algorithm $LL_2$ | 350 | 0.3629 | 71.50 | 98.08 | 82.70 | 79.49 |
| Algorithm of Suantai $L$ [20] | 350 | 0.3069 | 69.95 | 95.51 | 80.76 | 77.24 |
| Algorithm of Suantai $LL_1$ [20] | 350 | 0.3067 | 69.95 | 95.51 | 80.76 | 77.24 |
| Algorithm of Suantai $LL_2$ [20] | 350 | 0.3591 | 69.95 | 95.51 | 80.76 | 77.24 |
| **Test box 8** | | | | | | |
| Algorithm $L$ | 1477 | 1.2572 | 83.71 | 95.51 | 89.22 | 88.46 |
| Algorithm $LL_1$ | 1477 | 1.2445 | 83.71 | 95.51 | 89.22 | 88.46 |
| Algorithm $LL_2$ | 1477 | 1.4572 | 83.71 | 95.51 | 89.22 | 88.46 |
| Algorithm of Suantai $L$ [20] | 1477 | 1.2581 | 82.58 | 94.23 | 88.02 | 87.18 |
| Algorithm of Suantai $LL_1$ [20] | 1477 | 1.2520 | 82.58 | 94.23 | 88.02 | 87.18 |
| Algorithm of Suantai $LL_2$ [20] | 1477 | 1.4630 | 82.58 | 94.23 | 88.02 | 87.18 |

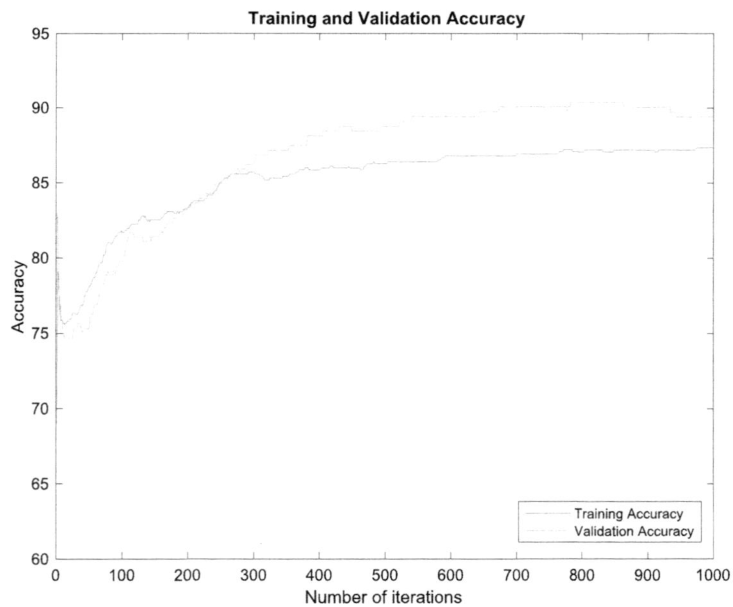


**FIGURE 9.2** The accuracy plots of training and validation of Algorithm $L$.

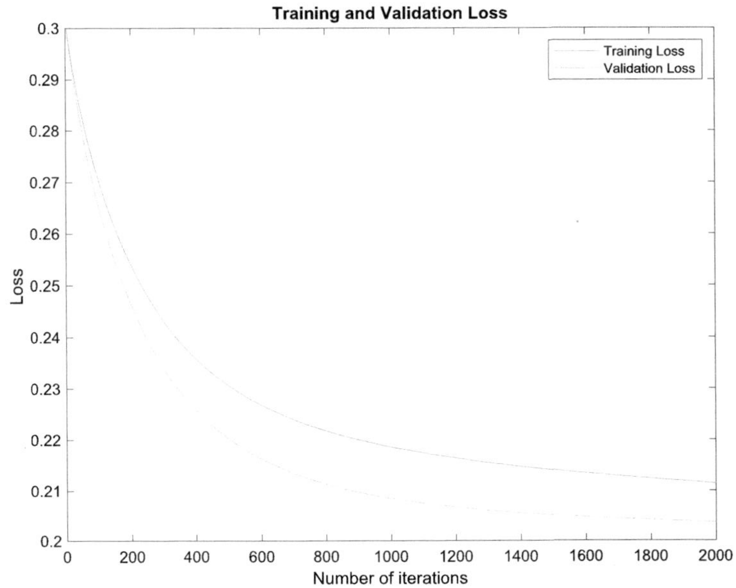


**FIGURE 9.3** The loss plots of training and validation of Algorithm $L$.

under suitable conditions in Hilbert space. Moreover, we show to choose parameters for three models of the least square problems in constrained sets considering in machine learning, as seen in Section 9.4. Finally, we found that the proposed algorithm can efficiently detect Parkinson's disease with an accuracy of 90.38%. The comparison shows that our algorithm in the ELM method is more efficient than in other machine learning methods.

# Data availability

The data are available in UCI (https://archive.ics.uci.edu/dataset/301/parkinson+speech+dataset+with+multiple+types+of+sound+recordings).

# Acknowledgment

The authors would like to thank the National Research Council of Thailand and University of Phayao (N42A650334) and Thailand Science Research and Innovation, University of Phayao (Fundamental Fund 2025).

# References

[1] R. Balestrino, A.H.V. Schapira, Parkinson disease, European Journal of Neurology 27 (1) (2020) 27–42.
[2] A. Benba, A. Jilbab, A. Hammouch, Analysis of multiple types of voice recordings in cepstral domain

using MFCC for discriminating between patients with Parkinson's disease and healthy people, International Journal of Speech Technology 19 (2016) 449–456.

[3] B.R. Bloem, M.S. Okun, C. Klein, Parkinson's disease, The Lancet 397 (10291) (2021) 2284–2303.

[4] E. Blum, From optimization and variational inequalities to equilibrium problems, Mathematics Student 63 (1994) 123–145.

[5] Y. Censor, T. Elfving, A multiprojection algorithm using Bregman projections in a product space, Numerical Algorithms 8 (1994) 221–239.

[6] E. Celik, S.I. Omurca, Improving Parkinson's disease diagnosis with machine learning methods, in: 2019 Scientific Meeting on Electrical-Electronics & Biomedical Engineering and Computer Science (EBBT), IEEE, 2019, April, pp. 1–4.

[7] W. Cholamjiak, S. Das, A modified projective forward-backward splitting algorithm for variational inclusion problems to predict Parkinson's disease, Applied Mathematics in Science and Engineering 32 (1) (2024) 2314650.

[8] P.L. Combettes, S.A. Hirstoaga, Equilibrium programming in Hilbert spaces, Journal of Nonlinear and Convex Analysis 6 (1) (2005) 117–136.

[9] K. Dashtipour, A. Tafreshi, J. Lee, B. Crawley, Speech disorders in Parkinson's disease: pathophysiology, medical management and surgical approaches, Neurodegenerative Disease Management 8 (5) (2018) 337–348.

[10] Ö. Eskıdere, A. Karatutlu, C. Ünal, Detection of Parkinson's disease from vocal features using random subspace classifier ensemble, in: 2015 Twelve International Conference on Electronics Computer and Computation (ICECCO), IEEE, 2015, September, pp. 1–4.

[11] K. Fan, A minimax inequality and applications, in: O. Shisha (Ed.), Inequalities III, Academic Press, San Diego, 1972.

[12] A. Gaenslen, I. Swid, I. Liepelt-Scarfone, J. Godau, D. Berg, The patients' perception of prodromal symptoms before the initial diagnosis of Parkinson's disease, Movement Disorders 26 (4) (2011) 653–658.

[13] Z. He, The split equilibrium problem and its convergence algorithms, Journal of Inequalities and Applications 2012 (2012) 1–15.

[14] G.B. Huang, Q.Y. Zhu, C.K. Siew, Extreme learning machine: theory and applications, Neurocomputing 70 (1–3) (2006) 489–501.

[15] Y. Li, C. Zhang, Y. Jia, P. Wang, X. Zhang, T. Xie, Simultaneous learning of speech feature and segment for classification of Parkinson disease, in: 2017 IEEE 19th International Conference on e-Health Networking, Applications and Services (Healthcom), IEEE, 2017, October, pp. 1–6.

[16] W.R. Mann, Mean value methods in iteration, Proceedings of the American Mathematical Society 4 (3) (1953) 506–510.

[17] A. Moudafi, Split monotone variational inclusions, Journal of Optimization Theory and Applications 150 (2011) 275–283.

[18] B.T. Polyak, Some methods of speeding up the convergence of iteration methods, U.S.S.R. Computational Mathematics and Mathematical Physics 4 (1964) 1–17.

[19] B.E. Sakar, M.E. Isenkul, C.O. Sakar, A. Sertbas, F. Gurgen, S. Delil, et al., Collection and analysis of a Parkinson speech dataset with multiple types of sound recordings, IEEE Journal of Biomedical and Health Informatics 17 (4) (2013) 828–834.

[20] S. Suantai, P. Cholamjiak, Y.J. Cho, W. Cholamjiak, On solving split equilibrium problems and fixed point problems of nonspreading multi-valued mappings in Hilbert spaces, Fixed Point Theory and Applications 2016 (2016) 1–16.

[21] S. Suantai, Weak and strong convergence criteria of Noor iterations for asymptotically nonexpansive mappings, Journal of Mathematical Analysis and Applications 311 (2) (2005) 506–517.

[22] C. Stewart, L. Winfield, A. Hunt, S.B. Bressman, S. Fahn, A. Blitzer, M.F. Brin, Speech dysfunction in early Parkinson's disease, Movement Disorders 10 (5) (1995) 562–565.

[23] K.K. Tan, H.K. Xu, Approximating fixed points of non-expansive mappings by the Ishikawa iteration process, Journal of Mathematical Analysis and Applications 178 (1993) 301.

[24] T. Thomas, N. Pradhan, V.S. Dhaka, Comparative analysis to predict breast cancer using machine learning algorithms: a survey, in: 2020 International Conference on Inventive Computation Technologies (ICICT), IEEE, 2020, February, pp. 192–196.

[25] R. Tibshirani, Regression shrinkage and selection via the lasso, Journal of the Royal Statistical Society, Series B, Statistical Methodology 58 (1996) 267–288.
[26] D. Van Hieu, Two hybrid algorithms for solving split equilibrium problems, International Journal of Computer Mathematics 95 (3) (2018) 561–583.
[27] S.S. Zhang, J.H. Lee, C.K. Chan, Algorithms of common solutions to quasi variational inclusion and fixed point problems, Applied Mathematics and Mechanics 29 (5) (2008) 571–581.

# Index

## L



## M



## N

9780443154454